Projective Processes and Neuroscience in Art and Design

Rachel Zuanon
Anhembi Morumbi University, Brazil

A volume in the Advances in Media, Entertainment, and the Arts (AMEA) Book Series

Information Science REFERENCE
An Imprint of IGI Global

Published in the United States of America by
Information Science Reference (an imprint of IGI Global)
701 E. Chocolate Avenue
Hershey PA, USA 17033
Tel: 717-533-8845
Fax: 717-533-8661
E-mail: cust@igi-global.com
Web site: http://www.igi-global.com

Library of Congress Cataloging-in-Publication Data

Names: Zuanon, Rachel, 1974- editor.
Title: Projective processes and neuroscience in art and design / Rachel Zuanon, editor.
Description: Hershey : Information Science Reference, 2016. | Includes bibliographical references and index.
Identifiers: LCCN 2016012759| ISBN 9781522505105 (hardcover) | ISBN 9781522505112 (ebook)
Subjects: LCSH: Design--Psychological aspects. | Neurosciences and the arts.
Classification: LCC NK1520 .P76 2016 | DDC 704.9/496128--dc23 LC record available at http://lccn.loc.gov/2016012759

This book is published in the IGI Global book series Advances in Media, Entertainment, and the Arts (AMEA) (ISSN: Pending; eISSN: pending)

British Cataloguing in Publication Data
A Cataloguing in Publication record for this book is available from the British Library.

For electronic access to this publication, please contact: eresources@igi-global.com.

Advances in Media, Entertainment, and the Arts (AMEA) Book Series

Giuseppe Amoruso
Politecnico di Milano, Italy

ISSN: Pending
EISSN: pending

Mission

Throughout time, technical and artistic cultures have integrated creative expression and innovation into industrial and craft processes. Art, entertainment and the media have provided means for societal self-expression and for economic and technical growth through creative processes.

The **Advances in Media, Entertainment, and the Arts (AMEA)** book series aims to explore current academic research in the field of artistic and design methodologies, applied arts, music, film, television, and news industries, as well as popular culture. Encompassing titles which focus on the latest research surrounding different design areas, services and strategies for communication and social innovation, cultural heritage, digital and print media, journalism, data visualization, gaming, design representation, television and film, as well as both the fine applied and performing arts, the AMEA book series is ideally suited for researchers, students, cultural theorists, and media professionals.

Coverage

- Cross-Media Studies
- Humanities Design
- Arts & Design
- Film & Television
- Drawing
- Color Studies
- Traditional Arts
- Print Media
- Fine Arts
- Geometry & Design

IGI Global is currently accepting manuscripts for publication within this series. To submit a proposal for a volume in this series, please contact our Acquisition Editors at Acquisitions@igi-global.com or visit: http://www.igi-global.com/publish/.

The Advances in Media, Entertainment, and the Arts (AMEA) Book Series (ISSN Pending) is published by IGI Global, 701 E. Chocolate Avenue, Hershey, PA 17033-1240, USA, www.igi-global.com. This series is composed of titles available for purchase individually; each title is edited to be contextually exclusive from any other title within the series. For pricing and ordering information please visit http://www.igi-global.com/book-series/advances-media-entertainment-arts/102257. Postmaster: Send all address changes to above address.

Titles in this Series

For a list of additional titles in this series, please visit: www.igi-global.com

Handbook of Research on Visual Computing and Emerging Geometrical Design Tools
Giuseppe Amoruso (Politecnico di Milano, Italy)
Information Science Reference • copyright 2016 • 924pp • H/C (ISBN: 9781522500292) • US $450.00 (our price)

Creative Technologies for Multidisciplinary Applications
Andy M. Connor (Auckland University of Technology, New Zealand) and Stefan Marks (Auckland University of Technology, New Zealand)
Information Science Reference • copyright 2016 • 467pp • H/C (ISBN: 9781522500162) • US $205.00 (our price)

Global Perspectives on Media Events in Contemporary Society
Andrew Fox (University of Huddersfield, UK)
Information Science Reference • copyright 2016 • 306pp • H/C (ISBN: 9781466699670) • US $165.00 (our price)

Political Influence of the Media in Developing Countries
Lynete Lusike Mukhongo (Moi University, Kenya) and Juliet Wambui Macharia (Moi University, Kenya)
Information Science Reference • copyright 2016 • 303pp • H/C (ISBN: 9781466696136) • US $200.00 (our price)

Impact of Communication and the Media on Ethnic Conflict
Steven Gibson (Northcentral University, USA) and Agnes Lucy Lando (Daystar University, Kenya)
Information Science Reference • copyright 2016 • 344pp • H/C (ISBN: 9781466697287) • US $185.00 (our price)

Handbook of Research on Media Literacy in the Digital Age
Melda N. Yildiz (Walden University, USA & Unite to Educate, USA) and Jared Keengwe (University of North Dakota, USA)
Information Science Reference • copyright 2016 • 532pp • H/C (ISBN: 9781466696679) • US $295.00 (our price)

Analyzing Art, Culture, and Design in the Digital Age
Gianluca Mura (Politecnico di Milano University, Italy)
Information Science Reference • copyright 2015 • 329pp • H/C (ISBN: 9781466686793) • US $185.00 (our price)

Handbook of Research on the Societal Impact of Digital Media
Barbara Guzzetti (Arizona State University, USA) and Mellinee Lesley (Texas Tech University, USA)
Information Science Reference • copyright 2016 • 789pp • H/C (ISBN: 9781466683105) • US $350.00 (our price)

IGI GLOBAL
DISSEMINATOR OF KNOWLEDGE
www.igi-global.com

701 E. Chocolate Ave., Hershey, PA 17033
Order online at www.igi-global.com or call 717-533-8845 x100
To place a standing order for titles released in this series, contact: cust@igi-global.com
Mon-Fri 8:00 am - 5:00 pm (est) or fax 24 hours a day 717-533-8661

Editorial Advisory Board

Table of Contents

Detailed Table of Contents

This chapter discusses issues related to two fields of knowledge: neuroesthetics and cognitive neuroscience of art. These two fields represent areas that link historically dichotomic instances: nature and culture. In the first section, the author introduces a brief discussion on this dichotomy, reified here as science and art/aesthetics. Based on a preliminary analysis of these fields, as well as potential interfaces and articulations, the author then situates neuroesthetics and cognitive science of art. In both cases, the main definitions, usual criticisms, and comments on potential expectations regarding the future of these two areas will be presented.

The aesthetic experience has been discussed throughout the history of mankind by philosophers and art historians, becoming a universal part of human experience, which leads us to some great interdisciplinary questions. It has been the subject of study by neuroscientists and neuro-psychologists since the 2000s. This recent evolution of neurology studies in the field of art, is due to in vivo brain imaging techniques, especially functional neuroimaging. Furthermore, recent research has provided evidence of cognitive interaction during the perception of an artwork indicating that the perceptual experience of art is not merely a passive one. This article reviews important studies in neuroaesthectics of visual art that point out that the aesthetic experience is related to the distribution in the neural architecture, suggesting the involvement of sensory-motor areas, emotional centers, reward system, memory and language.

The search for the source of beauty has exercised the speculation of philosophers and writers throughout the ages. The classical conception is that beauty consists of an arrangement of integral parts into a coherent whole, according to proportion, harmony, symmetry, and similar notions. The opposing view is that art is a fully subjective enterprise and our preferences are shaped by our values and experiences. To study beauty, neuroscience evokes the juxtaposition of concepts such as symmetry, geometry, proportion, judgment, affection, grotesque, erotic, sensitive with certain loci in the human brain. However, to explain and understand the aesthetic experiences at the network level, neuroscience have to incorporate the ambiguity in their research field. And with the continuous increasing of knowledge and comprehension of how beauty is perceived by the human brain, the world in its wholeness will be highly impacted, since we will start to live in pleasant societies.

The ownership feeling of our body occurs mainly due to feedback responses in real-time from environment stimuli to our own body. These constant feedbacks induce a neuronal arrangement, generating a representative map and, consequently, an ownership and unity feeling, named as a body schema. Although there is a relative well knowing of the sensorial mapping about each part of our body, there are still several gaps about how the integration of all this representation is indeed accomplished. Many researchers have shown high rates of prosthesis non-acceptance due to different reasons. Here, the authors discuss an experimental protocol to induce optimally the ownership feeling associated with upper limb prosthesis, by means of a crossmodal vibro-tactile stimulation over the individual's body. The main hypothesis is that through this procedure the participant will extend their proprioception and achieve an ownership feeling of the prosthesis.

This chapter is concerned with the teaching and learning of modelling in fashion design courses. Following a series of observations, it was found that fashion design students, with normal sight, have difficulties in fully understanding how an item of clothing can be transposed to a modelling display bust, which represents the body of the wearer. The same obstacle affects visually-handicapped students. This study seeks to explore ways of overcoming this problem. It involves seeking to introduce features into teaching that can allow a comprehensive learning program to be taught and in particular, to concentrate on certain key factors - cognition, constancy and abstraction - with regard to the information on fashion projects that can be found in the surrounding learning environment.

This chapter aims to present mental representations by seven audiovisual specialists on Sirena, the antagonist of a noir fictional story in a multilinear interactive short film script. To this end, a reception study was created with in-depth interviews with specialists after reading the script to discuss its several aspects. This approach is of qualitative character. The study was conducted between June and October 2015 in São Paulo, Brazil. It is possible to perceive that participants have interpreted the character differently in spite of its consistent build. We could verify that the mental representation of the antagonist was aided by the world-view and experience of each interviewed specialist, which has led to unique readings of the character.

Foreword

Projective Processes and Neuroscience in Art and Design arrives at a particularly cogent moment in the history of human culture. If the last scientific revolution focused on the digital, this one is centred on the technologies of mind. There is a perfect "brain storm": cognitive computing follows and amplifies the logic of the human mind at amazing speed; advances in neuroscience reveal and potentially manipulate brain function, and precision in data gathering and analytics allow us to express our behaviors and actions in as a quantified self, including the outputs of our minds. The moment offers the opportunity of deploying these new technological capacities to better study and understand what it is to be human, while altering the very nature of humanness.

Projective Processes and Neuroscience in Art and Design is an important contribution to this new era. As several writers suggest artists and designers tend to be early adopters of emerging technologies, providing critical and heretical insights, adapting technologies such as neuroimaging as a new material, inventing new tools, and creating new forms of expression. Science and technology studies remind us that scientific discovery and its applications are driven by structures of power that reside within and outside the apparatus of scientific cultures and effect the ways that science is technologized and socialized. Artists and designers ask fundamental questions about where human agency is positioned and should be positioned as powerful technologies, such as affordable virtual reality systems emerge. In the case of neuroscience there is a pas de deux.

As artists and designers have turned their gaze upon neuroscience, so has neuroscience increasingly turned its gaze onto art history, artists and design, seeking understandings of the ways that perception, pleasure, attention through the study of neuroaesthetics, and seeking to understand the neurological processes that accompany the production of art or design products. Studying the neurological processes of creative individuals can help analytics to understand how creativity per se operates and what the optimal conditions might be for these expressions.

In the research and applications described in this book collaboration between the arts and sciences, design and engineering are essential. These experiments now exit from a peripheral position and move to the centre. The stakes are high. Societies are beginning to realize that it is of critical importance to place human needs and desires at the centre of emerging technologies. STEM disciplines are increasingly dependent upon the mediation of design and art, on one hand, to create new applications, as discussed in this book.

One of the significant challenges for neuroscience in the field of contemporary art and design is to understand the complex weaving of cultural influences, context, the accrued language of a highly global yet equally localized art world side by side with the perceptual neurological processes of art-viewing. While some fields such as data visualization have long engaged with cognitive science, others such as conceptual art rely heavily on linguistic and psychoanalytic interpretation at odds with a neurological analysis than can reduce art experience to pleasure. This book takes the opportunity to weave together colliding philosophical traditions, interpretations of cognitive science and neurobiology and creative practices.

Sara Diamond
OCAD University, Canada

Sara Diamond *is the President of OCAD University, Canada's "university of the imagination". She holds a PhD in Computing, Information Technology and Engineering from the University of East London, a Masters in Digital Media theory from the University of Arts London and Honours Bachelors of Arts in History and Communications from Simon Fraser University. Diamond is a data visualization, wearable technology and mobile media researcher, artist and designer. She developed www.codezebra.net, a pioneering social media data visualization and sentiment analysis software. Diamond was founding Chair of the Mobile Experience Innovation Centre and was co-chair (with Deloitte). She is theme leader for the i-City project, a multi-million dollar urban planning initiative and was co-principal investigator on the Centre for Information Visualization/Data Driven Design and is the holder of funding from the Social Science and Humanities Research Council, Canada Foundation for Innovation, MITACS, the National Sciences and Engineering Research Council and Ontario Centres for Excellence. She is an appointee of the Order of Ontario and the Royal Canadian Society of Artists and a recipient of the Queen's Diamond Jubilee Medal for service to Canada. She is the winner of the 2013 GRAND NCE Digital Media Pioneer Award. Diamond was the Artistic Director of Media and Visual Art and Director of Research at the Banff Centre, where she created the Banff New Media Institute (BNMI) in 1995 and led it until 2005.*

Preface

This book is the result of the meeting of transdisciplinary, questioning, restless and greedy minds to explore territories and build trails, paths, bridges and roads to investigations that go beyong the limits of their individual fields of research. This path naturally - which does not imply simplicity neither ease - has designed dialogues, reverberations, synergies - sometimes complex - to the still unwary senses. The extended fields of art and design through the veins of the Cognitive Sciences, and especially neuroscience, reconfigure thoughts, break paradigms and transform gestures into attitudes that move the world towards the point that puts us as human beings, at the inseparable confluence between nature and culture, between what circumscribes us as 'I' to create awareness of the 'other', and on everything we create, design, develop, produce and deliver to the society.

The approach of art-science dates back centuries and announces its first signals from prehistory throughout the Paleolithic age, in the cave paintings considered the first illustrations of scientific visualization that arise as a result of careful observation of psychology and animal behavior carried out by painters in this period. This was also brings significant contributions to the chemistry history with the identification of different metals and their respective characteristics reflected in manually manufactured parts and famous for their aesthetic strength. Or even for their carefully arranged monoliths to indicate the exact positions of the heavenly bodies at different times of the year (Wilson, 2002). Centuries later, the historical period called as the Renaissance introduces what is known as the most significant phase of art-science synergy and elects the main representative of this integration in the world, Leonardo da Vinci. He was an artist who synthesized in his paintings his profound mastery of human anatomy, and his inventions revealed elaborate studies in mechanical engineering. To think the art separated from science, and vice versa, was not even considered an option in this context in which art and science were understood as complementary fields of knowledge and mutual contribution. Artists and scientists mutually united by interests that both have demonstrated for the area of knowledge of each other.

More centuries later, especially in the early twentieth century, the contributions of Gestalt laws to the aesthetic apprehension extend the collaboration between the fields of art and science, what we call as laws of perception. And although the Gestalt psychologists regarded these "laws as byproducts of some undiscovered physical principles such as electrical fields in the brain they didn't correctly answer why such laws may have evolved or how they came to be enshrined in the neural architecture of the brain" (Ramachandran, 2011, p. 200).

From the last decade of the last century, the advances of research in Cognitive Sciences, especially the field of Artificial Intelligence, with an impressive variety of neural networks and connectionist models - which have characteristics which are associated with human intelligence in its ability for reasoning, learning and problem solving – has opened significant prospects for the projective creation and development of artists and designers in deep synergy and exchange of experiences with engineers and computer scientists. The results of these cooperations are shown in the most diverse interactive intelligent and creative digital systems designed and implemented since then, involving speech recognition, image recognition, data analysis, autonomous robotics, game playing, among others.

In the context of Neuroscience – that investigates cognition from the observation of brain properties (Varela et al., 2003) and deals with the development, chemistry, structure, function and pathology of the nervous system (Lundy-Ekman, 2004) - art and science intensify the transdisciplinarity between their scopes with contributions of Semir Zeki, especially in proposing and structuring the concept of neuroesthetics (1994, 1999, 2000), focused on the cognitive neuroscience of visual aesthetics, especially in the explanations of how artworks function as perceptual stimuli and can serve as the foundation for a biological theory of aesthetics; and V. S. Ramanchandran (2011) with the formulation of nine universal laws of aesthetics, from which he proposes the construction of a neurobiological view of aesthetics, dealing with different levels of descriptions that complement both fields rather than contradict each other, with the identification of certain systematic principles that lie behind the various artistic styles variations and that cut across cultural barriers.

This very brief overview obviously does not exhaust the many existing and possible influences between the fields of art, design and science. Far from it, our intention here is to cast a glance at such cooperation over time, and how they contribute to the approach of what we are, by telling us more about ourselves and our experiences.

In this sense, reflecting and discussing the projective processes in art and design from the perspective of the mechanisms that are structured in the brain and define us as human beings - perception, attention, proprioception, cognition, memory, emotions and feelings, sensorimotor, among many others - has been my challenge over uninterrupted nineteen years and drives me to create and design digital systems and intelligent products focused on the neurophysiological system of the user and dedicated to the welfare and improvement of the individual's quality of life. Zizel (2006) summarizes this challenge in a unique way by stating that if you understand how people's brain and mind develop and function in different situations, and how they have evolved over time to respond to physical environments, then environments designed to support these capabilities as well as tasks, activities, and user's needs will contribute to people's quality of life, creativity and survival. Blass (2005) reinforces this thought by emphasizing that designers and researchers should take advantage of the brain's natural creative process to achieve their ends and that the understanding of brain function can provide insight into the nature of research, design and creativity. Therefore, exploring the brain even further must improve design and research in practice, once changes in brain structure imply changes in behavior and experience, and consequently in the interaction with artworks and products designed by design.

With my experience, this motivation overflows the boundaries of the personal level and identified resonances that here materialize in significant contributions expressed by the twenty-two authors invited to share the results of their transdisciplinary research of art-design-cognitive sciences interfaces, distributed in sessions (1) 'Multidimensional Brain: when Art meets Neuroscience meets Art'; (2) Aesthetic Experience and Emotion in Art and Design: Crossing Domains in Philosophy, Psychology and Neuroscience; and (3) 'Mapping and Designing Cognition: Artificial and Natural Perspectives of Learning and Memory in Design and iCinema'.

The correlations undertaken in this book assume that body, brain and mind are inseparable manifestations of a living organism (Damásio, 2004). And that the human brain seems to have evolved to be creative - to imagine new ideas, to put into practice what they invent and to critically analyze the results of human actions particularly suited to design things - concepts, tools, languages and places. In this sense, the brain creatively builds senses to its context, as well as artists and designers create artworks and products that respectively reconfigure these contexts and stimulate the brain to reconstruct the existing senses or to build new ones (Zizel, 2006).

More specific, sensitive detectors distributed by the human body participate in the construction of neural patterns that map the multidimensional interactions of the organism with any object, whether it is the result of an artistic creation or a design project. Such interactions result in images that the mind develops and preserves in different spatiotemporal scales. In other words, the human perception of any object or situation, whether coming from the inside or outside of the organism; the ability to respond to a stimulus, automatically or deliberately; as well as to anticipate and plan the future are always mediated by images of visual, auditory, tactile, olfactory, gustatory nature - built in the relationship of the body with its exterior - or by images from inside the body, such as pain and proprioception. Without mental images, the body would not be able to perform effectively the integration of internal and external information that are necessary for the survival and well-being. In summary, the regulatory operations of the brain in complex organisms depend on the creation and manipulation of mental images - ideas or thoughts - the process known as mind, a phenomenon that emerges from the cooperation of several brain regions. In the mental level, there is the integration of images from the current perception with those images from the memory. And these integrations respond directly by numerous manipulations of these images, which are indispensable to the solution of new problems and to creativity (Damásio, 2004). In this regard, Zizel (2006) proposes three main instances of brain articulation that constitute the creative process in any individual, not only on artists or designers: (1) Imaging, related to work of the brain's interpreter; (2) Presenting, related to work of the brain's actor; and (3) Testing, related to work of the brain's comparator. Each of these parts work together in the configuration of creativity. In this structure, (1) the brain interpreter is the responsible for creating explanatory models and concepts, by making up stories, myths, and concepts to explain experience, "playing a major role in research and design imaging and in the greater creative process" (Gazzaniga, 1998). In other words, the brain-interpreter helps us generate images and see connections between elements that already exist in new ways. That is what Changeux (1985) calls "pre-representations", which comprises the set of images sketched out by mind before making a final drawing. While (2) the brain's actor accounts for the combination of three levels of actions aimed at mental coherence on a certain situation: (a) sensor; (b) perceptor; and (c) motor responder, which involve mainly and respectively the brain's occipital lobe, parietal lobe and frontal lobe. As a sensor, the brain's actor employs the body senses like sight, touch and hearing for obtaining data from and about the environment, which will be organized by the mind of the individual in order to give them meaning and sense. At the time that these stimuli from the environment are perceived and identified more clearly and finally recognized, the brain's actor turns into the perceptor. And by instructing the body on how to act on the environment in response to previous instances of sensory and perceptual interpretation, the brain's actor assumes his role as motor responder, linking feelings and executive function in the brain, which will result in emotional actions of the individual. And finally, (3) the brain's comparator, in connection with (1) the brain's interpreter and the (2) brain's actor, comparing the assumptions made by the first and the perceptions and motor responses delivered by the second, in order to identify possible discontinuities that provide the brain the opportunity to modify and improve this interpretation. "After the comparator modifies the holistic concept the interpreter has generated, a re-entry is made in the interpreter, and a new cycle of refinement and further interpretation begins" (Zizel, 2006, p. 151). That is, the brain does not absorb environmental information and save them as a camera or recorder for later retrieval. Everything we think or remember is continually being modified by new learning, resulting from constant changes of connections between nerve cells of the brain (Freeman, 2003). And the brain, in order to promote reflection and action, subsidizes the desire for organization to attribute meaning to things in their respective contexts, as well as underpinning the planning and taking decisions intrinsic to the art and design. This interaction and coherence also corroborate the definition of ourself and of who we are in relation to the context. The self guides the mental process of planning in order to meet the needs of self-preservation of the organism. Without the self, and the feelings that constitute it, the integration of mental information on a large scale could not be oriented to the problems in life, especially to the problems of survival and well-being (Damasio, 2004). The world is highly complex and any brain knows only what they can create within themselves from their experiences.

From all these perspectives, this book aims to provide the meeting of transdisciplinary issues, still underexplored and widespread in the academic and scientific context, involving the dialogues between Art, Design and Cognitive Sciences, especially Neuroscience and mainly focused on contributions to the projective processes. Moreover, this publication aims especially to contribute with research centers in Art, Design and Cognitive Sciences in order to strengthen and expand this area of research by bringing positive contributions to the creation and development of products, environments, and services from the identification of a neuroscientific basis able to support and open innovative perspectives for projective methodologies oriented to sensory-perceptual and cognitive processes as well as behavioral characteristics of users. It is a promising field of research and projective development by indicating the way to an increasingly natural interaction between the human body and all objects we invent and create. At the same time, it broadens the role and responsibility of artists and designers in front of the artworks and products designed by them. Thus, this book is especially oriented and dedicated to all individuals, researchers, artists, designers and cognitive scientists used to the challenges that are naturally reflected from thoughts, research and transdisciplinary nature of practices and available to the endless search for answers made by the unceasing human desire to understand everything that is part of the existence.

DESCRIPTION OF THE BOOK STRUCTURE AND CONTRIBUTING CHAPTERS

This book is divided into three sessions. The first, 'Multidimensional Brain: when Art meets Neuroscience meets Art' is dedicated to reflection of the meeting of creative and projective processes of art in close cooperation with the fields of neuroscience and neurobiology. In this session, three chapters express this collaborative convergence. Rosangella Leote (Brazil) argues in Chapter 1 'Perceptual Processes and Multisensoriality: Understanding Multimodal Art From Neuroscientific Concepts' about the perceptual processes of a multisensory nature that take place in people with severe motor and vocal disabilities, as well as those without these restrictions, when they are in relationship with the artworks, which contain multimodal stimuli, promoted by physical and digital interfaces of an assistive nature. Maria Manuela Lopes (Portugal) in Chapter 2 'Inside/Out: Looking Back into the Future' explores ways of expanding concepts of memory, as functioning, loss and enhancement, bringing art as an important tool to wide perspectives in discussions about technologies for memory preservation and human body's improvements. In Chapter 3 'Cells, Organisms and the Living Brain as New Media for Art: A Pursuit in Art Research', Marta Menezes (Portugal) addresses how awareness about biological sciences and biotechnology has been having an increasing influence over artists, and how it reflects on the current position of the artist as an art researcher within the academic environment and the growing art-science community, in opposition to the position of the artist as an obsolete passive observer of the scientific research.

The second session 'Aesthetic Experience and Emotion in Art and Design: Crossing Domains in Philosophy, Psychology and Neuroscience' comprises five chapters which map, interrelate and discuss important contributions to thought and creative and design practice in art and design resulting from transdisciplinary articulations with the fields of philosophy, cognitive psychology and neuroscience. In Chapter 4, 'Neuroesthetics - Perspectives and Reflections', Fernando L. Fogliano and Hosana Celeste Oliveira (Brazil and Helsinki) from a neuroaesthetic approach they address the aesthetic experience as a key aspect of subjectivity and interactions in the world beyond the field of art. Alexandre Siqueira de Freitas (Brazil), in chapter 5 'A Study on the Interface between Arts and Sciences: Neuroesthetics and Cognitive Neuroscience of Art' discusses the dichotomy of nature and culture in the context of science and art/aesthetics spheres to reflect, on current and future perspective, how these spheres articulate interfaces in the fields of neuroesthetics and cognitive neuroscience of art. In Chapter 6, 'Neuroaesthetics: Insights into the Aesthetic Experience of Visual Art', Ana Teresa Ribeiro Contier and Laila Brito Torres (Brazil), from studies in neuroaesthectics of visual art, address the aesthetic experience in relation to intrinsic neural architecture of the human being, involving sensory-motor areas, emotional centers, reward system, memory and language. The authors also discuss some important findings in this research context from three artworks of the world famous Brazilian modernist painter, Candido Portinari, (1903-1962): the drawing "Crib" (1931); "Denise with the cat" (1961); and "Portrait of Maria Jordão Arruda and her daughter" (1926). In Chapter 7, 'Understanding the Interdisciplinary Meaning of Beauty to Neuroscience: Designing Beauty to Neuroscience', Bruno H. S. Araujo and Ibrahim Elias Nasseh (Brazil), from the classical view of beauty, provided by the philosophy of proportion, harmony and symmetry notions in opposition to the subjective construction of the concept of beauty by art. Thus, they discuss the multiplicities and ambiguities of this concept from the perspective of neuroscience. And Paula Landim (Brazil) in Chapter 8 'Design and Emotion: Contributions to the Emotional Design' presents cultural aspects and economic situations surrounding the design concept in contemporaneity to discuss the interrelation of design and human emotion and its contribution to the role of designers in industrialization and globalization scenarios.

Finally, the third session 'Mapping and Designing Cognition: Artificial and Natural Perspectives of Learning and Memory in Design and iCinema' creates the collaboration between neuroscience and distinct projective development scenarios in six chapters. Leila Reinert (Brazil), in Chapter 9 'Qualia and Extended Field of Contemporary Design', builts the association between the qualia concept, as a fundamental aspect for a conscious mind, and the contemporary design to discuss and review the relations between artifact, user, and projected environment, understanding this association as a necessary articulation to improve the contemporary surroundings. In Chapter 10, 'Conceptual Experiments in Automated Designing', Petr Ivanovich Sosnin (Russia) brings the artificial intelligence approach and the analogies between the designers and scientists work when conducting experiments describing a conceptual experimenting in automated designing objects and systems circumscribed as a Software Intensive System in which the objects are units of designers' behavior focused on solving project tasks in conceptual design. In Chapter 11, 'Can We Induce a Cognitive Representation of a Prosthetic Arm by Means of Crossmodal Stimuli?', Mateus Sousa Franco, Bruno Almeida Salles, Tiago V. Ortiz, Henrique Alves Amorim and Jean Faber bring contributions to the product design, from the concept of extended proprioception sense and the ownership feeling of the prosthesis, discussing an experimental protocol focused on optimally inducing a cognitive representation of an upper limb prosthesis through a crossmodal vibro-tactile stimulation applied in the user's body. Geraldo Coelho Lima Júnior, in Chapter 12 'Cognitive Processes in Fashion Design-designing of modelling projects for the visually handicapped', approaches in the fields of fashion and neuro-education design, specifically the aspects of cognition, constancy and abstraction, to propose elements to the teaching-learning process of a sewing course that contribute to overcoming the difficulties faced by students who can see and students with visual impairments in understanding the transposition of a flat pattern of a garment to a mannequin bust. In Chapter 13, 'Cognitive Processes in the Reception of Interactive Short Film Script: Mental Representations by Audiovisual Specialists', Patricia Bieging and Raul Inácio Busarello approach the fields of interactive cinema and cognition, especially with respect to mental representations raised by audiovisual specialists by the antagonist character of a noir fictional story. In this context, the authors verify the impact that individual and distinct life experiences, and subsequent judgments, generate on the interpretation of the same character. In Chapter 14 'Understanding How The Mind Works: The Neuroscience of Perception, Behavior and Creativity', Claudia Feitosa-Santana emphasizes the importance of neuroscience knowledge in the Art and Design curriculum and brings #TheDress viral Internet phenomenon to discuss how context and experience influence human perception and how #TheDress Meme gave the opportunity to the popularization of neuroscience knowledge. Moreover, she discusses the neural basis of creativity and how the work environment could foster creativity, as well as how emotions are intrinsic related to the human choices.

In conclusion, our aim is that this varied and extensive mapping scenario of interrelations, reflections and discussions about the creative and projective processes of Art and Design in deep dialogue and intersections with the fields of Cognitive Sciences may in fact result in contributions able to be replicated and widely disseminated in order to help the contemporary society to better prepare the paths to be followed and expanded by the society of the future.

REFERENCES

Changeux, J. P. (1985). *Neuronal man*. Princeton, NJ: Princeton University Press.

Damasio, A. R. (2004). *Em busca de Espinosa: Prazer e dor na ciência dos sentimentos*. São Paulo: Companhia das Letras.

Freeman, W. J. (2003). O cérebro solitário. In *O livro de ouro da mente*. Rio de Janeiro: Ediouro.

Gazzaniga, M. S. (1998). *The mind's past*. Berkeley, CA: University of California Press.

Lundy-Ekman, L. (2004). *Neurociência: Fundamentos para reabilitação*. Rio de Janeiro: Elsevier.

Ramachandran, V. S. (2011). *The tell-tale brain: A neuroscientist's quest for what make us human*. New York: Norton.

Varela, F. J., Thompson, E., & Rosch, E. (2003). *A mente incorporada: Ciências cognitivas e a experiência humana*. Porto Alegre: Artmed.

Wilson, S. (2002). *Information arts: Intersections of art, science, and technology*. The MIT Press.

Zeisel, J. (2006). *Inquiry by design: Environment/behavior/neuroscience in architecture, interiors, landscape, and planning*. New York: Norton.

Zeki, S. (1999). Art and the brain. *Journal of Consciousness Studies: Art and the Brain*, *6*, 76–96.

Zeki, S. (2000). *Inner vision*. New York: Oxford University Press.

Zeki, S., & Lamb, M. (1994). The neurology of kinetic art. *Brain*, *117*(3), 607–636. doi:10.1093/brain/117.3.607 PMID:8032869

Acknowledgment

The editor of this book wishes to thank many individuals for their significant contributions to this book:

First, I would like to thank the members of the IGI Global publishing team: Lia Kambouroglos for inviting me to work on the book; Kayla Wolfe and Lindsay Johnston for kindly forwarding this proposal beside the Editorial Content Department; Jan Travers for all legal assistance; Brandon Carbaugh, Katherine Shearer, Keith Greenberg, Christina Henning, Amanda Smith, Henry Ulrich and Deanna Zombro for their kind and attentive support along this editing process.

I would like to thank Sara Diamond, the President of OCAD University for writing a Foreword, and all the co-authors who contributed to this book by writing chapters and providing images illustrating their researches and projects, for their cooperation, in order as they appear in the book: Rosangella Leote, Maria Manuela Lopes, Marta Menezes, Fernando Fogliano, Hosana Celeste Oliveira, Alexandre Siqueira de Freitas, Ana Teresa Contier, Laila Torres, Bruno H. S. Araujo, Ibrahim Elias Nasseh, Paula da Cruz Landim, Leila Reinert, Petr Ivanovich Sosnin, Mateus Franco, Tiago V. Ortiz, Henrique A. Amorim, Jean Faber, Geraldo Coelho Lima Júnior, Patricia Bieging, Raul Inácio Busarello, Claudia Feitosa-Santana. Also many thanks go to my colleagues and students at Anhembi Morumbi University who daily contribute to deepen my research.

I wish also to thank the members of the book's Advisory Board for their impressive cooperation to make this book become real.

Warmest thanks to my marvelous parents, João José Dias e Maria Zuanon Dias; my beloved godmother, Zilda Rosa Baptista and our angels; my heart's mother, Iracy Conceição da Silva; my sweet brother, Sidarta Zuanon Dias; and all my friends who bring happiness to my life.

Rachel Zuanon
Anhembi Morumbi University, Brazil

Section 1

Multidimensional Brain:
When Art Meets Neuroscience Meets Art

Chapter 1
Perceptual Processes and Multisensoriality: Understanding Multimodal Art from Neuroscientific Concepts

Rosangella Leote
Universidade Estadual Paulista (UNESP), Brazil

ABSTRACT

Perception seems to be an overly discussed subject in theories of Art, giving us the impression that there is nothing new to add. So much research and so many conceptions have been developed on the subject. Nevertheless, many holes can be perceived in these theories with regard to the mental process that operates in the perception phenomena. We have chosen to look to neuroscience for possible answers to these holes. In this paper, based on the knowledge of Vilayanur S. Ramachandran and António Damásio, among others, we focus on that which we are emphasizing as "perceptual processes". We will restrict ourselves herein to the perceptual processes of a multisensory nature that take place in the perceivers' relationship with the artworks, which contain multimodal stimuli, promoted by physical and digital interfaces of an assistive nature. Said perceivers, however, are both people with severe motor and vocal disabilities as well as those without these restrictions.

WHY?

In this paper we present the evolution of a part of our research that is being developed within the research group under our coordination - GIIP[1] in one of its lines of research, namely "Physical and digital interfaces for the arts: from diffusion to inclusion"[2].

The approach assumes the existence of poetic, not merely artistic, produced and/or perceived by people with special needs, but also by those who are "exempt" from this condition.

We believe that the perceptual experience involves both the enjoyment and the making of Art by any person, despite the diversity of physical, emotional and cognitive qualities that individuals possess. Based on that, we demonstrate that it is the relationship between self-perception, stimulus and sensing that qualifies the "perceptual process" [3].

DOI: 10.4018/978-1-5225-0510-5.ch001

Aspects of our perception are clearly important for the construction of our consciousness and hence of our world and understanding of Art.

As artists, our understanding of the perception phenomenon is important in many ways and for different reasons. To understand it, artist-researchers who objectively seek understanding come across and develop several methodologies, generally empirical in nature. Like them, we have followed a similar path based on a variety of fundamentals and reasons. Until now, our strategy has been to develop low-cost assistive interfaces that can be used by disabled and non-disabled people to develop and/or teach Arts.

The poetics, in our study, are those that present multisensoriality and multimodality. However, we seek to analyse and substantiate them through contributions from neuroscience. This contribution does not propose the exclusion of other analysis models. On the contrary, it aims to verify which contributions from neuroscience lead to a deeper understanding of perception, and can be associated with some of the better known theories for this purpose, used by the arts in general, namely, Semiotics and the Theory of Complex Systems. We have added phenomenology and Gestalt psychology to the scope of our work. Nevertheless, the relationship between theories is not discussed in this text, and the reason for this omission will be dealt with in this text.

In this study we favour those works qualifiable as artistic, i.e. aesthetic manifestations, in diverse poetics and produced by a multimodality of stimuli.

It became necessary for us to study neuroscience when we began trying to understand the processes taking place in the individual's mind when his or her perception is triggered, so that they can be verified through brain mapping technologies such as, for example, magnetic resonance (fMRI) and tomography (PET scan).

Although no scientific field has as yet been able to shed light on the cerebral operation whose occurrence generates these mental processes, is it through neuroscience that we can come closest to the quality of these processes.

It is clear to us that multimodality leads to multisensoriality, but the latter varies according to the acuity of the perceiver's natural sensors, or in consonance with the physical nature of the multimodal stimuli. We will later deal with the multimodal amplitude[4].

The myth that external or internal stimulus is only understood, qualitatively, by only one specific meaning according to the nature of the channel, has been weakened. The perceptual process experience is a continuous one. We live because we perceive. By identifying certain events that we give emphasis to during a significant experience in our daily living, we are revealing world events that we were forced or decided to experiment, in varying intensities.

The stimuli provide elements that trigger consciousness of sensations, as explained by Damásio (2000), captured by the various senses in a simultaneous operation, in the construction of the brain map that enables the stimulus that is highlighted by the perceiver to be identified in each perceptual experience.

All perceptual processes occur selectively. But those whose selectivity can be easily tested in humans, through several assessment tests, are hearing, sight, smell and taste.

We are selective up to the limit of our core consciousness (Damásio: 2000, 2004 e 2011). There are many other perceptual states that occur in a manner that is unintelligible to the core consciousness, i.e. they are in the extended consciousness. Neuropsychology has endeavoured to shed light on this fact. However, we still know more about the reasons that control what we can see even if by optical illusion or another psychophysical conditioning, than what we cannot see, even though it is in our perceptual field. It is highly unlikely that we will develop instruments or our extended consciousness enough, to the point of being able to take in perceptual events in their entirety. If this occurred, it would be an anomaly. Most of the perceptual choices we make, i.e. selecting from among the stimuli offered to us, become lodged forever or for quite some time in the less accessible layers of our consciousness (in the extended consciousness).

Despite being choices, our measuring instruments are not able to present the mind process that experiences the event intensely and makes such choices. We know about the existence of parts "that have been suppressed by the memory" when, for whatever reason, unintentionally or induced (i.e. therapy, hypnosis, medications and other drugs) we re-examine the events and retrieve certain suppressed memories of things we did not realize at the time of the experience. The nature of trauma is like this, and also of subliminal messages

Hi-Tech has made it possible to capture information with significant details about the areas of the brain[5] that are activated depending on the stimulus provided. However, this information, of a physiological order, is not by itself efficient at showing WHAT thoughts, feelings and emotions are and hence the perception of the world and HOW they are built. Science understands a large part of the "WHERE" this takes place. For now, we will approach the first two because it is likely that a long time will pass before scientists can take us to the innermost recesses of the human mind.

EXPOUNDING ON THE WHAT

The concept of multisensoriality is explained by the simultaneous activation of multiple natural sensors in the interactor's various levels of perception. This concept evokes that of multimodality.

This term generated our concept of "interactive multimodality" used to describe artistic and non-artistic works, where multiple modes of stimuli, together, trigger this perceiver/interactor's various natural sensors within the testing space, so that the perceiver is conscious of this multisensoriality.

We can extend the notion of multisensoriality to comprise both natural and artificial sensors, (which are extensions of the natural ones). In any case, consciousness is at the same time a filter and organizer of multisensory states.

Although multisensoriality is present in any event experienced in our daily lives, the examination of the event, within an exhibition space, requires from the artist or proponent of the work, a better understanding of how it happens. It is also necessary to guide the user's cognition towards an interaction that is more coherent with the Multisensoriality model that the work intends to achieve.

When, in the perceptual event involving a work of art, it does not occur to us to pay attention to the technological device, or even to the scientific knowledge used, and these take second place, it is the poetic that is highlighted. This is what the majority of artists that produce in this field desires. If this happens it is often said that the technology became invisible. Although technologically impossible in our era, it is a plausible qualification. Even if our technology becomes literally invisible, there will still be the perceptual processing that attracts our attention, including the technological structure that contributes, with its share, to the work.

We would venture to say that there is a straight and horizontal line leading multimodality to multisensoriality. But the same cannot be said between stimuli and sensing. This is because one stimulus may trigger several sensors simultaneously.

Our focus is purely a resource to understand how multimodals operate and how multisensors respond.

In fact, scientists are increasingly unravelling how human senses function and have been convinced for a long time that the division into five senses, based on Aristotle, is limited. For present-day Sciences, and depending on the theoretical line, more than 30 senses may be listed. However, if we were to take into account the natural sensors that we know of, including those of animals, the number would increase greatly (i.e. thermal, photo, mechanical, chemical, etc.). And, we haven't even touched on the so-called "extra-sensory senses." If we accept the thinking of António Damásio (2000, 2011), although he did not discuss this matter, we infer that perception of an "extrasensory" nature is also the result of the organization of the brain maps, which reflect different states of consciousness, therefore, everyday occurrences in the brain that the individual may be made more or less aware of. But this is a controversial issue, as we cannot measure the stimulus that generates the extra-sensory perceptual processes.

We, of course, agree with the idea that the division into any number of senses is limited, but we do not have sufficient knowledge to concur with any specific number of human senses or, likewise, to discuss a subdivision about which not even scientists have come to a consensus.

As this is not our aim, what interests us, for now, is to examine the multisensoriality that reveals itself to perception at some heightened level of consciousness, produced by various modes of stimuli.

Multimodality can be understood as an immeasurable abundance of stimuli, which are offered by the world we create, to our consciousness.

Nevertheless, it occurs to us that as the worldview and the construction of reality is the work of each individual, in line with his or her Umwelt[6], it would be impossible to quantify both multimodality and multisensoriality.

However, as a work methodology, we reduced the breadth of meaning of multimodal stimuli to focus on those who are both easier to be recreated and also perceived with our natural sensors. Similarly, we study the natural sensors that can be simulated by the technologies available in our contemporaneity.

Thus, we distinguish stimuli related to certain natural sensors, with or without the intervention of artificial ones. The choice is linked to the modal range we can observe in the hybridized field of Art/Science/Technology. Therefore, to us sight, hearing, smell, touch, taste, proprioception and kinaesthesia are relevant.

In this part of the research, we highlight sound and visual perceptions.

The Deaf Perceive Sound

Several studies and devices that were created to replace senses that are barely or not in the least functional have been around for a long time. We have been acquainted with some of these devices for so long that we barely pay attention to their purpose linked to perception, such as eyeglasses and hearing aids.

Although our society is "moved" by images, it is sounds that reach us, in most cases, without invitation. So that we do not hear them we need to carry out an effort of psychic disconnection that is greater than the rapid action of closing our eyes, which we do so as not to see.

Sound information is a type of signal that is converted by ear canals into a auditory brain map and captured by our consciousness as a code that can only be decoded in greater detail by those who hear it. But it is actually received in greater intensity by the sense of touch than by the sense of hearing. This invalidates the notion of deafness, for the deaf perceive sound.

The range of sound awareness that people have is greatly reduced in comparison to the sound frequencies that we know through physics, mathematics and chemistry. Each person develops his or her own abilities to perceive in the space of frequencies combined with amplitude, between the bass (about 20 Hz) and treble frequencies (about 20,000 Hz) that are not solely responsible for our understanding of sound.

The condition of the environment where the sound source and the perceiver are located, contributes both to the sound's qualification, and to a higher acuity of its perception. By the route taken by the sound to reach our body we can identify its course and origin, as well as balance and locate ourselves in the space in relation to it. Sound can be perceived without the person's ear being able to catch it. Sound frequencies are received competing with other stimuli witnessed and activating the natural sensors available, promoting understanding of "the existence of sound", although without qualification and importance beyond the auditory fact. It is the ear that allows the body to balance itself.

Therefore, a malfunctioning of the ear may affect an individual's proprioception, the body's notion of balance, without this having a direct bearing on his or her ability to hear. We say direct because, as we have seen, other aspects cause sound to be perceived.

We have heard about deaf people's ability to develop rhythmic sensitivity to dance or music, through the tactile perception of sound (SACKS, 2003). On a large-scale, everyone perceives sound's vibration in his or her body, when it is high enough. Tolerance to pitch also varies from individual to individual.

Sound establishes a path that touches the whole body by the impact of the sound wave. But it only becomes relevant when this impact reaches the most appropriate channel for its conversion into a auditory brain map, which is the ear. We are barely aware, however, of the tactile role that the hearing process plays, with the simple reverberation of sound in the eardrum.

This is one of the reasons why some people tend to listen to music with the headphone volume turned up too high, despite the decibel control system installed in the sound reproducing equipment. People, when using headphones, fail to perceive the sound's reverberation throughout their entire body, therefore they tend to compensate for the lack of this tactile experience by increasing the volume.

Research and artistic work done by the Spaniard Josep Cerdá[7] use what he calls inaudible sounds to show other sound frequencies, not perceived by our senses. He argues that no sound is repeatable, however controlled the conditions are for its repetition. That is, the frequencies, both those inaudible to human senses and those not perceived by our attention at the time of the experience, but which are constructive parts of the listening experience, underlie the sound event. Indeed, we agree that the soundscape underlying any perceptual sound experience will possess, unconditionally, unique and competent matrices to make the landscape specific. Here we are using two concepts jointly.

The first concerning soundscape in the way identified by Raymond Murray Schaefer (1992; 1997), can be summarized as the specificities of the sound conditions that can be found in that specific location, the space surrounding the perceiver. The idea comes from the theory of Acoustic Ecology, founded by him and his research group in the 70s.

The second refers to the concept of sound matrix, developed by Lucia Santaella (2001), who identifies it through semiotics in firstness state. According to her, matrices are elements that, by semiosis and hybridizations, will become sounds, per se, and may become the language of sound (musical or otherwise). The same would happen with the visual and verbal languages. All these matrices are potentially capable of developing into hypermedia language.

The reception of auditory information, in a different way from those who are not able to hear, only occurs after pressure occurs in the eardrum with each pulse of the captured frequency. Only after this pressure/tact information travels the entire path by a mechanical vibration process of the ossicles does it reach the cochlea, where it is converted into electrical signals that are directed through the auditory nerve to the brain. Other physical conditions of the perceiver exert pressure before this moment, but those shall be left aside for the purpose of simplifying the concept. Next, the encoded sound information becomes perceivable by the so-called auditory brain, the brain mapping area that predominantly interprets and brings the auditory perception to the core consciousness. To reinforce the previous idea, we could say that we would be "hearing with the ears".

But everyone's auditory acuity is not identical. Each individual following his or her own auditory route will always be "a little deaf" to certain sound conditions. Perceptual acuity changes with the variations of our bodily state that is directly influenced by external factors such as altitude, air pressure, temperature, number of people etc. Internal factors create interferences that are often determinant to the rest of a person's life.

There are also the transitory factors such as pain, fever, allergies, flu and emotional states that also alter not only our auditory perception, but also interfere with all our natural sensors.

This collection of characteristics that is formed individually and varies from moment to moment, is what we use to build our world and our aesthetic sense. With it we also complete the meaning of the works of art, in our case, the multimodal ones.

The soundscape is only one part of the multimodal spectrum. It is the soundscape's specificity and originality that induces the perceiver to multisensoriality. He or she is immersed in the event, even if unconsciously in part. It is in this immersion that we locate the qualification of the relationship between multimodality and multisensoriality. Note that there will never be control over how or how many of the interactive modes will affect the perceiver, in order to imprint on him or her the consciousness of multisensoriality.

We Reinterpret the Image

The "visual brain," as described by António Damásio (2000) and Margareth Livingstone (2008)[8] builds the intelligibility of the visual stimuli that reach the retina in the form of light pulses. Part of the process is optical; another is mechanical and the other neurological, until the image map is formed in the mind. This investigation also focuses on the visual map. Just as with sound, it is not only what is "available" to sight that is interpreted in the layer of core consciousness. This visual map resides only in the mind. In fact, most of what is absorbed in the perceptual process is unknown to the interpreter.

Our vision is very selective. The area that we see is limited to the arc that the eye draws when scanning the environment and to our ability to approach the viewed object. This arc varies according the anatomy of the perceiver's eye and face.

Apart from the perceiver's physical conditions, all the properties of odour, taste, volume, temperature and light reflected from the object that are available to sight, come together to build the perceptual dynamic. But it is not enough to perceive the physical information to establish judgement about them, as this occurs in the core consciousness. So we say that the individual is conscious of the perception made. If there is any perceived light, a visual perceptual process is going on, even if it does not allow the person to identify the general qualities of the stimuli at work in this sliver of light. Ultimately total blindness is the extreme inability to perceive light.

NÃO SE APROXIME / DO NOT COME CLOSER

We have aimed for transdisciplinarity, by using theoretical bases and studying our artistic production.

We are interested in how interactors deal with contemporary art installations. This issue is crucial to us and demands constant investigation that includes analysis of the relationship between art, science and technology and the perception of these interactors in these transduced and contaminating fields.

In order to demonstrate a part of what we perceive and use in the field we shall describe the interactive installation called "Do not come closer" (Figure 1), which we developed in 2013. It is a work that stemmed both from conceptual research and artistic production. The work aims to promote interaction with two video clips that overlap on the same projection area, The progression and fusion between them is computer instructed by the contact of the interactor's hand on a surface sensorized by ultrasounds. We used low-cost technology and open source programming. Although we are the authors of the work, it was developed alongside some of our research partners[9].

We changed the project in 2014, when minor alterations were made to the program and in the arrangement of the ultrasound sensors used to detect the interactor's presence and actions.

The performer's (this author) image is projected on video, on a narrow and vertical screen. There a fusion occurs between the clothed body that invites interaction and the naked body, whose hands are suspended and cuffed. In the two videos the body turns 360 degrees, slowly ever looping. The plot entails a performer wearing a long black dress, who with gestures invites the interactor to come closer. If the interactor does not move closer, the image is modified using ultrasound sensors applied to a demarcated area at the entrance of the installation, where the interactor places his or her hand. Otherwise, the video image of the clothed body is maintained. Therefore, it is only at this spot, away from the screen, that the interactor visualizes the body's transformation when disrobing, in the fusion programmed between the two videos. In other words, when trying to get closer to the screen, i.e., near the naked body, the interactor is obliged to remove his or her hand from the sensorized area, causing the first video to keep on playing. In summary, drawing closer prevents the vision of the naked form.

Symbolically, the interactor's bold proximity reduces the possibility of the shackled creature's release. Any approach results in the destruction of the original meaning, creating ambiguity about the installation's intent. The fruition zone[10] is not limited to the voyeur aspect.

The approach that "Do not Come Closer" introduces is in the context involving the interactor's participation at the installation, where our gaze is turned to the problem of visual and sound perception.

Figure 1. Não se aproxime (Do not Come Closer)

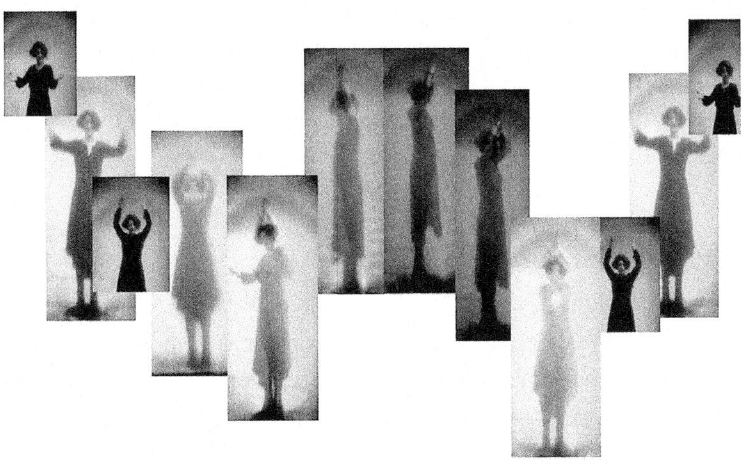

Obviously, the engagement of the visual sensors is the most prominent mode of stimulation. Without any sound included, the work leads to immersion and sound perception, both because of the "silence", and the space. It encourages spatial perception, due to the size of the dark hallway (where, at the end, the bluish image is projected) and the invitation to the visitor's journey to the screen; olfactory perception regarding the materials (wood, fabric, paint, metals, electronic components) and bodily odours of other interactors and/or the interactors themselves; tactile perception, resulting from the necessary action of touching the sensorized area, as well as other perceptions whose nature depends on the interactor.

The installation "Do not Come Closer" is, in fact, soundless. The purpose is to make the interactor immerse himself in the image, while at the same time realizing that the lack of a sound source is not silence. On the contrary, on noting the lack of a sound source in the work's syntax, the auditory sense's perception is highlighted; the impression of space and the relationship with the image is changed. Despite the prominence that the visuality has in this work, the image is not treated as the main component. What interests us is the relationship that the individual who perceives it has with the image, the space, the silence and the prompting to a particular type of interaction, even if it does not occur (i.e. placing a hand on the sensorized area).

With the continuous reduction of equipment and components, investment in these researches can be more rewarding, from a marketing perspective, accelerating the accessibility of these devices to the needy and interest parties. The price of the technology developed by large corporations is still very high, but the bottom up movement, permitted by this miniaturization and, similarly, by the popularity of open source, has reverted their knowledge to another field of application, one with lower cost equipment. The proposed production of devices for technology focused on the arts that we are developing fits the second (but not only) opportunity. With open source and hardware with low-cost, easily customizable components, we can reach a wider and less privileged social group in economic and social terms (Figure 1).

As in this work, an important feature that we perceive in the artworks using emerging media is the ability to conceive an expanded constitution of the work's identity. We describe this at another time (2008) when we discuss the expansion of the work's identity in greater detail. However, to summarize the thinking, we say that the work that has technological aesthetic characteristics can have its body or structure expanded to the point of involving electronic spaces (virtual) and physical spaces (lived). By experiencing theses spaces (virtual and lived) we immerse ourselves in them.

We must differentiate these spaces (virtual and lived) and consider that while together they compose the same sensation of lived space, relative to places that we hold dear, separately they represent different levels in the form of experiencing. With Bollnow (2008) we learn that living and experiencing a space are different things. This difference is very important because it is experience that promotes the understanding of the concept of place. The more we experience a space, the closer it comes to the concept of place.

We believe that a work moves toward concretization when it can direct the interactor into making the installation space a place of his or her own.

ABOUT THE ANSWERS

Therefore, what is perception? We do not need to criticize or propose another referral to the logical model of interpretation of perceptual phenomenon made by Peirce's semiotics. We are fully confident that it is competent and even spectacular. As for Gestalt psychology (Wertheimer, Koffka, Köhler) although very coherent in its structure, it poses certain questions about the verification models of the experiments that led to the conclusions that gave rise to the theory. These experiments are still used today. Our criticisms dwell precisely in the somewhat hazy interstitial space between the modal stimulus and the brain map created. The theory was developed in the early 20th century; long before the technological systems of brain reading achieved the breakdown of the "concrete" brain operations available nowadays. While such devices are still far from providing full understanding of perception - and we will probably never reach that level with the kinds of technologies we know about - they can already give us indications of certain brain action specificitics that cannot be measured because they occur only in the mind.

Many of the conclusions from Gestalt analyses were obtained more through inference than the organization of solid data that could be placed in isolated conditions with the possibility of test reproduction and whose analysis would be predominantly objective.

Another factor that compels us to consider the need to carefully examine Gestalt theory is the concept of consciousness. Arno Engelmann (2002) argues about the existence of two types of consciousness: "immediate-consciousness" (true fact) and the "other's-mediate-consciousness" (probabilistic). This idea that there is an outside and an inside world does not reconcile with what we understand as consciousness based on our neuroscience studies.

Another incompatibility with our scientific basis of both semiotics and neuroscience is the concept of percept, introduced by some Gestaltists, which is understood as "characterizing the type of perceptual content"[11] of the consciousness (Engelmann, 2002).

According to our understanding, the difference between percept and perception in Gesltalt literature is not clear. We take percept as firstness while percept, as extracted from Gestalt Theory, would be a complete perceptual process.

We hope this research will help us to seek clarity and the connection between what we have already learned from the theories we use in art, especially the two aforementioned theories (Semiotics and Neuroscience).

As we have seen, in standard perceptual conditions, the senses, interdependently, provide our core consciousness with the assurance that we exist in full connection with our surroundings and create our world. But what is the difference in the amplitude of this guarantee in people who possess limited conditions of perception? Although at this stage of our research we are far from answering this question, we dare to propose that the very awareness of the lack of acuity or severe limitation of some sense guarantees the creation of one's own world, thanks to the action of the core consciousness, based precisely on these limitations.

Silent films, that strictly speaking never were so, can exemplify this. In the space of meaning, between the image and the lack of sound, there were endless audible perceptions coming from stimuli existing in the environment, not only from the orchestral music which accompanied the film, in most of the screenings. The soundscape and also the impression of sound created in the viewers' minds, coming from the sounds promised or imagined in the actions and dialogues in text form, built their listening experience. The film was not heard. The viewers knew, through memories and cognitive experiences, that in the film would reside sounds whose quality was intuited by each individual. In other words, each viewer created auditory brain maps, based on memory.

We understand that it is in this "intuitive" capacity, for lack of another word, that resides one of the origins of the qualia of sound, its absolute firstness. If we listen to a dialogue without images, preferably within a cinema's spatial condition, the same will occur with the image: we will create it, through brain maps, in relation to that sound. Touch, smell and taste are conducted in the same manner. Proprioception, on the other hand, receives less interference from external stimuli of this nature.

People who have completely lost the tactile perception of their body, or parts of it, due to trauma, may not feel it when one of their limbs or their entire body is touched, but be aware of the fact of the contact. Memory will retrieve the *qualia* in the experience. It can be said, then, that the existence of the contact is perceived to some degree. Touch, by its very nature, depends on the functioning of specific receptors (corpuscles and terminations).[12]

But perception, at least to the core consciousness, will always be "mistaken". We hope that António Damásio and Oliver Sacks can agree with this. As perception is selective, we can either not perceive or overperceive. This applies to the perception of pain or touch in limbs that have been amputated. These are the so-called phantom limbs, discussed at length by Sacks (2007, 1998) and Ramachandran (2005, 2012).

A perceptual experience can be carried out with people whose arms have a fully functioning tactile capacity, even though they haven't suffered an amputation.

Children usually do it as a game. Someone is asked to close his or her eyes tight and to stretch out either arm to the person applying the test. The tester slowly travels up the arm with his or her fingers, as if walking, gently, from the middle of the hand towards the anterior elbow crease (it can also be done with the tip of a pencil). The person whose eyes are closed must inform the other when his or her fingers reach the crease line. Interestingly most people indicate the contact on the crease long before or after it is reached. This has to do with the tactile receptors in this area because, on the forearm, relative to other areas of the arm, the number of these receptors is lower and its distribution is more dispersed. The back of the knee is even less sensitive than this area of the arm. In this case, identification of the touch on the crease depends not only on tactile perception. At the very least, temporal, spatial, auditory and also visual perception contribute to the experience. This is because, even with his or her eyes closed, a person can sense the amount of light received, i.e. the shadow of the person applying the test can interfere with the emission of light rays, indicating the other's approach; both the opening of the fingers and this "walk" can intuit the path travelled. The duration of the movement will also supply information about this path and the warmth of another person's hand provides indicators to the senses. But all this only takes place if the perceiver's consciousness of the limb corresponds to its physical quality and if the perceiver's cognition of such stimuli is profound. Now some of these perceptual details may not reach the core consciousness, and even so, the perceiver is able to locate the exact point of the touch on the crook of his arm. This means that the expanded consciousness has perceived the qualias of the experience.

But if there is no cognition, based on the body's memories, not even these qualias will be perceived. The body's lack of cognition occurs under special conditions, where the person has been deprived, even before birth, of said cognitive experiences involving these perceptions. The reasons are several, but those we have paid most attention to are related to congenital development problems and paralysis.

How We Can Conclude This?

In short, we are always ascending from one perceptual state to another (or interweaving, transducing them) until reaching the core consciousness.

That which is learned about the perceptual states (which occurs though imprecise states of involvement with what is being perceived) is what an individual uses to build his or her world, which is always under construction.

For now, semiotics provides us with us the logical fundamentals for the subject.

Both neuroscience, and neurotechnology and biotechnology present results that can be understood as predominantly "verifiable", regarding the problem of perception. As we know that none of these fields of study can as yet give us the definitive answer to what we could consider an understanding of the perceptual phenomenon, in the current phase of our work we chose to study perception using theoretical instruments from neuroscience. A future step will be to organize these results in comparison with the theories indicated above and, if possible, focus our work on the development of an analytical model combining all of them. This is still a hypothesis.

REFERENCES

Bollnow, O. F. (2008). *O homem e o espaço*. Curitiba: Editora UFPR.

Cairns-Smith, A. G. (1996). *Evolving the mind on the nature of matter and the origin of consciousness*. Cambridge, UK: Cambridge University Press.

Damásio, A. (1996). *O erro de Descartes*. São Paulo: Cia das Letras.

Damásio, A. (2000). *O mistério da consciência*. São Paulo: Cia das Letras.

Damásio, A. (2004). *Em Busca de Espinosa: prazer e dor na ciência dos sentimentos*. São Paulo: Companhia da Letras.

Damásio, A. (2011). *E o cérebro criou o homem*. São Paulo: Companhia das Letras.

Engelmann, A. (2012). A Psicologia da Gestalt e a ciência empírica contemporânea. *Psicologia: Teoria e Pesquisa, 18*(1). 10.1590/S0102-37722002000100002

Koffka, K. (1957). *Princípios da Psicologia da Gestalt*. São Paulo: Cultrix.

Köhler, W. (1980). Psicologia da Gestalt. Belo Horizonte: Itatiaia.

Livingstone, M. (2002). Vision and art: The biology of seeing. New York: HNA Ed.

Peirce, C. S. (1977). *Semiótica*. São Paulo: Editora Perspectiva.

Ramachandran, V.S., & Blakeslee, S. (2012). *Fantasmas da mente: uma investigação dos mistérios da mente humana*. Editora Record.

Ramachandran, V. S., & Hubbard, E. M. (2005). The emergence of the human mind: Some clues from synesthesia. In *Synesthesia - Perspectives from Cognitive Neuroscience. Oxford-NewYork* (pp. 147–190). Oxford University Press.

Sacks, O. (1996). *An Anthropologist on Mars*. New York: Vintage.

Sacks, O. (1998). *The Man Who Mistook His Wife for a Hat: and other Clinical Tales*. New York: Touchstone.

Sacks, O. (2007). *Alucinações Musicais*. São Paulo: Companhia das Letras.

Santaella, L. (2001). *Matrizes da Linguagem e Pensamento*. São Paulo: Iluminuras.

Schafer, M. R. (2001). *A Afinação do Mundo*. São Paulo: UNESP.

Schafer, M. R. (2003). *O Ouvido Pensante*. São Paulo: UNESP.

Uexküll, T. (2014). A teoria da *Umwelt* de Jakob von Uexküll. *Revista Galáxia, 7*, 19-48. Retrieved from: http://revistas.pucsp.br/index. php/galaxia/article/viewFile/1369/852

Wertheimer, M. (1971). O fenômeno Phi como um exemplo de nativismo na percepção. In *Textos básicos de História da Psicologia*. São Paulo: Herder e EDUSP.

KEY TERMS AND DEFINITIONS

Art and Neuroscience: Connection between art and neuroscience that involves artistic and/or scientific production and study, whose benefits can be shared between the two areas of knowledge even if only one is emphasized.

Multimodality: Simultaneous stimuli that are given to the individual, and that can lead to perceptual processes.

Multisensoriality: Simultaneous perception, through more than one of the individual's sensory receptors, whose reception may not be conscious in its totality.

Perception: Selective and adaptive capacity of beings endowed with neuronal functions, to gather information from the surrounding environment and the organism's internal functions through nerve endings. This phenomenon depends on the mental and biophysical qualities of the organism, when applicable.

Perceptual Processes: Event during which the organism's internal or external signals are configured as a perception.

Open Source: Open software technology, which allows different individuals to modify significant parts of it, without onus in relation to law and patent software. However, this type of distribution can establish copyright, on key parts of the software, which can not be changed without granting of rights from the author.

Qualia: First instant of the perceptual process that is triggered by mental processes which qualify in firstness, the emergence of a perception.

ENDNOTES

1 GIIP (International and Inter-institutional Research Group on Convergences between Art, Science and Technology) is composed of Research Group Leaders and the researchers linked to them as collaborators. Currently, 36 people participate in the group including researchers, students and technicians, all registered in the CNPq system, in addition to students that are teaching assistants and Scientific Initiation (IC) students without a scholarship. Researchers both nationally and internationally are also part of the network. GIIP's lines of research are 1 - Convergence between

Art, Science and Technology; 2 - Physical and Digital interfaces for the arts: from diffusion to inclusion; 3 - Digital and Mobile Technologies and their Applications; 4 - Nanoart. The group is certified by the UNESP and by the CNPq.

2 This line of research has five sub-projects. These results refer to this author's subproject entitled "Neuroscience and Perception: Multisensoriality and Multimodality'. It was developed as a Post doctorate research at the Universidade Aberta (Open University) in Lisbon, Portugal, under the supervision of Dr. Adérito Marcos, and research grant from the Fundação de Amparo à Pesquisa do Estado de São Paulo - FAPESP (Research Support Foundation of the State of São Paulo) in Brazil.

3 As explained in our paper/publication "Multisensoriality and Synesthesia: Possible Poetics?" In: Proceedings of the 6th International Conference on Digital Arts, ARTECH. Lisbon: 2012 and in the Revista Ars. Ano 2, nº 24, 43-61. São Paulo: USP.

4 By modalities we understand types of stimuli.

5 Here there is no concept of separation between brain and body. Our approach follows that of Damásio (2000: 61), for whom the brain is part of an integrated, live and complex organism.

6 Jacob Von Uexküll, a German biologist, developed this concept in 1909. He created the building blocks for the development of the theory of biosemiotics.

7 Cerdá is one of our research collaborators, a GIIP member. The artist's website: www.artesonoro.org. Access in January 2014.

8 The reader may notice that we are quoting Margaret Livingstone instead of Semir Zeki. We suggest consulting our publication "Multisensoriality and Synesthesia: Possible Poetics?" In: Proceedings of the 6th International Conference on Digital Arts, ARTECH. Lisbon: 2012, where we criticized the author's text and, as explained in another publication, although we accept his statements about the qualification of the visual brain, we disagree with some of his theoretical applications in the context of art.

9 Team: Rosangella Leote - Project, design and development; Rec(O)rganize (Rodrigo Rezende and Fernanda Duarte) - hardware, software, programming and video editing; Miguel Alonso - Installation Assembly Assistant. We used ultrasonic sensors, an Arduino microcontroller and Pure Data software on the Windows system.

10 "Zonas de Compensação" is both the name of the extension project and also of the annual exhibition held by the same project. It is organized by the GIIP, under our coordination, and is where the work was shown.

11 Emphasis added

12 Expanding the concept, Somaesthesia can be treated as a perceptual group where the submodalities known as touch, pain, temperature and pressure are found. Somaesthetic perception occurs thanks to receptors that are known as Meissner's corpuscles, Krause's corpuscle, Pacinian corpuscle, Merkel's corpuscle, hair follicle endings, Ruffini endings and Free Nerve Endings. This is a scientific classification, which we are not concerned with here, understanding that so far it has seemed, first of all, unnecessary to invest in this breakdown into minute detail, since we prefer to focus on multisensoriality with stimulus modalities that can be simulated by technologies. Secondly, because the research sources we have encountered up to now have not provided us with enough

clarity to get into this issue, as we have noticed excessive incompatibilities, especially in the classification of the senses. It seems, however, that there is consistency in the recognition of three kinds of perception: the exteroception (stimuli from outside the body), proprioception (the body's own stimuli - also called kinaesthetic) and visceroception, also known as interoception (stimuli of the viscera), although the latter is also treated as a specific sensing. The somaesthetic system as a whole includes sight and hearing (containing the information receptors that are outside the body). Despite smell and taste also carrying information from outside, these are not considered part of exteroception but rather visceroception, which once again, puts us in an uncomfortable position to accept this or that classification.

Chapter 2
Inside/Out:
Looking Back into the Future

Maria Manuela Lopes
Universidade de Aveiro, Portugal

ABSTRACT

Technologies, for memory preservation and enhancement of our humane bodies, are developing at a fast pace, and the corresponding dystopic and utopic future scenarios are presented in speculative news reports, science research studies and popular culture such as science fiction films. Lopes' artistic/research projects explore ways of expanding concepts of memory, body, and representation/mediation, using art as a tool to enhance public awareness of several anxieties and technologies, raising ethical dilemmas and questioning norms of behavior and normality. This chapter is an exploration of the issues raised on the development of several artwork projects during the course of the author's PhD and present Postdoc research, when in residency at several Medical and Scientific Research Institutes, dealing with distinct studies on memory (functioning, loss and enhancement).

INTRODUCTION

The author's PhD research – *Representational Strategies on Alzheimer's disease Studies: A Practice-Based Arts Research in a Neuroscience Laboratory* - was in the Instituto de Medicina Molecular and Hospital Santa Maria in Lisbon. This provided a unique fieldwork and complementary scientific/laboratory/artist studio in which to develop a progressive and situated practice-based inquiry into how memory loss (e.g. Alzheimer's Disease) is scientifically researched. Science and health philosophers Marc Berg and Annemarie Mol (1998) claim that medical and scientific investigation is based on practices of construction, assemblage and incorporation of techniques, habits, ideas, images and conversations. This represents L. concept of an archive, envisioned as a network of four intertwined 'archival spaces' (Figure 1):

- The patient;
- The art studio;

DOI: 10.4018/978-1-5225-0510-5.ch002

Figure 1. 'The four virtual archival spaces'
(L., 2019) Source: L., 2009

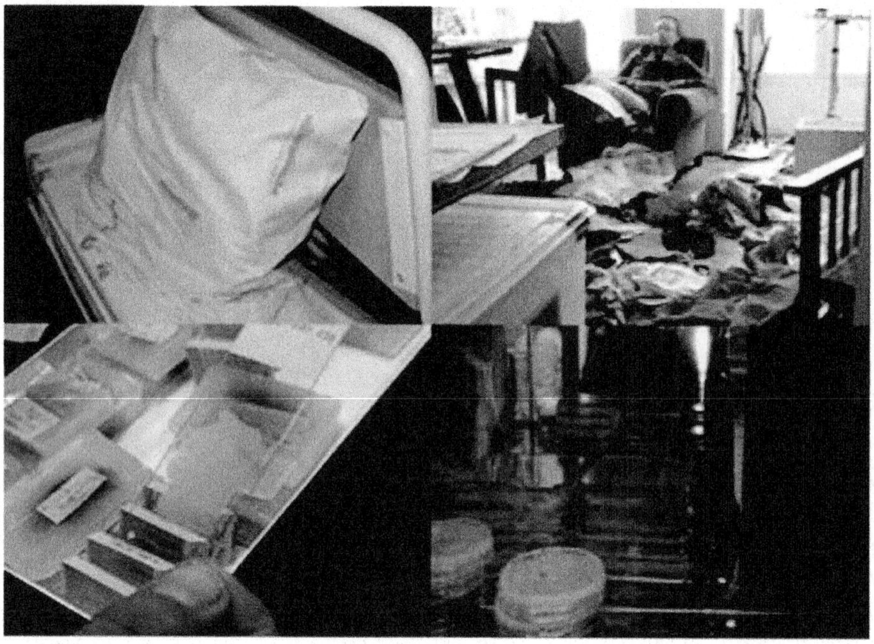

- The Alzheimer's research laboratory;
- The cellular and molecular research laboratory[1]

In this process the author documented the assessment and categorization of Alzheimer's patients and their exposure to various therapies. XX also documented molecular and cellular research laboratory procedures and collected laboratory materials. In order to maintain a common thread when approaching these spaces, methods have focused on neuroscientific diagnostic images (MRI and PET scan), graphs, diagrams, notes, models, photographs, instruments, physical space conditions and hybrid forms of these, and establishing a critical link to art practice. In laboratory studies these visual approaches are designated representational strategies or practices (Lynch & Woolgar, 1990: 1-18).

The human brain is a kind of an archive (L., 2013) that retains traces of our experience of the past, yet is often unreliable and arbitrary when we attempt to gain access to it. The images and experiences we retain, the places we have known and the feelings we have experienced tend to become fragmented and reorganised when our brain attempts to recall them. The result is a collection of subjective adapted recollections as, in the process of evoking past experience, our memory reconstructs, converts and updates it.

Aims of the Investigations

In the Doctorate period L. was aiming to develop strategies for evoking the nature of autobiographical memory within a contemporary art practice, by generating installations using a period of residency at Lisbon' University's Instituto de Medicina Molecular (Institute of Molecular Medicine); to explore the representational strategies of Alzheimer's studies in the scientific laboratory through art practice;

to explore and construct connections between visual art practice and scientific research in the field of neuroscience studies on Alzheimer's; and to further locate the author' art practice in relation to other artists conveying activities with regard to natural sciences laboratories and topics that overlap with autobiographical memory.

The purpose of both the laboratories and L. art practice is very much bound into an idea of montage, juxtaposition and circulation, where the images are assembled and transferred between locations as diagnostic, inscriptions, suggestions, or metaphors according to the transporter and distribution context.

L.' wonder whether in the stage following memory creation and preceding its transmission, memory does exist. Does a memory endure in matter, when a letter has not been read or an artwork is not experienced? What may then be the matter of memory? At present in a Post Doctorate research L.' is undergoing another endeavour that takes the form of a collaborative art/science research model. It is discursively located within the paradigms of memory, life sciences and visual arts. Within the process it allows the boundaries between aesthetics, philosophic and scientific disciplines to become significant in a relational network associated with the themes of remembering/forgetting, presence/absence, and concepts of self-identity/enhancement.

When in the neuroscience laboratories the researcher found visual significance, narrative and metaphorical richness in the neuropsychological assessment materials and in mobilized laboratory representational strategies. These representational strategies allow translations, and simultaneously evoke absence and presence (blots and gels instead of cells, cells instead of brains, diagnostic images instead of the patient's body).

Further the researcher' concluded from the previous research project that on an epistemological level, the bridging and reciprocity between science and art and interdisciplinary practice-based research projects may contribute to the introduction of a more open research paradigm in contemporary research. Nevertheless, the practicalities are complex and also depend on financial factors such as the need for support through scientific research grants compounded by the disadvantages of under-funded artists attempting to develop research collaborations.

L.' present inquiry (i.e. *Art Making with Memory Matters*) intends to extend her research approach and timeframe into a broadening field of study within the life sciences, with various outcomes exploring, interpreting and disseminating previous research outcomes and new findings into a more varied, enriched and non-academic audience. Thus this research is sustained by robust methods founded in her PhD practice in addition to unique ethnographic approaches within a molecular biology laboratory (Instituto de Biologia Molecular e Celular – Unversidade do Porto), botanical gardens and museums, and neuroscience clinical and pathology contexts (e.g. Newcastle University Brain Bank). The research is undertaken in ID+ (Instituto de Design, Media e Cultura – Univerdidade de Aveiro + Universidade do Porto) and IBMC (Instituto de Biologia Molecular e Celular – Unversidade do Porto).

Art Making with Memory Matters

Explores/expands on the representational strategies of memory studies in the scientific laboratory through art practice, alluding to the complex, fluid and multiple nature of memory and the studies on functioning/dys-functioning and enhancement/preservation of memory. This project is anchored in an International Research Project - NERRI[2] (Neuro Enhancement Responsible Research and innovation) an European research group on the introduction of Responsible Research and Innovation in Neuro-Enhancement in the European Research Area contributing to the shaping of norms underpinnings the ruling of neuroenhancement technologies.

Meaning and knowledge, sometimes as memory, seem to originate in a place that consciousness would like to disown. Thoughts, feelings and memories do not always present themselves as helpful to the daily projects of life or science, so are disregarded. If it were possible to archive and explore rather than to discard these random connections and impulses laying in matter, they might offer an interesting insight into other ways of conceiving memory and identity: as glimpsed fragments of how things could be and as images that touch us briefly and find an understanding and reaction.

L.' artistic installations and this chapter looks at the way in which artistic practice overturns traditional theories of technology as purely 'instrumental', theories of the subject in which identity is tied to the body, and the assumption that the will to self enhance is a typically humane condition. The researcher emphasizes that the produced and exhibited installation work embodies the power to communicate complex scientific and human concerns through art, and demonstrates the strength of using aesthetic strategies to present complex notions of disease, memory, archiving, clinic diagnostic, treatment or therapies, identity and the body, and also technology. L.' explores how we take for granted human enhancement methods from education or coffee, to the eyeglasses or cell phones, but than tremble when hear words such as nanorobot, gene-therapy, brain computer interface or cyborg.

BACKGROUND

Recent neuroscientific research, centered on the neurophysiological case studies of patients with brain injuries as well as cognitive studies on healthy subjects, suggests that memory is in fact a multiple memory system (Tulving, 1972; 1985). Memory seems to be not a single entity, but a group of different abilities mediated by distinct brain systems, such as the sensory cortex, the dorsal striatal, the amygdala and the hippocampal region. Each neural system holds differing connections and receives different information from distinct brain structures. The working of these neural systems is understood as both independent and cooperative, which explains dissociations in people with brain lesions, as well as the unified experience of memory in a normal person.

In result of its complexity, memory is researched in many different fields. Hence memory studies are a field of interdisciplinary inquiry that harmonizes intellectual strands from many domains, from anthropology, to philosophy and psychology. Recent discussions (L., 2013) focus on distinctions between 'types' of memory; representation, materiality and the mind; and the dialectical relationship of memory and forgetting. L. research contributes to memory studies agenda, as art became an important agency to the memory-work required by contemporary life and culture (Gibbons, 2007) by challenging ontological assumptions that underpin much recent research on memory, including understandings of biological memory, and the role of it's enhancement.

Recent discussions (L., 2013) on memory studies focus on distinctions between 'types' of memory; representation, materiality and the mind; and the dialectical relationship of memory and forgetting. Memory is understood to be fundamental to the constitution of the self /perception and understanding of the world (Nunes, 2008) working in tandem with external archives and aids of memory ranging from personal mnemonic objects to collective, social and public memory (e.g. gardens, museums).

In neuropsychology, when a test is performed on a patient's memory, the meaning associated with memory is, according to Tulving and Craik (2000), a neurocognitive capacity described as a three-stage process of encoding, storage and retrieval. The process of encoding the links to the initial absorption of information into the memory leaves traces in the nervous system that preserve the effects of experience across time. Storage is the maintenance of these traces. The act of remembering in neural terms is described as a network of connections among nerve cells (Kandel, 2006). Information arriving from the world is stored in the limited-capacity process of short-term memory, before being lost or transferred to a separate long-term memory system.

Autobiographical memory links us to places, people, sensory-perceptual details of events (episodic memory) and facts (semantic memory) such as when we were born, or where we studied. Cognitive psychologist Daniel Schacter (2001) views such information as personal interpretation rather than truthful record, asserting that the awareness and reconstruction of the individual's personal experiences depends on other mental capacities:

[…] retrieval of past episodes usually recodes, or updates, the information. Thus, active reconstruction, rather than mere retrieval, appears to be essential to episodic memory, and this necessitates the involvement of certain cognitive faculties […]. (2001: xx)

Psychologist David Rubin claims that many systems play a role in autobiographical memory:

[…] individual senses (especially vision, hearing, and smell); a multimodal spatial system, which notes the location of objects and people; emotion; language; a narrative system that keeps track of causal relations […] and an explicit memory system that coordinates or binds information from the other systems. Each of these systems has its own processes, forms of organization, and roles with respect to memory. (2005: 79).

L. proposes that this process in a multiplicity of systems implies a tendency to assemble, and may be evoked through montage, juxtaposition and video installation. The notion of linear time dominates our daily lives and is intertwined with narrative convention and construction of the self. Narratives allow us to link past, present and future, and memory, due to their dynamic functioning around encoding, storage and retrieval, implying an assemblage capacity that binds fragments to experience and imagination. In Alzheimer's disease communication within this memory system seems to be disrupted; the assembling capacity therefore seems to be shattered.

The late or scarce physical symptoms of Alzheimer's disease prompt L.' interest in how the intermediation of the technological image substitutes for some clinical sensorial examination of the body, and how certain laboratory practices turn the biological materiality of the loss of memory into a visible phenomenon. Analysis of the researcher' working methods identified the need for some studies on the status of images, visuality and representation, due to the interdependency of my ethnographic exploration on vision and technology. This reflection served to critically consider the author's voyeuristic position, using the artistic gaze on others' practices and diseases to question the roles of observing, archiving and reconstructing. As an example, Jonathan Crary (1992) focuses on the embodied observer's ways of looking; examining how optical devices shaped practices, including the *camera obscura*[3], panoramas, magic lanterns, kaleidoscopes and stereoscopes, thus ascribing supposed neutrality to the technologies leading to codified modes of viewing. The concepts of 'visuality' and 'scopic regime' are often used in discourses around image and vision and, according to Gillian Rose (2007: 2), refer to ways in which both what is seen, and how it is seen, are culturally constructed.

Reflecting on the idea of 'looking at' in institutions, French philosopher Michel Foucault explores the ways in which institutions such as schools, prisons, hospitals or asylums depend on several forms of surveillance. In *The Birth of the Clinic* Foucault argues that the mechanical-clinical gaze transforms our bodies in sites where organs and eyes meet (2003: xiv), detaching the body from the person, or 'externalizing the internal'. He explains that this gaze allows a passage from small-scale practice to an archival kind of functioning, substituting body presence for large manipulation of coded records permanently present. This approval (Struken and Cartwright, 2002: 281) of the instrumental image as a provider of access to invisible truths is linked to an increased use of institutional regulation and categorization or archiving. This allows the clinical gaze to extend from the individual body to the wider population, and a stronger weaving of technology, culture, medicine and society.

In clinical practice, perspectives similar to those adopted by Crary or Cartwright assume a supposed neutrality of the technologies and ascribe objectivity, methodological consistency and precision to the images[4]. Representation is the substitution of symptoms and behaviors reported by the patient. Images are seen as essential to systems of power[5] and concepts of knowledge and self-regulation. Foucault uses a powerful metaphor to explain this idea: the panoptic prison[6] functioning as a reminder of the permanent exposure to visibility to which one is from then on subjected.

The panoptic mechanism arranges spatial unities that make it possible to see constantly and to recognize immediately. (1995: 200)

As Foucault suggests, scientific and medical understanding of the body focuses on surveillance using practices of assessing, recording and archiving to classify and define the boundaries between illness and health. An apparent organization of the sensorial perception of space and time thus articulates power relationships. According to Struken and Cartwright (2002) this looking is linked to other activities that ascribe meaning to what vision uncovers, such as experimenting, measuring, analyzing and ordering. The panoptical gaze surveys the human body, which for the purposes of the clinical investigation becomes transparent; the invisible disease becomes accessible; volumes, spaces and time become visible in representational strategies, both portable and juxtaposable in a translation format. Vision prevails but is dependent on other sensory and cognitive processes, and also on context.

The most radical example of this practice was the development of the X-ray as a medical representational strategy. The opening of the clinical gaze to a wider audience is explored in several studies that focus on transparency and approach of the body's interior, an experience only possible via X-rays and medical imaging techniques such as ultrasound, MRI and PET scans. Joseph Dumit[7] (2004) examines how identity is constructed visually through PET scans, arguing for a direct link between a person's behavior and symptoms and the image representing it. Dumit's study focuses on the shifting meanings and uses of these images when they move from the laboratory to contexts such as magazines or scientific journals. In a parallel methodology L.'s research also explores and re-frames the representational strategies on scientific studies on memory outside their usual confined paths.

In *Practices of Looking* (2002) Cartwright and Sturken explore how images are observed and by whom. They present a social and cultural association between truth, knowledge and vision intertwined in the scientific representational strategies[8]:

The idea that the truth lies beneath the surface, and needs to be seen to be fully understood, has predominated in Western culture since of the Greeks. It is a common sign in contemporary culture to use the image of looking inside someone as seeing their 'true' identity. (2002: 298)

Several studies have focused on visual strategies common to science, medicine and art, questioning the externalization of the invisible body and the relationship between seeing and knowing. These include Barbara Stafford's (1991) Enlightenment Era survey on pictorial common practices such as abstraction and marking; Bettyann Kevles's (1997) study on the social and cultural impact of X-ray technology; José Van Dijck's (2005) exploration on the myth of the transparent body undamaged and ready for improvement; and Robert Zwijnenberg's (2009) reflection on how scientific knowledge is transformed in art, questioning integration of these new image technologies in the construction of identity.

The relationship between technology, science and nature is explored in Bruno Latour and Steve Woolgar's seminal study of 'laboratorial life', and Latour's studies around action in science[9]. Latour suggests that science is a process of addition of inscriptions, which primarily works through study of images, graphs, maps, tables and text data rather than the real phenomena. He emphasizes the advantage lent by such images and inscriptions over scientific claims, with a focus on the whole process of mobilization surrounding their production. The invisible truth may be sustained by showing an image:

We are so used to this world of print and images, that we can hardly think of what it is to know something without indexes, bibliographies, dictionaries, papers with references, tables, columns, photographs, peaks, spots, bands. (1986: 13)

Latour's analysis suggests that observing means constructing and bringing into being. This applies particularly to technologies such as MRI and PET scans, in which observation by construction is mediated and functions at the very boundary of what can be conceived as visual[10].

Latour (1987) emphasizes the central role played by images such as inscriptions and material artifacts in consolidating the networks established between laboratory and the outside community. He calls these delegates 'immutable mobiles', which should provide important information of the places and people concerned.

Experience in the clinical and scientific spaces is oriented to observation, collection and video and photographic captures of repeated gestures, sentences, materials produced, archival strategies; this includes recording the neuropsychological assessment sessions and part of the clinical research and aftercare, as well as digitalizing the material outputs of their direct performance (drawings and handwritten narratives). This process is considered an enactment of the archive.

L.' has recorded herself undergoing self-assessment sessions in the studio space (comparable to the ones patients undergo on their clinical trials), which involves the production of drawings, hand-written sentences, sound and gestures. These sessions included revisiting recordings of the laboratories and patients, as well as visual materials produced, and contrasting them with the recordings of myself, and their respective outcomes. This process is considered a re-enactment of the archive.

Decisions about editing and juxtaposition were informed by reading neuroscientific studies on autobiographical memory and Alzheimer's disease, which gave clues to editing rules. This re-enactment of the archive was intended to establish a critical gap between the original scientific coordinated performance and the parallel cultural and subject-driven practice.

Currently the author is exploring studies on enhancement strategies (such as traditional mnemonics practices explored in diverse workshops[11] and also collaborating with scientific research groups that investigate on drugs and technologies and its relations to potentiation of cognitive functions. Enhancement is not a new subject, according to Vincent Clark and Raja Parasuraman (2014), Humans have long used several cognitive enhancement methods to expand the competence and range of the various mental activities we perform. Enhancement is not limited to physical functions and qualities; scientists are using neuroprosthetics to mend (cognitive) deficits from dementia, stroke and other brain injuries. Nor is it necessarily limited to enhancing biological human functions and abilities. In terms of cognitive enhancement or, Neuroenhancement it

[...] describes the use of neuroscience-based techniques for enhancing cognitive function by acting directly on the human brain and nervous system, altering its properties to increase performance (....) This includes transcranial electromagnetic stimulation methods, such as transcranial direct current stimulation (tDCS) and transcranial magnetic stimulation (TMS), along with deep brain stimulation, neurofeedback, behavioral training techniques, and these and other techniques in conjunction with neuroimaging. (Clark & Parasuraman 2014, 889)

In this view, neuroenhancement implies operating directly on our nervous systems, to improve our cognitive function and increase our capabilities. Clark and Parasuraman in a recent scientific review on the subject conclude that we are living in a period where scientific research has provided theories and methods to transform our dreams in reality. The methods mentioned above and others described by Eduard deBono Foundation, such as lateral thinking or creativity or many drugs promoted by pharmaceutical companies can be used to improve, attention, perception, memory, creativity and other forms of cognition or even reduce the duration and costs of treatment in mental health patients.

Hub Zwart (2014), resumes the scholarly debate around the themes of neuroehnancement referring post-humanism, bio- conservatism and the work of Peter Sloterdijk as the most important examples. Zwart claims that until recently the debate was polarized between the authors in favour of a complete technological remake of our human capacities (trans and post humanists), and the opposed defenders of the natural evolution (bio conservatives). Zwart calls attention to Francis Fukuyama (2002) concerned evaluation of the possible alterations of the brain as matter and the human personality by the intake of the future genomically targeted neuropharmaceutics.

This position echoes Brem et al. (2014, mentioned in Clark and Parasuraman, 2014) suggestion that cognitive enhancements might also come with some costs, leading to broader, long lasting changes in cognition and brain physiology. Science fiction and current research on soldiers and other practices, such as 'smart-drugs' taken by University students (e.g. Ritalin, methylphenidate and modafinil), are suggesting that instead of addressing a lost function, neural interfaces could allow the healthy to gain some new function. Visible signs are in the public domain since adverts are marketing in-home 'trans cranial direct-current stimulation', which supposedly has a cognitive-enhancing effect.

While some researchers are investigating (Wilson, 2012) the possibility of increasing human intelligence by surgically implanting genetically engineered tissue into our brains, employing sophisticated brain to computer interfacing technologies (i.e. BCI) and genetically engineering human embryos, others are exploring immortality through 'mind uploading' – the process of transferring the mind from a human brain into a digital, analogue or quantum computer.

Peter Sloterdijk (2009, cited in Zwart[12]) places the debate in a different platform considering the will to self enhance as typically humane condition and analyses this history of 'self-improvement' in a broader historical perspective extending Nietzsche and Foucault's views on our dependence of self improvement. In line with his review of Sloterdijk' self-enhancement as human trait Zwart (2014) calls attention to our neurological plasticity and the way our brain networks are "sculpted by our socio- technical environments.

To sum up, L. (2013) claims that scientific and medical investigation is based on practices of assemblage and incorporation of techniques, habits, ideas, images and conversations (i.e. representational strategies). Representation is the substitution of symptoms and behaviors. Life science studies focuses on surveillance using practices of assessing, recording and archiving. Looking is linked to other activities that ascribe meaning to what vision uncovers, such as experimenting, measuring, analyzing and ordering. Observing means constructing and bringing into being. The researcher demonstrates that this process may be evoked through montage, juxtaposition and video in art installations. Truth, knowledge and vision are intertwined in the scientific representational strategies.

Hence, if concepts such as truth, vision and knowledge are already complex in science and art fields, they seem to be contested in the future scenario when vision may be altered towards a more cyborg like performance (with x ray or infrared possibilities). In the near future, neural implants could improve our ability to perform physically and mentally, at present researches are exploring ways to improve athletes' performance with gene doping (i.e. enhancing performance by adding or modifying genes), creative surgical enhancements (e.g. using skin grafts to create webbing between a competitive swimmer's fingers and toes) and mechanical prosthetics (e.g., the prosthetic legs used by double-amputee, athlete and top model Aimee Mullins). We were recently presented with the exoskeleton that allowed the paraplegic to kick the ball at the beginning of the Brazil world cup in 2014; the BCI suit was developed by more than 150 researchers led by Brazilian neuroscientist Miguel Nicolelis.

MAIN FOCUS OF THE CHAPTER

Art Making with Memory Matter

We are facing a moment in time when our curiosity and technical possibilities (alongside some anxiety and fears) are propelling us to move from using the created techniques to restore our bodies to "normal" functioning, to using them to increase our innate abilities and, perhaps, add new body and mind capacities. L.' believes that, our use of enhancement techniques will increase in frequency, in the sophistication of the technologies and in the purposes for which they are currently used.

It's difficult for humans to live in the present; we are mind wired to be anxious about the future, and to dwell upon the good things of the past. These instincts allow us to carry on, since overestimating negative future possibilities better prepares us for them, and having nostalgia for the past suppresses memory of pain, hardship, making life overall seem better and more worth investing in. As the author investigates more about emerging technologies, alongside the experience of the Doctorate research in connection with the Dementia patients (that showed the fragility of memory and cognition), L. located what was looking for in terms of ethics, philosophy, aesthetics and the ability to have tangible tools for bettering the world. The author instills the spectator's natural fire in them to drive innovation, cooperation and life preservation, The provocation and complexity presented in the installation art works intend to ensure innovation gets pushed forward, defending empathy, better communication and outreach for these fields. These works are ultimately about the fragile and contingent nature of memory and human futures; they invite the audiences to ponder the different dimensions, costs and unintended consequences of enhancement. L.' work extends and challenges this entwined relationship.

The unique method/structure of the research presents an innovative holistic model for practice-led research: negotiating the interests of the involved institutions and NERRY (Neuro Enhancement Responsible Research and Innovation) project, schedules, bureaucracies, funding, and the public dissemination of its results in educational workshops, conferences, performances and exhibitions. It brings together understandings of the institutional and cultural framing of visual strategies, archiving and therapies; interrogating the possible application of scientific practices/inscriptions in subjective/visual discourse on memory but also assessing how are the novel technologies and practices (or the knowledge of their existence) for neuroenhancement reaching the general public.

The installation art projects [*The Approach* (Figure 2); *The Assessment* (Figure 3); *The Archive* (Figure 4), and *The Therapy* (Figure 5)] developed at the previous research period, were concerned with the referred aims and confined to the universe of neuroscientific study on memory loss (Alzheimer's disease studies); while the following art projects (*Platonia; Box, Shadows and Definitions*, and *Enhancing the Mind's I*) correspond to a broader universe of research on different areas of memory, from technologies and techniques for memory enhancement, to strategies of acquisition (astrophysics imaging), preservation (archiving, recording, sampling - Botany) and communication (drawing, writing). Testing the boundaries, mixing old methods (such as mnemonics) with the current new technologies (such as Brain Computer Interface or neurofeedback) L. makes the audiences wonder whether we should enhance ourselves, or seek to modify our children? What is being modified? There is no ground zero once many of the dimensions of human enhancement are evolving with the species subtly and pervasively.

Figure 2. The Approach. Exhibition A Arte do Cérebro, Centro Cultural da Figueira da Foz (L., 2009) Source: L., 2009

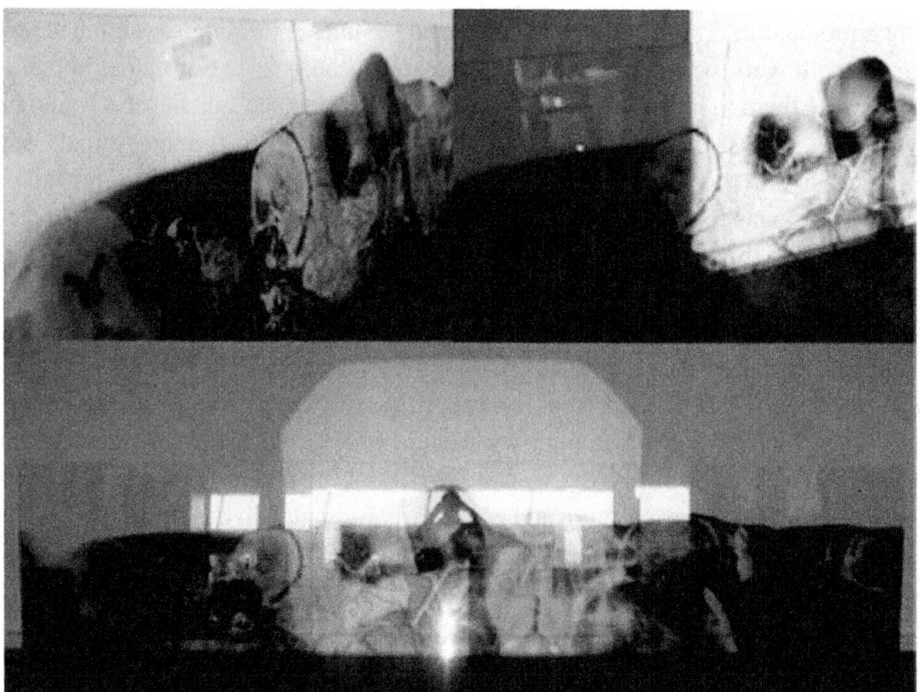

Figure 3. The Assessment. Installation trailer. Video stills
(L., 2011/12) Source: L., 2012

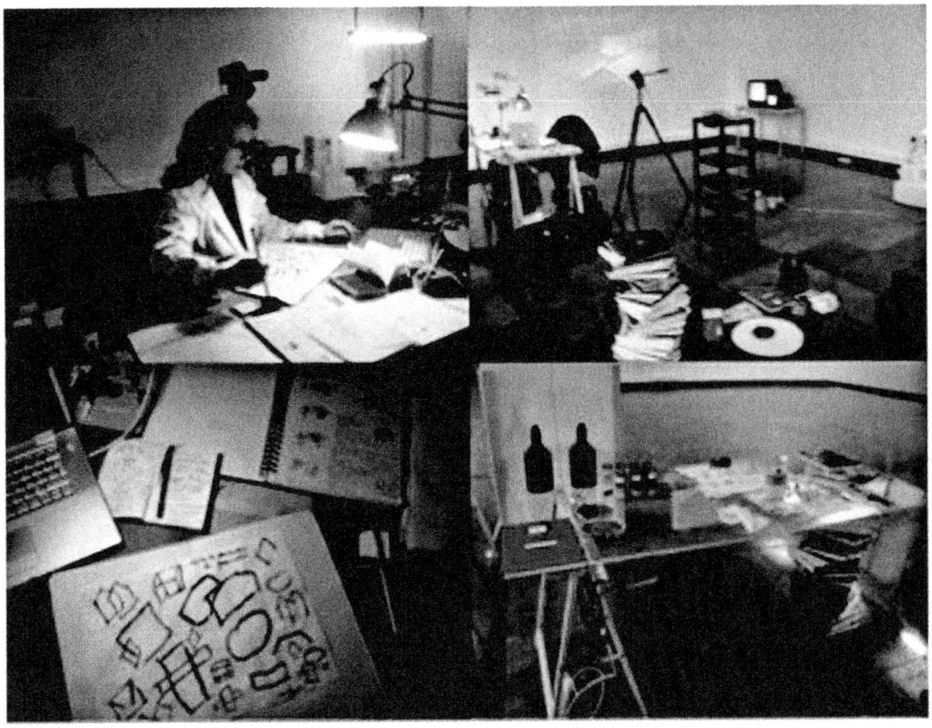

Figure 4. The Archive: Enactment/Re-Enactment of the Archive. Performance at IMM, 18–25 March
(L., 2011) Source: L., 2011

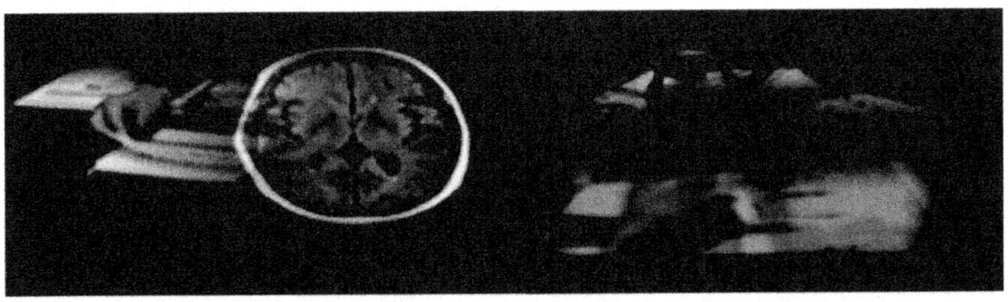

Art Projects

Platonia

Platonia (Figure 6) is a multimedia installation produced in 2014 for the Biennale *Touch Me Festival* in Zagreb (Croatia) in collaboration with Paulo Simeão Carvalho (Professor at Faculdade de Ciências, Departamento de Física e Astronomia, Universidade do Porto). It was exhibited in *Kontejner* gallery, from 15th November to 07th December. The installation comprises a stool, a telescope, and a black mirror on the outside of the main room; On the inside space it has a video projection of a montage film (a light cone) that 'enters' a jar of water and disappears through a gap in a black hanging screen; the cone

Figure 5. The Therapy. Exhibition Uma coisa entre muitas. Museu Nacional de Historia Natural e da Ciência, Lisbon
(L., 2012) Source: L., 2012

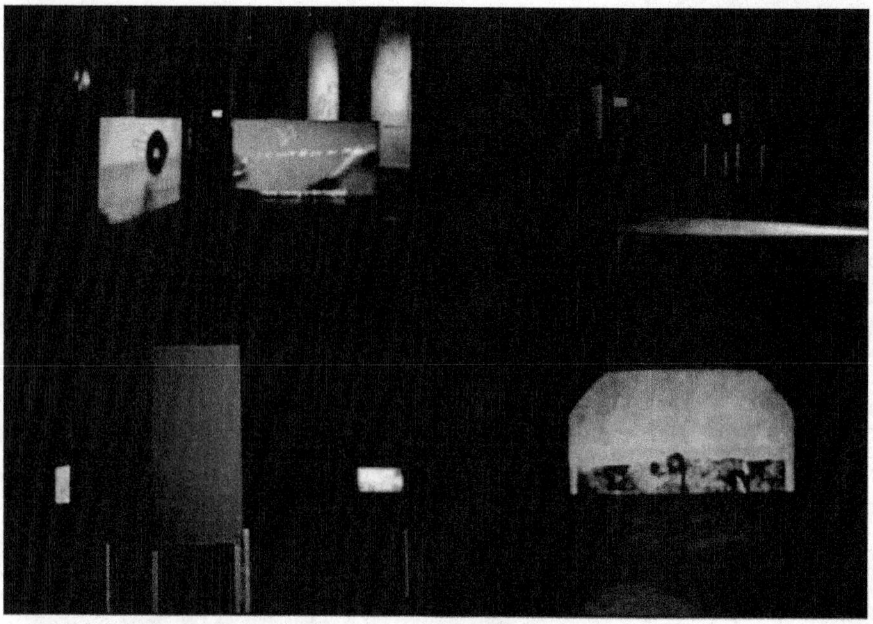

Figure 6. Platonia. Detail; Exhibition Touch Me Festival, Kontejner gallery, Zagreb Croatia Portugal
(L., 2014) Source: L., 2014 photo credit Toni Mijac

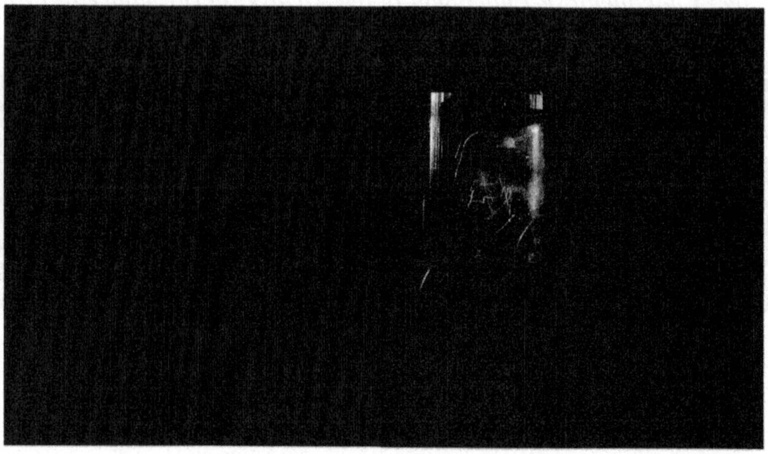

of light is reassembled into two distinct images (through the use of four glass half cylinders sections) and the film it contains is projected in a spectral way into two white suspended screens. Towards the end of the room the wall exhibits a large video projection with a very slow progression of images (that look like nature photographs of mountains, glaciers or clouds to the eyes of the distracted public); each image sequence fades to white and that moment is timed with a white strobe Light that fills the room momentarily blinding the viewers.

The title is based on a temporal interpretation of the functioning of time passage as dependent on memory and the experience of a 'timeless realm' as in Julian Barbour's (2000) notion of *Platonia*. When face with the challenge to comment on notions of time, L. decided on a reflexion on time related to the speed/duration of a lifetime, of the Earth and the Universe and its relation to memory in the scientific and personal realm.

We live in an unsettling position between the need and possibility of archiving a lot of our experiences digitally and the unnerving control of the archive, seduced by the possibility of a better future performance (of life extension or an un-faulty memory) with the enhancement promises. Memory and time are intangible and unexplainable experiences crossing borders from personal to social, and touching many disciplines. Humans have always been augmenting their senses, physical powers and cognitive abilities through ingenious tools and technologies. As seen x rays, the telescope, functional magnetic resonance imaging and atomic force microscopes can be viewed as extensions of the senses and simultaneously help construct new synaptic brain connections shaping our senses. Humans are continually developing new ways to see the invisible, to understand when the world started, to share knowledge and manage our social lives remotely and simultaneously how to extend life.

We depend on memory to be able to learn and interpret the world around us; Simultaneously our memories are subtle, faulty and subject to change. The seduction of the recording devices and the frail nature of the human matter allied to the reconstructive capacity and plasticity of the brain, elicits a desire to explore intangible understandings of experience and perception. For Henry Bergson (1990), the term duration reflects the inner temporality of a naturally occurring process as against the imposed temporality of external markers fixed by calendars and clocks. Similarly the time based nature (looping) and light dependence of the installation deals with intensity, duration and the completion as key elements in establishing the quality of experiences and the subjective/scientific understanding of time.

Spaces, places and objects hold presences, experiences, wishes and memories that are constantly reshaped. By interweaving metaphors of memory (such as Proust's telescope) along history, with personal archival elements and public natural imagery, as well as optical instruments and physics notions of time understanding, my work responds into creating a suspended moment were simultaneously images of different scales and contexts are both creating and evanescing. The public is invited to climb a little stool and look through a whole (i.e a portion of the telescope that is crossing the wall) in the entrance wall to the installation room. The viewer is rewarded with a drawing of a rabbit holding a pocket watch (a cartoon illustration figure from Louis Carol novel *Alice in Wonderland*). When entering the room the figure is nearly invisible (not just covered by the other in between screens as upside down on a wall distancing seven meters).

By presenting images of the family archive (extracts from the author' childhood *super 8* movie clips) the work hints at a shared personal passage of time. Thus by interlacing those with small clips from specific scientific movies on nature and other peculiar everyday life recordings, *Platonia* responds to the impalpable presence within a given moment. The personal and mundane becomes something that is interweaved with the social and natural world to the point of becoming something different, suspended, repeated … as time. The black mirror (lining on the entrance wall, a few steps from the stool and telescope, facing the entering spectator), reflects the exhibition space, sculpting and distorting all its stories, while the apparent linearity of the projection room is often broken by the strong strobe light and the consequent imperceptibility of the surrounding space; almost as a need to reset and form (new) memories of the already passed moments and experiences.

Boxes, Shadows, and Definitions

This project has currently two distinct versions; the first (*Boxes, Shadows and Definitions*) was produced for the exhibition *Sul Sal*, held in Odemira (Portugal), from the 20th of December 2014 to the 17th of January 2015 (Figure 7). The second one *(Boxes, Shadows and Definitions – Nostalgia)* was produced for the exhibition *Praxis e Poiesis 1.2.9,* held in Museu Santa Joana Princesa, Aveiro (Portugal) from 4 to 14th June 2015 (Figure 8).

If *Platonia* was based on a sophisticated notion of truth that overcomes the Platonic dualism between truth and appearance, *Box, Shadows and Definitions*, challenge the narrative of crisis predominant in 1990s theories of new media, which considered the digital image in terms of a profound loss of memory and historicity, moreover it explores the idea of projection as a mediated critique to experiencing life and constructing myths and narratives and even nostalgia. The past does not appear as a sentimental and escapist fantasy, but on the contrary, as a strategy to reflect critically on the present and re-imagine the future.

In the first version *Box Shadows and Definitions* (a projection onto a semi circular screen made of bleached leaves on glass jars) proposes an inter-medial approach, describing a field of contemporary scientific concerns with powerful art-historical resonances, harbouring essential questions about human subjectivity and memory in the face of the digital replacement of nature/landscape and experience with simulation, simultaneity and depthless surfaces (i.e. photography, TV, laboratory inscriptions, cell phones or computer screen, cinema, etc.). Concurrently it evokes the future scenario of a different body and an extended possible capacity to perceive in a different way and apprehend what lacks in the gaps of the screen.

If our memories are constructed, fragile and usually distorted, might our beliefs about our lives and ourselves be intrinsically inaccurate? If we enhance our memory and cognition will we loose nostalgia?

Figure 7. Boxes, Shadows and Definitions. Detail; Exhibition Sul Sal, Biblioteca de Odemira gallery, Odemira, Portugal
(L., 2014) Source: L., 2014

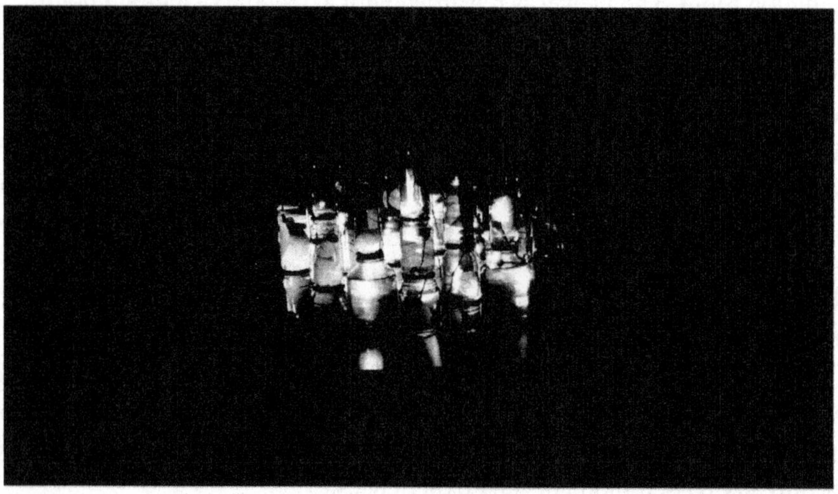

Figure 8. Boxes, Shadows and Definitions: Nostalgia. Detail; Exhibition Praxis e Poiesis 1.2.9, Museu Santa Joana Princesa, Aveiro, Portugal
(L., 2015) Source: L., 2015

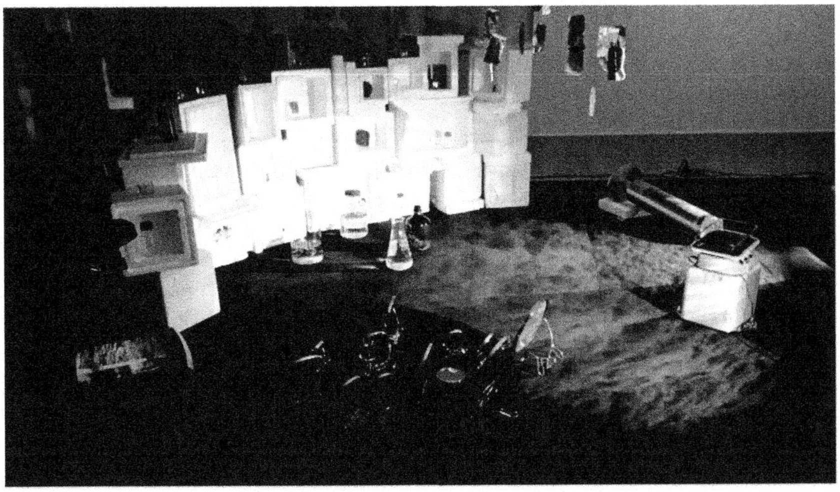

What is the matter and place of/for memory? These and many other questions arise from and also ignite the present artistic research; thus scientific culture, news, literature and cinema (mainly science fiction) are feeding contexts for L. production and reflection.

Recent neuro scientific proposals echoes the understanding that memories are constructed by combining bits of information from different levels of autobiographical memory (events and concepts) like a jigsaw puzzle assembled from many pieces. Further current studies on memory contend that it is influenced by our present desires, needs and future goals.

These considerations lead L. to question if we are unable to see or understand our own lives and are only confronted with portions of experiences (boxes), uncertain images from our time/place position (shadows) and constructed facts, myths and life stories (definitions). For the practical exploration of the subject L. uses the potential of imagery to evoke a language and its deceiving status in traces such as, writing, drawings, tree leaves, containers (domestic and laboratorial flasks), water and light in the form of video projection (in the first version and transparency and slide projection on the second version). This project aims to address the textures, shadows and discontinuities, processes and methods, techniques and technologies that circulate in different scientific fields in order to build a 'perfect memory' (Bowker, 2005) of the past, one that is preserved in order to construct definitions.

As memory is often wrongly perceived as (only) a functional archive and a process of calling data to mind, remembering is considered as an interpretative action of traces of the past. Thus L. articulates the conception of memory as an experience and explores the memory practices circulating in scientific laboratories. The author plays with strategies, such as deceived visualization, to allow the viewer to actively contribute to the construction of the work, interrogating the possible application of scientific practices/inscriptions in subjective/visual discourse on memory.

Boxes, Shadows and Definitions deals with both, the desire for freedom of form and greater cognitive capacities promised by *transhumanism* and neuro-enhancent technologies (More & Vita-More, 2013), as with the loss of coherence of the human body, exploring visually when the locus of definition and subjectivity is dispersed across the surface of the multiple screens (*i.e.* tree leaves suspended in water jars).

Recognizable animation pictures and scientific data, presented as disappearing images in projected videos that display their tenuous connection to the reality they index, all appeal to a preserved fleeting shadow and loss. As matter and image physically breaks down, meaning relies in a diffusion of the viewer's gaze across the surface of the moving image in an encounter with the emotional tension of the experience. Once a recollection can be lodged anywhere, in any artefact or a process driven by conscious activity – a tree leave, a glass jar, or a song, perpetuating a gap between the fragments of images and the use of familiar glass jars as containers, suggests practices of appropriation, archiving strategies and constant rebuild of our definitions and perceptions as with the flux of memory and the experience of duration.

When a fragment of nature is invested with a human being's conscious and wilful input we have a memory embedded in that fragment of modified matter. This (embodied) memory will be communicated to those who will meet the object, to those who will recognise it as culture and knowingly use it as intended by its producer. When placed in sequence and presented together, the fragments of video establish previously invisible links that approach the separate actions and construct bridges, (absent/present, nature/culture) transforming what was an invisible link into a necessary narrative. The multidimensionality of the installation and the fragmented experience of the viewer becomes the representation of multiple layers of memory and meaning, in a constant interrelation between the external experience and its internal reality. Stories, narratives, memories and the meanings of things must exist in some dormant, invisible form between production and communication; as a text written but not yet read, memories must contribute to some level of objectivity that is not maintained by the presence of minds (Figure 8).

The most recent form of this installation *Boxes, Shadows and Definitions - Nostalgia* is a multiple screen projection of three simultaneous image narratives of the sea (fragments juxtaposed) and a slide projection of familial images on a multidimensional screen made with stereo foam laboratorial boxes and many other objects collected both in scientific laboratories and the art studio (sketches, raw materials or souvenirs), in a condensation of an extended exploration of different actions, moments, locations, and materials. The installation comprises also a scientific oscillator used in molecular biology laboratories with a glass container laying on it and oscillating (containing salted water and familiar photographs on film taking the place of protein gels which are developed in molecular biology laboratories); the movement of oscillation mimics the sea motion. Further the viewer's motion also triggers a fan lying on the floor that when rolling on surface makes the little pieces of foil, hanging from the ceiling, producing a sound similar to waves touching the sand.

As Henry Bergson (1990, p.9) has observed, material objects do not merely exist in themselves, unrevealed and undetectable. They can be seen, smelt, heard, and felt. Thus, physical bodies, boxes or matter are "an aggregate of images". For Bergson an image stands between a representation (a private mental process) and an object (a thing). In a challenging perspective Bergson (2010) proposes that memory is formed alongside perception and not after it. The ontological duality of the real as something that exists objectively and can be perceived as an *aggregate of images*, in itself entails the tendency towards the separation of the epistemological process into perception and memory. This notion of memory in the present and matter as a palimpsest of images was helpful to clarify the author' opinion on the formation of nostalgia; thus the conception of the new version of the work.

Will Humans be nostalgic of our present frail condition, with the propensity to error, our ability to forget and entailed and constant reconstruction of ourselves? This work explores visually these worries and affect, in non-fixed images, changing according to movement of spectators in space (sensors detect spectator's presence and trigger the several devices of the work constructing deferring view points and perspectives each time it is visited or depending on the cooperation of several simultaneous visitors).

In this multiform/multi-location installation, L. explores the disjunction between the fragment and the whole by using a structure in which disparate meanings are simultaneously possible. Interweaving projected image, objects from both the studio and different laboratories allied to the development of a critical visual montage practice, provides hints into the complex territory of identity and memory (personal or cultural) using the image from the familial but and also the public domain (the sea images extracted from a digital public archive) to bridge the differences that are constructed at distinct margins of the oceans (each with differing political and historical backgrounds, language, laws, etc...) seeking to enrich the discourse surrounding the way memory is researched in the sciences and through the visual arts reconciled through the visual and sensuous play.

Enhancing the Mind's I

If the idea we have of ourselves is to be projected onto a material to be assessed by others (as in neuropsychology assessment) what is being judged? The communication skills, the capacity to ly and perform as expected by the viewer, the speed on which the task is performed, the imagination or the embedded memory? L. on participating as witness in several clinical neurological assessments, found visual plasticity and conceptual challenge on the relationship established between the complexity of the functioning of memory and the personal and constant construction of identity and also the strategies used by different fields of research to explore this entwined relationship. As an artist the researcher shares some tools and technologies with scientists, such as drawing, human-computer interaction and techniques of biotechnology; similarly L. is interested in understanding the materiality of memory, its functioning and the extension of remembrance, thoughts and personality from personal to collective engagement. If the preserved memory is intended to be passed on to others, it needs to be extracted from one's own mind and presented in a way that would make it accessible for others to perceive. The author explores whether is it possible to translate ones emotions directly into an object as a memory of a certain moment.

The result is the opening of a critical gap between the way sciences produce knowledge about the subject and the affect produced by the experience of the viewer on the installation art works. The idea behind *Enhancing the Mind's I*[13] (Figure 9), is to design with the mind (both metaphorically and literally!) multiplying ways in which memories become embodied and externalized, by using L. brainwaves as tools for a new humanistic arts based-approach. The *I* is metaphor for an allegorical inverted Cartesian Theatre with neurons/brain waves illuminating a search for the self.

L.' explores unpretentious ideas of enhancement, such as the invention of the writing itself, as a development that simultaneously extended and impaired human memory, by providing an externalised written record but diminishing people's ability to memorise by removing the necessity of learning by heart. Andreas Husseyn (2000) inquires about the consequences of writing for human memory and further on the role technologies impact on our natural skills providing examples for contemporary discussions around human enhancement through technology and how external memory aids and other new technologies such as brain computer interfaces, are always double-edged, extending or fixing certain powers while eroding traditional skills.

Although the use of Brain Computer Interfaces (BCIs) in the arts originated in the 1960s, with the pioneering performances of Alvin Lucier, there is a limited number of known applications in the context of real-time artistic performances and accordingly the knowledge base of this area has not been developed sufficiently. Among the reasons are the difficulties and the unknown parameters involved in the design and implementation of the BCIs.

Figure 9. Enhancing the Mind's I. Detail; Performance, Brave New World, Casa da Música, Porto, (Portugal)
(L., in collaboration with Bastos, Marques & Teixeira 2015) Source: L., 2015

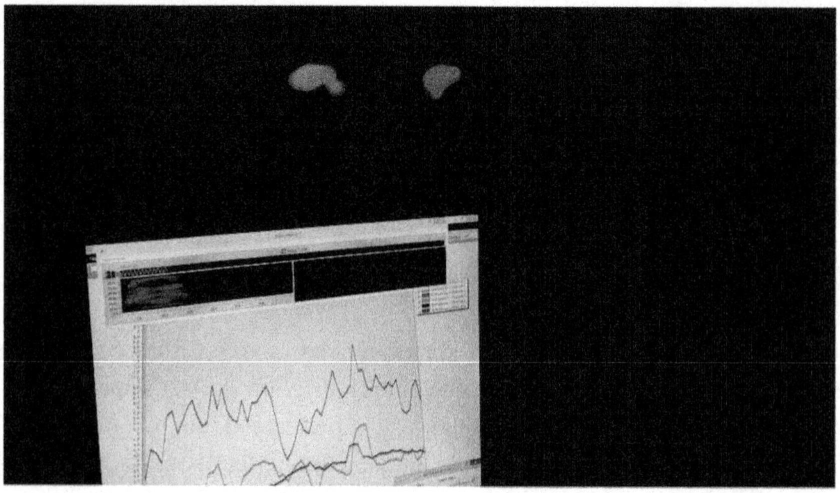

However today, with the dissemination of the new wireless devices, the field is rapidly growing and changing, and it is visible that artists work are harnessing the electroencephalogram (EEG) signals, bridging various methods and technologies and arising questions both in art/technology and in human experience. This project aims to develop cross-disciplinary relationships and encompass their evolving research to interrogate societal, artistic and ethical attitudes to neuroenhancement, memory and the preservation of the body, exploring the potential of neuro-stimulation/neurofeedback and EEG for artistic research.

A BCI is a system that captures the brain electrical activity in the form of EEG signals; further it translates those specific features of the signal that represents the intent (or unconscious desire) of the user into computer readable commands, Allowing its user to control a machine (e.g. a computer, an artificial limb, or any other machine) solely with brain activity rather than the peripheral nervous system. A typical BCI combines neurophysiological measurement technology with machine learning software to automatically detect patterns of brain activity that relate to this specific mental task. Control with a BCI is originated when a user performs a specific mental task.

Exploring 'drawing with the mind' (through the body action and possibly literally through thought and memory) by using the performer'/artist's memory brainwaves as tools for a new humanistic arts approach, is driven by several questions, such as;

- Is it possible to measure something intangible as memory and creativity?
- What can we learn from what brainwaves show us?
- Will wearing an EEG device influence the drawing process?
- Will clinical neuro-enhancing through neurofeedback expand the drawing capacity?
- Can I translate my emotions/memories directly into an object as a memory of a certain moment/place?

In the performances L. produces a series of Drawings following a neuroscientific/ psychological guideline for psychological assessment on identity and sense of self (i.e. The TST – twenty statements test protocol is to complete the 20 times in a row the sentence 'I am …'). The drawings are constructed by writing, in graphite, sentences starting by 'I am'. After 20 sentences there is an interruption in the flow and the artist erases partially the resulted drawing. Immediately after that effacing action L. counteracts enacting the procedure of writing another set of 20 sentences. Each event is performed after sessions of neurofeedback (Figure 10) for cognitive enhancement (memory and creativity stimulation) and while having the author's brain being sensed for the different waves it produces. EEG measures frequencies of L. brain activity (Alpha, Beta, Delta, Gamma, Theta) relating to her state of 'consciousness' while wearing it. The data collected from EEG is translated in real-time to the computer that uses software to detect the brain waves and then transforms that information into data sending it to another computer which performs further actions (such as sending information into Processing or Arduino, which is linked with Max/MSP to receive data and generate sound or video or even controlling several printing and embroidering machines that produce further drawings). The EEG records the 'drawings' that L. brain is producing while thinking 'who she is' (i.e. who 'am I') and the production of the drawing is recorded in real time video. That video file is feed into the computer that edits it accordingly with the coordinates dictated by the performer brainwaves while drawing. The audience as access to the performance scenario and also to the video images of the brain waves captured by the first computer and further the projection of the final edited video. The drawings will endure until exhaustion of the material surface, her memory or other unpredictable enforcement.

Spaces, places and objects hold presences, experiences, wishes and memories that are constantly reshaped. We conceptualise our memories, verbalise them and confine them to the boundaries of a narrative (or several and in distinct moments, contexts and materials). Is art able to hold memories without deforming them by rationalization? Is BCI able to confirm that if as an artist L. is able to externalise her

Figure 10. Enhancing the Mind's I. Detail of Neurofeedback session performed at Neurobios, Clinica Professor Doutor Marques Teixeira, Porto, (Portugal)
Source: L., 2015

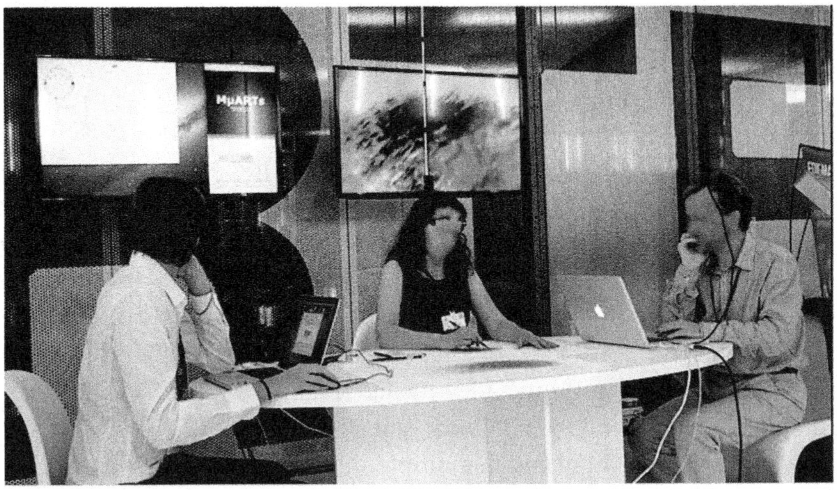

memories whilst remaining faithful to their own fleeting emotions, paradoxical, liminal and conflicting feelings, sorrows and joys, all those irreducible inner events that constitute the nature and identity of a person.

CONCLUSION

Memories are formed and informed by their social and cultural context and further by their technological and medial frameworks (Rigney, 2005). Accordingly as Marita Sturken claims in *Tangled Memories* (1997) the illusion of a transparent individual memory is in fact the outcome of complex technologies that produce cultural memory, through objects such as texts, icons images or monuments. The installations unsettle conventional representations of memory, drawing and also unsettle viewing relationships. The performative audiovisual work offers ways to negotiate individual memory enhancement and loss, and explore scientific practices for investigating this process and also construction of Identity.

The images and objects in the works are a presence, representing something beyond what is ascribed by the authority of science. In sculptural montages of the appropriated images/objects the disjunction of the fragment (present signifier) and whole (absent signifier) are manipulated in a way that gives priority to no specific meaning, but endless possibilities of interpretation. The different values attributed to the fragments of scientific research in the studio or exhibition context, coexist in no hierarchical order, none claiming a higher degree of truth than another, in the heterogeneous space created by the works.

The display of objects from the laboratories used in various ways in the installations emphasises a need for visual spatial order, and also provokes excitement about the idea of visiting and exploring the unknown space. These objects refer directly to the workings of autobiographical memory, and thus resemble mnemonic traces of a lived experience. They are also part of the on-going archive. In exploring the archive and the scientific strategies to test, understand and expand memory, L. proposes to challenge the meaning of the scientific representational strategies, extending their possible readings by transforming them, colliding and juxtaposing them to familial objects and images and bringing them into the context of the exhibition. To expand the researcher' understanding on memory, Alzheimer's studies and patients were approached, and in a similar note it is those individuals traditionally classified as "disabled" who are currently at the vanguard of human enhancement technologies. The impaired are the pioneers of the testing and improvement of assistive technologies and potentiation drugs, blurring the threshold between a perceived "lack" and an unfair advantage in relation to the general population. Within L. research, the recognition of time and memory and the role these play in the construction of definitions of memory workings or loss and cognitive enhancement have come to the fore. Her explorative works intervene within the scientific/popular discourse to contribute to expanded ways of thinking about and looking at construction and validity of memory and identity, body and normality, and representation versus mediation, using art as a tool to enhance public awareness of several anxieties, disorders and technologies, raising ethical dilemmas and questioning norms of behaviour. The ambiguity of historical/personal time and the myth of authenticity are considered through an exploration of how the archive is assembled.

The chosen approach is a poetics of time; identity and fragility, which explores the past, present and future of the memory studies portrayed within the scientific archive. The structure of the practice as a temporal *briccolage* displays a fragmented, multiple, jumbled narrative, where chronology itself is disrupted. The fragmentary nature of the practice ensures that no complete meaning can be fixed. The interlocking of historical and personal time, and the conjunction of both facts and fiction, enables a plurality of voices to be heard, contesting any dominant scientific or historical linear narrative.

ACKNOWLEDGMENT

FCT Grant SFRH/BPD/98356/2013

REFERENCES

Barbour, J. B. (2000). *The End of Time: The Next Revolution in Our Understanding of the Universe*. London, UK: Phoenix-The Orion Publishing Group Ltd.

Berg, M., & Mol, A. (Eds.). (1998). *Differences in Medicine: Unravelling Practices, Techniques and Bodies*. Durham, NC: Duke University Press.

Bergson, H. (1990). Matter and Memory. New York, NY: Zone Book.

Bergson, H. (2010). *Mind-Energy: Lectures and Essays*. Kessinger Publishing.

Bowker, G. (2005). *Memory Practices in the Sciences*. Cambridge, MA: MIT Press.

Clark, V. P., & Raj, P. (2014). Neuroenhancement: Enhancing brain and mind in health and in disease. Neurolmage, 85, 889-894.

Crary, J. (1992). *Techniques of the Observer: On Vision and Modernity in the Nineteenth Century*. Cambridge, MA: MIT Press.

Dumit, J. (2004). *Picturing Personhood: Brain Scans and Biomedical Identity*. Princeton, NJ: Princeton University Press.

Edward de Bono Foundation. (2014). *Edward de Bono Foundation*. Retrieved April 9, 2014, from http://edwarddebonofoundation.com/

Foucault, M. (1995). *Discipline and Punish: The Birth of the Prison*. New York, NY: Vintage.

Foucault, M. (2003). *The Birth of the Clinic: Archaeology of Medical Perception*. Taylor and Francis e-Library.

Fukuyama, F. (2002). *Our Post human Future: Consequences of the Biotechnology Revolution*. New York, NY: Farrar, Straus, and Giroux.

Gibbons, J. (2007). *Contemporary Art and Memory: Images of Recollection and Remembrance*. London: IB Tauris.

Huyssen, A. (2000). Present Pasts: Media, politics, amnésia. *Public Culture, 12*(1), 21-38.

Kevles, B. H. (1997). *Naked to the Bone: Medical Imaging in the Twentieth Century*. New Brunswick, NJ: Rutgers University Press.

L., M. M. (2013). *Representational Strategies on Alzheimer's Disease: A PracticeBased Arts Research in a Neuroscience Laboratory*. (Unpublished doctoral dissertation). University of Brighton and University for the Creative Arts, Farnham, UK.

Latour, B., & Woolgar, S. (1986). *Laboratory Life: The Construction of Scientific Facts*. Princeton, NJ: Princeton University Press.

Latour, B. (1987). *Science in Action: How to Follow Scientists and Engineers through Society*. Cambridge, MA: Harvard University Press.

Lynch, M., & Woolgar, S. (1990). Introduction: Sociological Orientations to Representational Practice in Science. In Lynch & Woolgar (Eds.), *Representation in Scientific Practice* (pp. 1-18). Cambridge, MA: MIT Press.

More, M., & Vita-More, N. (Eds.). (2013). The Transhumanist Reader. Wiley-Blackwell doi:10.1002/9781118555927

Nunes, B. (Ed.). (2008). *Memória: Funcionamento, Perturbações e Treino*. Lisbon, Portugal: Lidel.

Rigney, A. (2005). Plenitude, Scarcity, and the Circulation of Cultural Memory. *Journal of European Studies*, *35*(1), 11–28. doi:10.1177/0047244105051158

Rose, G. (2007). *Visual Methodologies: An Introduction to the Interpretation of Visual Materials*. London, UK: Sage Publications.

Rubin, D. C. (2005). *A Basic-Systems Approach to Autobiographical Memory*. Retrieved February 12, 2015, from http://dukespace.lib.duke.edu/dspace/bitstream/handle/10161/10105/RubinBasicSystems.pdf?sequence=1

Schacter, D. L. (2001). The Seven Sins of Memory: How the Mind Forgets and Remembers. New York: Houghton Mifflin.

Stafford, B. (1991). *Body Criticism: Imaging the Unseen in Enlightenment Art and Medicine*. Cambridge, MA: MIT Press.

Sturken, M. (1997). *Tangled Memories: The Vietnam War, the AIDS Epidemic, and the Politics of Remembering*. Oakland, CA: University of California Press.

Sturken, M., & Cartwright, L. (2002). *Practices of Looking: An Introduction to Visual Culture* (2nd ed.). Oxford, UK: Oxford University Press.

Thompson, H. (2012). Performance enhancement: Superhuman athletes. *Nature News*. Retrieved May 19, 2014, from http://www.nature.com/news/performance-enhancement-superhuman-athletes-1.11029

Tulving, E. (1972). Episodic and Semantic Memory. In E. Tulving & W. Donaldson (Eds.), *Organization Of Memory* (pp. 381–403). New York, NY: Academic Press.

Tulving, E. (1985). How Many Memory Systems are There? *The American Psychologist, 40*(4), 385–398. doi:10.1037/0003-066X.40.4.385

Tulving, E., & Craik, F. (Eds.). (2000). *The Oxford Handbook of Memory*. Oxford, UK: Oxford University Press.

Van Dijck, J. (2005). *The Transparent Body: A Cultural Analysis of Medical Imaging (In Vivo: the Cultural Mediations of Biomedical Science)*. Seattle, WA: University of Washington Press.

Wilson, D. H. (2012). Bionic Brains and Beyond, *Wall Street Journal.* Retrieved May 19, 2014, from http://www.wsj.com/articles/SB10001424052702303640104577436601227923924

Zwart, H. (2014). *Limitless as a Neuro-pharmaceutical Experiment and as a Daseinsanalyse: on the use of fiction in preparatory debates on cognitive enhancement. Medicine, Health Care and Philosophy.* Dordrecht: Scientific Contribution; doi:10.1007/s11019- 013-9481-5

Zwijnenberg, R. (2009). *Brains*. Art, and the Humanities.

KEY TERMS AND DEFINITIONS

Alzheimer's Disease: A common form of dementia; a brain pathology that leads to a progressive cognitive decline in the ability to remember, learn, think and communicate; implying in its sufferers' a gradual need for navigation and translation of the world.

Archive: A fluid generation methodology, envisioned as a network of four virtual intertwined archival spaces depending on a complex negotiation of the space, time and actions between people, objects and concepts.

Brain Computer Interface (BCI): A system that combines neurophysiological measurement technology with machine learning software to automatically detect patterns of brain electrical activity that relate to a specific mental task, allowing its user to control a machine solely with brain activity rather than the peripheral nervous system.

Matter: Resonating instances fundamental to our personal process of becoming, remembering and forgetting; they transfer affect and allow us to perform 'who we are'.

Memory: Rehearsed performances of reconstruction, narrated and practiced. Multimodal complex cognitive capacity that constantly translates the world through the sensory details and images that are held in brain patterns comprising the feeling, context and meaning of experiences which become linked in the process of remembering to form memories, rather than existing innately as memories.

Neuroenhancement: Cognitive methods and neuroscience-based techniques to expand or increase the competence and range of the various mental or nervous system activities we perform.

Project-Led Research: A progressive and situated practice-based inquiry; sustained in theory/practice interaction and on self-reflexivity in which theoretical research, ethnographic exploration and studio practice all trigger the experimental process, as well as on developing and reflecting on the practice, elucidated by the methods and structure of the written work, which in turn fed into the practical decisions.

Representational Strategies: Catalogues or archives of categories or systems, by which life and diseases are ordered in the biological sciences. A way of translating the invisible under-structure of life to a visual domain of notation, drawing, instruments and hybrid forms of these.

ENDNOTES

1 The research network lead L. to the Caselas MRI clinic (Lisbon) where patients were scanned or the Memory Clinic (Lisbon) where patients received cognitive stimulation therapies.

2 For further information's please consult the URL http://www.nerri.eu/eng/home.aspx

3 The camera obscura is a common metaphor for memory. In *Metaphors of Memory: a History of Ideas About the Mind*, Douwe Draaisma presents an interesting case study of the way in which descriptions of memory, supported by the vague but continuous and dynamic images of the *camera obscura*, rapidly gave way in mid-19th-century psychology to the photography-inspired contradictory notion that memories are fixed and accurate but static.

4 *The notion of photographic truth hinges on the idea that the camera is an objective device for the capturing of reality, and that it renders this objectivity despite the subjective vision of the person using the camera.* (Sturken and Cartwright, 2002: 280)

5 Foucault coined the notion of *biopower* as a reflex on the body on the regulatory power institutions. The training, exercise and regulation of the body, categorized and registered through photography, is explained as the simultaneous rise of 19th-century social institutions and the emergence of photography. For expansion, see *Practices of Looking* (Marita Struken and Lisa Cartwright, 2002).

6 Foucault uses the design of a model prison conceived by reformist Jeremy Bentham (1748-1832). In Bentham's model: *at the periphery, an annular building; at the centre, a tower; this tower is pierced with wide windows that open onto the inner side of the ring; the peripheric building is divided into cells, each of which extends the whole width of the building; they have two windows, one on the inside, corresponding to the windows of the tower; the other, on the outside, allows the light to cross the cell from one end to the other. All that is needed, then, is to place a supervisor in a central tower and to shut up in each cell a madman, a patient, a condemned man, a worker or a schoolboy. By the effect of backlighting, one can observe from the tower, standing out precisely against the light, the small captive shadows in the cells of the periphery. They are like so many cages, so many small theatres, in which each actor is alone, perfectly individualized and constantly visible.* (Foucault, 1995: 200)

7 His work is important to reinforce the connections established between the biology of the body and of the cellular and molecular research on it and the cognitive impairment dementia that patients present.

8 Clearly reporting the influential approach taken by John Berger in *Ways of Seeing,* a seminal text on the relationship between the image and the viewer.

9 These are particularly useful and new insofar as they interweave discourses on biotechnology research and production of facts and knowledge using an ethnographical approach. The reader is driven through different construction sites of facts and machines staging science as an inscription-making practice, prompting discussion of issues around representation and inscription and science as a semiotic practice dealing with signs and symbols, presence and absence.

10 For example, MRI depends on the use of very strong magnetic fields to make the nuclei of atoms spin in a certain direction. This effect is fundamentally undetectable in any form by the human sens¬es, although the amount of noise might be misleading in this respect. In contrast to a microscope that magnifies visible light, the sophisticated measurement of electromagnetic energy, outside the scope of human sight, must be reconstructed in a pictographic way that may be interpreted by expert technicians and doctors. Accordingly, not only is the interpretation subjective, but also the

visual choices made in the graphical interpretation of the scan data for diagnostic purposes may obscure other readings; norms and consensus are under constant discussion in terms of choices of how to represent the scaled measurement of the magnetically spinning nuclei in a visual way.

11 L. developed creative workshops, with different lengths (2 to 8 hours), using traditional mnemonics interweaved with some recent technologies (Augmented Reality or EEG) and has been presenting them to different ages groups in academic conferences and high schools as part of her collaborating with the NERRY project).

12 Zwart, H (2014), was citing Sloterdijk, P. 2009. Du musst dein Leben ändern. Über Anthropotechnik. Frankfurt: Suhrkamp.

13 The work is in process and in collaboration between Maria Manuela L., Paulo Bernardino Bastos, Horácio Tomé Marques and Francisco Marques Teixeira (the last two founders of μARTs {should read muarts}).

Chapter 3
Cells, Organisms, and the Living Brain as New Media for Art: A Pursuit in Art Research

Marta Menezes
Cultivamos Cultura, Portugal

ABSTRACT

Throughout art history numerous artists have explored connections to science. In the society of today, the relationship between art and biology has been acquiring special visibility. Moreover, the current importance given to science and technology by today's public opinion directly drives an increased awareness about the relationship between art and science. The public eagerly follows any breakthroughs of scientific research that reach the public sphere, with mixed feelings: simultaneously awe, hope, and fear for potential misuse. Such awareness about biological sciences and biotechnology has been having an increasing influence over artists, with a strong emphasis on the biological sciences. Nowadays, the position of artists as mere interpreters and observers is obsolete, and is reminiscent of a time before the term "science" was created. Now the artist is no longer a passive observer of the scientific research, but does not quite fit the current definition of a basic science researcher. The artist is, rather, an art researcher. This particular position, has been developed within the academic environment and within a growing and trendy artscience community. It has led to the development of strategies to promote the exploration of research possibilities deriving from a cross-talk between artists and scientists. While art and science seems to be establishing itself as a new art practice, as a research field it has been rooted on the timeless investigation strategy of science (as interpreted by artists) in order to develop its own new methods of practice, new media and installation strategies, and new ways to manipulate the materials for artistic expression. Art and Science (and especially art and biology) is itself, in my view, a great example for the setting of brackets to the field of art research.

DOI: 10.4018/978-1-5225-0510-5.ch003

INTRODUCTION

In today's society it can be argued that biology has replaced the physical sciences – physics in particular – in becoming the scientific discipline with greater impact in the public opinion. In the first two-thirds of the twentieth century advances in nuclear physics and quantum dynamics represented for the society the most visible science in respect to fear of misuse (nuclear catastrophe) and also the promise of great benefits (such as clean and cheap nuclear energy, or space-exploration). However, the molecular revolution of biological sciences – made to a great extent by physicists like Delbruck or Pauling– led to landmark discoveries that placed biology and biotechnology in the centre stage of public visibility (Hudson, Horace & Maddox 1979). It is therefore no surprise that art has also developed a strong relation with the biological sciences, regardless of the breadth of the field.

In the frame of this context of Biology and Art, we can find several works, developed in recent and not so recent years, which represent this paradigm (Bulatov 2004). Any analysis of those works will necessarily lead to the conclusion that what is usually described as: "art and biology" (Pandilovski 2008) or "bio-art" (Bulatov 2004; Kac 2007; Pandilovski 2008) is a very heterogeneous group of very different approaches and outcomes. As it has been clear to me from the beginning of my research, it becomes an extremely difficult and lengthy endeavor to attempt the inclusion of all strategies of both parts of the artistic practice (research and production) in the same analysis in the perspective of artistic research. I will therefore focus on my own methods and experiments, with some examples from other practitioners, to provide contrast and/or reinforcement of a point. Through this approach I will explain my position in this vast field of research and, most importantly, I will present the argument for the connection between art research and knowledge production in art practice.

I am aware, as a practitioner, that the process I use in my research and art practice, in the field of art and biology, is influenced by the way scientists conduct their own research. My research, and its similarities and differences from the scientific research (frequently both occurring in a shared laboratory, or outside it in the studio or even in the field) is a crucial point in the outcome of my projects and their formalization. I believe this attitude and strategy is also an important issue in the engagement of my fellow scientific collaborators or art colleagues and in our personal relationship to eventually create the artwork itself.

Neurons, Representation and Knowledge

In my research and practice I not only try to make some sense of the concepts I use and develop for my artworks, but I also aim to experiment with new material, live, pulsing, changing material to express these concepts. This aspect is fundamental to what I do as an artist. It is my conviction that both aspects of my strategy in the studio/laboratory on the development of the research for an artwork and the way I adapt and tackle the ideas and questions in the conceptualization of the artwork's production, present new explorations and new knowledge in the art research field, with the added point of exploring a living material. In the project *Tree of Knowledge* I decided to explore interactions between art and science by using novel cell imaging and tissue culture technologies in order to create live sculptures. When creating a sculpture, the choice of medium is critical as it directly has an effect on how the sculpture is experienced. For example, the same object made of wood, stone or steel can have a different meaning, and can drive a different response in the beholder. When considering what would be the most adequate medium to represent the 3D structure of neurons, I concluded that I ought to use neurons themselves. By covering a scaffold with live neurons, or by filling glass tubes with these cells, one can achieve a representation of the delicate structure, maintaining the dynamic nature of a neuron: always changing, establishing new connections, eliminating old ones, growing, and living. It is important to note that this strategy is not an attempt to represent reality as accurately as possible, but simply to explore the material that seems more adequate to represent such a living object. Such sculptures are a form of art that is literally alive.

But also those cells are the representation of the questions I have while working with living materials: what is the meaning of making a live sculpture and what does it mean to say it is a living sculpture? The concept I question here is our idea of what we consider alive or not alive. There is a second layer of complexity concerning our perception of what is an object and an organism. I have not yet been able to get a satisfactory answer for those questions, and I do not expect to get it any time soon (this is not the objective). But the tension emerging from the reasoning to try to understand why we consider something alive or not alive, is an important concept in our society and that is the relevant issue to rise in the artwork itself (Figures 1 and 2).

For art there is, of course, yet another fundamental question that is raised by my work "*Tree of Knowledge*". Are the neuron sculptures, and therefore the work, a representation? Is it really a representation of something, or rather a presentation of something? Can life itself ever be considered merely a representation? I believe live can indeed be a representation, especially if one takes life as a malleable plastic material (Catts & Zurr 2002). The neurons are part of the artwork, being manipulated and presented as a vehicle for questioning a concept. They are not there to present life itself, they are there to represent the question of what we may or may not consider considered a tree, or more importantly a network of connecting cells, this is why the cells are grown into a sculpture. A similar argument could be made concerning my work "*Nature?*" where I manipulated the wing pattern of live butterflies, addressing what is natural and artificial (Menezes 2011). That work is also presented as a live butterfly installation and not a series of images or preserved butterfly specimens. That is also the reason for the performance of

Figure 1. "Tree of Knowledge" Sculpture made of live neurons. Installation at the exhibition Hybrid in TEKS Gallery, Trondheim, Norway
Menezes, Marta.

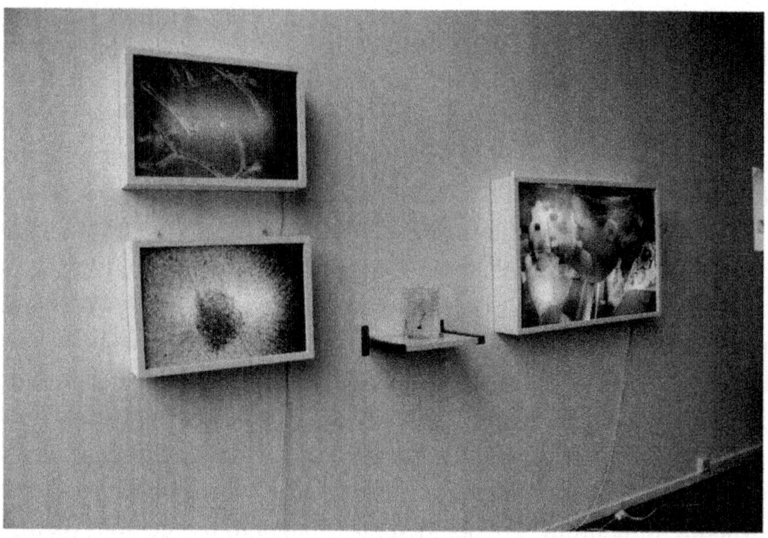

Figure 2. "Tree of Knowledge" Sculpture made of live neurons. Close up of sculpture
Menezes, Marta.

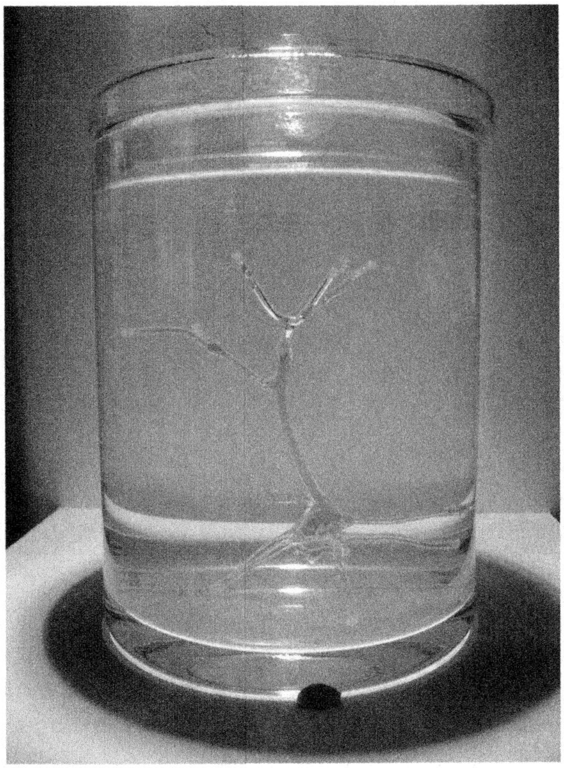

manipulations of the pupa within the gallery space. The outcome is considerable different from what would be achieved with the exhibition of photos from past acts. It is the distinction between what is made real and tangible by their unequivocally living presence and "photoshopped" digital products (Figure 3).

In this example of work, there is a clear and marked contribution to the production of knowledge, through the practice of art research, required for its development. The argument I make here relates to the fact that the concept of the artwork itself is important to address in the art field, though it may be argued that any concept and idea has its place in its relevant context. Art is, much to my personal satisfaction, very inclusive in terms of relevant ideas to express and develop. The concept of nature, and of what is considered natural or non-natural, is not an unique or obscure framework to be explored in art. Indeed, it is very well explained in art that nothing arises from nothing, and while we rethink old concepts constantly we re-render them in new perspectives and even, hopefully, making some other people rethink their own ideas taking advantage of that new perspective. There are many examples, from pre-historical art to the present, that illustrate this point while addressing the definition of nature and natural. Some of the most recent examples have also used living material and biological research techniques, both in their production and public presentation, which allows me to argue that the use of living tissue as a plastic material for art production is becoming an increasingly more common media, constituting a real addition to the known techniques and methodologies of artistic practice (Miller 2014; Bulatov 2004). More importantly, its development as a tool and strategy is a very striking example of a practice in art that irrefutably shows a "progression line" leading to addition to the previously known

Figure 3. "Nature?" Live Butterflies with modified wing patterns. Byciclus Anynana butterfly Menezes, Marta.

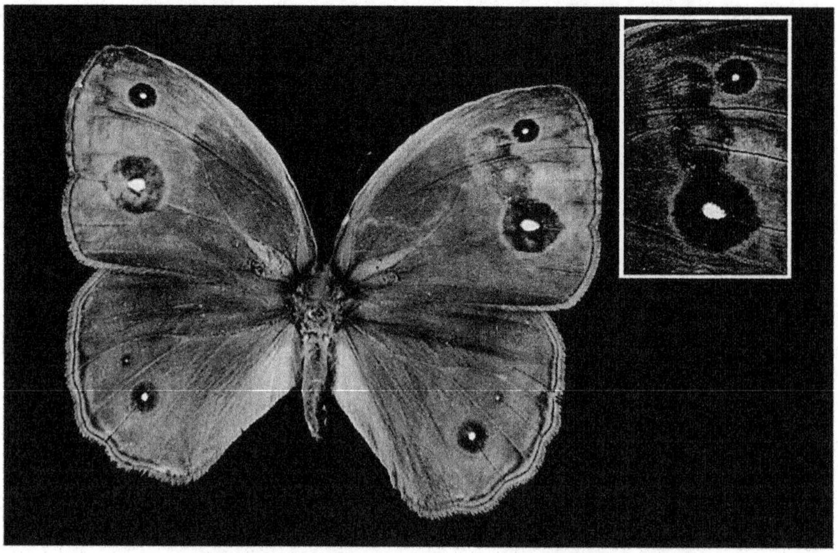

toolkit for the production and development of artworks, much like digital art or web art. What is this adoption of new methods and materials if not the outcome of knowledge-producing art research? If such contributions to knowledge are clear, it is remarkable I often still find myself arguing this point within fellow artists and academics.

There is only one deduction to be made. The discussion of practice-based research, namely in the framework of doctoral thesis in the arts is often centered on the issue of production of knowledge. Although production of knowledge is important, one should not neglect some other issues such as how is that knowledge recorded and how it is made available to the wider community. The general rule is that the knowledge is acquired as the artwork is researched and produced. However, that knowledge is not necessarily apparent in the artwork itself, as the artwork is not (generally) a clear description of the developments and insights gained during its production. Therefore, a systematic record of the research leading to an artwork is usually absent unless a deliberate effort is in place to clarify methodologies for the benefit of the artistic community at large (and to the public).

The added difficulty, some may think, is that art is, by its very nature, a subjective endeavor and pursuit. The art practice is full of opinions and questions, and with very few answers and statements of fact. Art is a field that loves and lives from the different and the unique, from the new and the unexplored, and from the freedom of choice and divergent perspectives. This is as much a choice for its practitioners and viewers as an essential part of what art is. Without these characteristics art would become another knowledge pursuing endeavor of humanity, but it would definitely cease to be art (Menenzes 2011).

Returning to the issue at hand, how do we make a field of research that already exists informally (and which can be even defined by this informality) into a knowledge producing field, that shares the knowledge it produces in order for it to be re-used by others. One must set clear guidelines, even if rules and guidelines may sound counter to the spirit of art. In fact, such guidelines are not meant to govern the process of artistic creation, production, or even art research. However some common ground is needed for the communication and sharing of knowledge generated through the process of art research, while not necessarily for the making of art. Mostly, I think this task is important because the knowledge being produced is extremely valuable and not the property of any one person. I suspect this is the "spot of bother" for most artists: While the research that we conduct as artists, for the benefit of making our artworks, is a personal and individual endeavor and part of our intellectual property, there are aspects of it that derive from knowledge produced by others, and aspects that lead to the development of new knowledge. It is that new knowledge that transcends the artwork and should not be bound by the artist but rather made available for the benefit of the community. That is the knowledge that has to be recorded and shared. And the guidelines I suggest are critical not for the creation of the artwork (that should remain without boundaries) but for the sharing of knowledge. Such guidelines can be seen as "rules of engagement" to allow for precise recording, preservation, and clear communication in a simple form.

This process of recording, sharing and curating knowledge, is something in line with the activities of academic institutions. In fact, many other disciplines have their specific knowledge is managed by the academia, mostly and foremost because that knowledge is key for academic activities namely teaching. It is within the academia such knowledge has the most impact by being available to emerging new practitioners. Therefore the schools of art are the privileged place for the storage and dissemination of knowledge in art research.

Two Cultures with Similar Approaches

My usual answer to the common question scientists asked first when I was developing "Nature?" at Paul Brakefield's laboratory in Leiden: "why do you claim to do art and not science, when daily in the lab you do the same things we do?" is in fact rooted in the ideas expressed above. Actually, although apparently we perform the same procedures in the laboratory, we are in fact doing quite different things. We simply happen to use the same technique and knowledge to approach a very different set of questions.

It appears to me that when most people consider art practice, they never think of it as the result of research. Most frequently this vision is correct, as perhaps the most visible side of Art does not require significant research. However, several examples can be found where art is a direct result of research in new methodologies, new media, and new practices. In many cases those advances do not just emerge from the work and research of only one artist, but rather from many artists, working around the world on similar concepts.

It should be noted, however, that not only this art research is mainly invisible to most people, but it is also frequently conducted without the support of an infrastructure (namely academic) by several individuals who pursue their own projects and have a variable impact on the field of art.

Art research and practice have always moved forward through the refusal to submit to control and "domestication", and art has always been proud to break the "canon". As I am an artist who has researched about art and science, I believe science, and biology in particular, offers good examples from where artists can learn to establish a new field of art research independent and at the same time complementary to art theory.

It is my understanding that science, as art, has developed from an individual point. What I mean is that it progressed from a series of individuals dedicating themselves to their experiments and hypothesis without the organized support of institutions and with difficulties in sharing ideas in the way we do nowadays (almost instantly). These early scientists shared their discoveries, of course, by letter, personal publications or even meetings, but nothing like today, even if contemporary scientists still get frustrated with the slow review of their papers prior to publication. The way science is conducted today is very far away from the conditions one century ago, when research was being conducted in small personal laboratories, often in the basement of scientists' own houses with only one or two assistants. Although today science is no longer conducted as an individual endeavor, it started out like that. However, at a certain point in its history it could not continue to progress at an individual scale, and changed its process to function more as a global system of collaborative efforts and not individual discoveries. We find today's science being produced in buildings and institutions where many researchers work together in the most varied fields, sharing not only the infrastructure, equipment and services but most importantly the company of fellow researchers. If we think of people like Madame Curie, Einstein or Darwin we are considering individuals that conducted their research alone, or nearly alone, before presenting their ideas publicly. Sometime in its history, science could not continue to progress as efficiently as it would at the individual scale and changed its process to function more like an organization based on a system of collaborations and team efforts that involved several areas of science. It does not make much sense anymore to say that any country is more developed scientifically than any other when so many publications are the result of international collaborations. Science has no nationality any more.

Likewise, there has been a tendency in the art of today to move towards a more collaborative endeavor. In many cases, being presented in festivals and exhibitions of new media, it is necessary to bring together a team of people with different expertise for the completion of an artwork. However, unlike research institutions in science, it is still infrequent to find in the arts a clear system providing collaborative support to the process of contemporary art research. In science, it has been increasingly difficult to identify the individual genius behind a discovery (as demonstrated in recent years by the increasing number of authors per scientific communication or the rarity of Nobel prizes that are not shared). In the same way, the idea of a single artistic genius is a myth that has been artificially preserved by artists, gallery managers, collectors and buyers for a very long time.

The advances in art are also, like in science, being supported by the shoulders of giants who preceded the more recent creators. In spite of this, in science and in art, there are a few researchers who push more intensively the general knowledge of the field and acquire a more prominent notoriety.

Art Research

I have the strong conviction that, like art itself, artistic research should be mainly a practical research field, without it being reduced to a mere simple exercise in art production. This does not imply that the artwork cannot be criteria for evaluation in art research. The artwork is the real proof or demonstration of a complex process, with investigative components. However, an artwork is not a thesis in itself (in principle), or even a hypothesis. If anything it can be compared with an experiment that contributes to the demonstration of an hypothesis. As such, the argumentation about the process of conception, research, production and even the result of its exhibition can be fundamental to the clarification of an idea and its contribution to the field of art. The way by which all this process adds to an already existing body of knowledge should be the major criteria for the possible evaluation of artistic research. In other words, the relevance of a research project should not be determined by the quality of the artwork itself but rather by the novel contribution brought to the artistic field: what new media, new technology, or new paradigm, is provided by the artist/researcher that can benefit other artists that will use these knowledge for their own research and/or art practice.

The following diagram represents a schematic view of the process/method of the development of an art project (process/method which is not exclusive to the art world, nor it is universal). If we take a typical art project and divide it by stages of development we can distinguish (Figure 4):

This way, and following the arrows, one can observe that with very little or no investigation, an idea can move very quickly to being produced or constructed to become an artwork to be publically exhibited. However, the research developed can originate a set of ideas leading to new directions of research and even to the alteration of the final artwork to be produced.

In a similar way, the exhibition of a project while it is still in research stage (work in progress), or while the final shape/form is still just a prototype of the possible final product, can be subsequently reformulated in response to the public response.

I will again illustrate this process with an example from my own work.

The Brain Inside the Portrait

Artists have been trying to represent in portraits not only the physical appearance of the subject, but also characteristics of the personality, by using the pose, elements in the composition, and even the choice of technique. Science has developed powerful techniques to visualize what is hidden under the skin: and not only morphological characteristics, but also functional data. In this work I have been using functional magnetic resonance imaging (fMRI) of the brain, in order to visualize the regions of the brain that are active while a given task is being performed. With this visual information it becomes possible to create portraits – functional portraits – where besides the physical appearance of the subject, the function of its brain while performing a chosen task is represented. For example, I created a self-portrait incorporating images of my brain activity while drawing inside the fMRI scanner; a portrait of scientist Dr. Patricia Figueiredo allows the visualization of her brain activity while she was playing the piano (Figure 5).

Figure 4. Diagram
Menezes, Marta

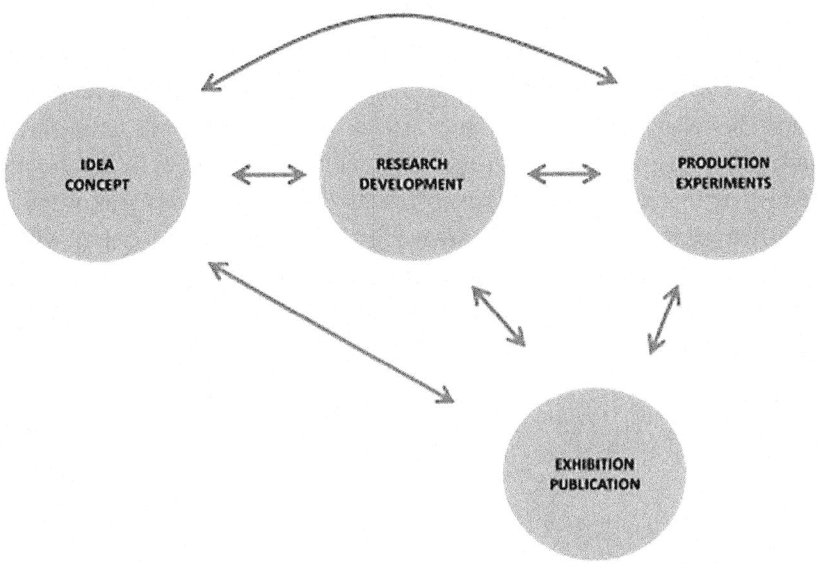

Figure 5. "Functional Portraits: Patricia Playing the piano" FMRi scans, photographs and sound
Installation image at Marta de Menezes BioArt exhibition in S. Miguel, Azores, Portugal

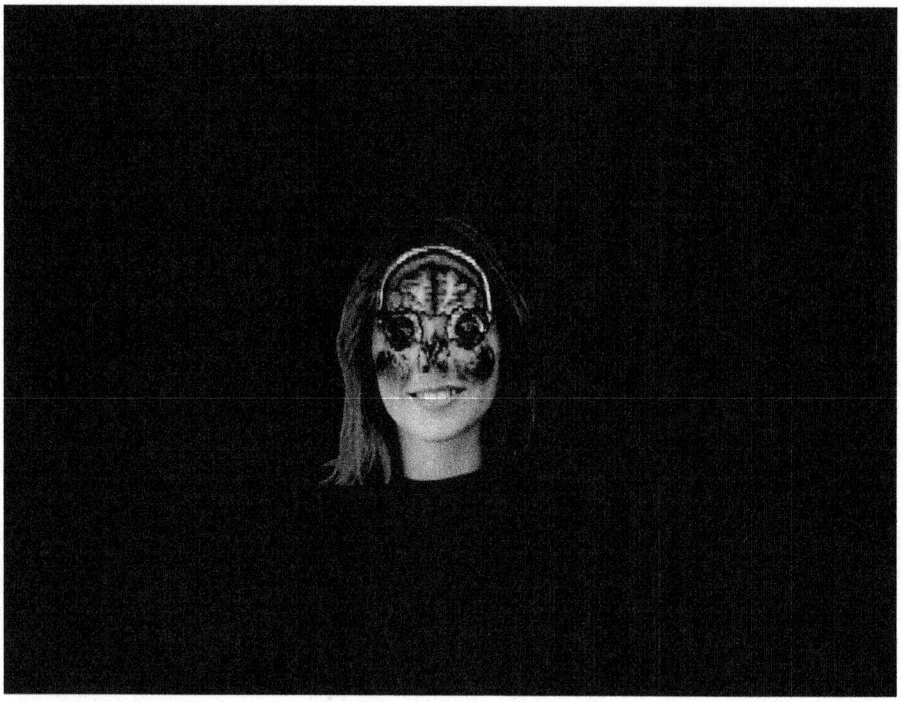

Like in "*Tree of Knowledge*", all phases of any given artistic project can and are being continuously re-evaluated and the progress is not unidirectional or even unique, leading necessarily to a single artwork. The most important conclusion is to understand that the art research happens on any of those steps and processes. This research can be solely referring to the artist making the artwork, but should always contribute to the common knowledge of the field of art in general, and being relevant to the art community. The research process described can even refute, in time, the scheme I have proposed.

In any of the possible outcomes, I make the case that it does not matter how personal the project may be, like a self-portrait. The key issue is that it must create precedence, experience, perspective useful for the author (of course) but also potentially for other artists, also looking for different strategies to develop their work, to think about their research and to proceed with their practice. In fact I am grateful to many different artists that contributed with their own art research to the development of my own art practice.

Therefore I reaffirm that the question in art research as a discipline is not whether we produce knowledge as art researchers, but how to sift that knowledge to separate the personal relevance from the information that is valuable to the peers. And, in addition, how to simply and eloquently present the knowledge while saving it for future use and further developments.

REFERENCES

Bulatov, D. (Ed.). (2004). *Biomediale. Contemporary Society and Genomic Culture*. Kaliningrad: The National Centre for Contemporary Art.

Catts, O., & Zurr, I. (2002). The Aesthetics of Care?. SymbioticA, School of Anatomy and Human Biology, UWA, Perth.

Eisner, E. W. (1981). On the Differences between Scientific and Artistic Approaches to Qualitative Research. *Educational Researcher*, *10*(4), 5–9. doi:10.3102/0013189X010004005

Gauch, H. G. Jr. (2003). *Scientific method in practice*. Cambridge, UK: Cambridge University Press.

Judson, H. F., & Maddox, J. (1979). *The Eight Day of Creation: Makers of the Revolution in Biology*. London: Jonathan Cape.

Kac, E. (Ed.). (2007). *Signs of Life – Bio Art and Beyond*. Cambridge, MA: MIT Press.

Lesage, D. (2009, Spring). "Who's Afraid of Artistic Research? On measuring artistic research output" (PDF). Art&Research - A Journal of Ideas. *Contexts and Methods*, *2*, 2.

Menezes, M. (2003). The Artificial Natural: Manipulating Butterfly Wing Patterns for Artistic Purpuses. *Leonardo*, *36*(1), 29–32. doi:10.1162/002409403321152257

Menezes, M. (2011). *"(Use)less is More! Is the artist a master of the useless?" in Useless, by Max Bruinsma and Federico Duarte for Experimenta Design 2011*. Lisbon: Babel.

Miller, A. (2014). *Colliding Worlds*. London: W.W. Northon and Company.

Pandilovski, M. (Ed.). (2008). *Art in the Biotech Era*. Experimental Art Foundation.

Shapiro, A. (2011). Of Art, Biotech and the Body in the World, by Claudio Cravero. *The Illusion Beyond Art*. Retrieved from http://www.alan-shapiro.com

Snow, C. P. (1998). *The Two Cultures*. Cambridge University Press.

Webster, M. (2011). *m-w.com*. Encyclopædia Britannica.

KEY TERMS AND DEFINITIONS

Art: Art is a diverse range of human activities in creating visual, auditory or performing artifacts – artworks, expressing the author's imaginative or technical skill, intended to be appreciated for their beauty or emotional power.

Art Researcher: One of the characteristics of artistic in opposition to scienctific research is that it must accept subjectivity and a relevant and fundamental factor in the discovery methodologies, something very contrary to the classical scientific methods. As such, it is similar to the social sciences in using qualitative research and intersubjectivity as tools to apply measurement and critical analysis.

Bioart: BioArt is an art practice where humans work with live tissues, bacteria, living organisms, and life processes. Using scientific processes such as biotechnology (including technologies such as genetic engineering, tissue culture, and cloning) the artworks are produced in laboratories, galleries, or artists' studios. The scope of BioArt is considered by some artists to be strictly limited to "living forms", while other artists would include art that uses the imagery of contemporary medicine and biological research, or require that it address a controversy or blind spot posed by the very character of the life sciences.

Biology: Biology is a natural science concerned with the study of life and living organisms, including their structure, function, growth, evolution, distribution, and taxonomy.

Knowledge: Knowledge is a familiarity, awareness or understanding of someone or something, such as facts, information, descriptions, or skills, which is acquired through experience or education by perceiving, discovering, or learning.

Research: Research can be defined as a studious inquiry or examination; especially investigation or experimentation aimed at the discovery and interpretation of facts, revision of accepted theories or laws in the light of new facts, or practical application of such new or revised theories or laws.

Science: Science is the systematic knowledge of the physical or material world gained through observation and experimentation.

Section 2

Aesthetic Experience and Emotion in Art and Design:

Crossing Domains in Philosophy, Psychology, and Neuroscience

Chapter 4
Neuroesthetics:
Perspectives and Reflections

Fernando Fogliano
Centro Universitário Senac, Brazil

Hosana Celeste Oliveira
Universidade Estadual Paulista (UNESP), Brazil

ABSTRACT

This paper presents some considerations regarding the aesthetic experience from a neuroaesthetic point of view. The approach proposed here can take the discussions beyond the field of art allowing the aesthetic experience to be considered as a key aspect of subjectivity and interactions in the world. A reflection on the roles of emotion and language as fundamental aspects of the aesthetic experience is presented. Based on this reflection, some assumptions that are at the core of culture are made and re-framed here considering the postmodern turn.

INTRODUCTION

The paper presents some considerations regarding the aesthetic experience from a darwinian scenario aiming at providing evidence that reviews important aspects of culture in postmodernity. In this scenario the aesthetic experience is seen as responsible for mediating and structuring forms of knowledge, assisting in organizing the essence of things, allocating attention and categorizing objects (Ramachandran, 2014; Brown & Dissanayake, 2009). The aesthetic experience is therefore considered an essential aspect of environmental and social relations, paving the way for the construction of memories and reality itself.

The text shows how some biological laws, incorporated during the evolutionary process and which are common to all living beings, build the aesthetic experience as well as being responsible for creating sensory categories and patterns that promote selective attention for decision-making and problem solving. Such laws cross cultural barriers and are experienced by living beings to a greater or lesser degree of sophistication. According to neuroscientific findings of recent decades, emotion would be the trigger for these laws to be activated in the brain.

DOI: 10.4018/978-1-5225-0510-5.ch004

In the darwinian scenario neuroesthetics appears as an emerging field of knowledge that brings, based on scientific advances arising from the neurosciences, important contributions not only to deepen the understanding of the artistic phenomenon but also of cultural practices in various fields of human activity. The studies in neuroesthetics by Ishizu and Zeki (2011) are given prominence. The theoretical model of the authors inspires the correlation between language and the aesthetics experience, and both are discussed in parallel in this paper because they are deeply interconnected, since it is through language that the aesthetic experience is achieved.

In addition to Ishizu and Zeki's neuroesthetics model, the one proposed by Brown and Dissanayake (2009) is also considered. These two authors assist in the comprehension of the role of emotion in the aesthetic experience and help to conceptualize it. As defined by Brown and Dissanayake, the aesthetic experience is understood as the emotional appreciation of an object (Brown & Dissanayake., 2009), be it of the physical world, a work of art or something abstract.

This scientific notion of aesthetics, underpinned by a theory of emotion, is not new and dates back to Darwin, who already foresaw the aesthetic experience as a class of emotions whose biological function is to generate the critical evaluation of the object's properties, and to extract from it important information for survival. Usually, emotional appreciation of an object is discussed by neuroesthetics based on issues in which the artwork is seen as a class of stimulus intentionally designed to induce a variety of affective, emotional, perceptual and cognitive responses in the subject (Seeley, 2011).

The aesthetic experience is presented in its relations with language, thought, conscious experience, attention and subjectivity. Language is emphasized because of its important role in the development of social constructs and cultural practices. Therefore in addition to the evolutionary point of view, cultural and social issues related to taste variations and individual preferences are also considered to be equally important aspects and components of the aesthetic experience.

SCIENCES AND HUMANITIES: A NECESSARY CONVERGENCE, NOT AN INEVITABLE CONFRONTATION

Before initiating the considerations on the aesthetic experience, it is necessary to discuss the importance of the convergence between sciences and humanities to rethink culture.

Placing distinct fields of knowledge side by side may still be surprising to some. This occurs even after the emergence of multi-disciplinary fields of knowledge such as the theories that explore complexity. Nevertheless, ecology, for example, and the integrated vision that it provides, has gained importance.

The distinction between the fields of knowledge is, in Wilson's view (1999), "artifact of erudition". The natural sciences provide the epistemological foundations to address a whole variety of issues about the universe around us. One consequence of this approach is the need to consider knowledge as a unique and complex body. The perception that the universe is a fragmented body made up of a compound of autonomous disciplines is illusory (Fogliano, 2002).

"Trust in consilience is the foundation of the natural sciences. For the material world at least, the momentum is overwhelmingly toward conceptual unity. Disciplinary boundaries within the natural sciences are disappearing, to be replaced by shifting hybrid domains in which consilience is implicit. These domains reach across many levels of complexity, from chemical physics and physical chemistry to molecular genetics, chemical ecology, and ecological genetics. None of the new specialties is considered more than a focus of research. Each is an industry of fresh ideas and advancing technology" (Wilson, 1999, pp. 9-10).

The idea of consilient knowledge is fundamental in areas beyond the natural sciences. This view of knowledge is a demand of contemporaneity. Historically, especially in recent decades, we have seen science refocus its instruments and methods of observation on questions previously only dealt with through syllogisms and the qualitative observation of phenomena.

Positron emission tomography, for example, has opened up enormous possibilities for brain studies. This imaging technique has enabled a new type of observational access to the brain's functional dynamics, providing an unprecedented accumulation of knowledge about the most complex organ in the human body. At the end of the 1990s, neuroscientists were able to expand lines of research, like those initiated by the Gestalt school in 1920, which later received important contributions such as those of Rudolf Arheim. In the twenty-first century, the cognitive sciences continued on their evolutionary path leading to a number of lines of research and giving rise to new fields of knowledge like Neuroesthetics (Onians, 2007).

New findings called into question certain long-standing concepts in many areas of knowledge. Lakoff and Johnson (1999) consider that more than two millennia of a priori philosophical speculation have met their end because of the accomplishments of cognitive science.

Due to scientific advances, philosophy will never be the same again. According to Gibbs (2007, pp. 1-13) the separation between body and mind established in traditional philosophy severely limits the academic studies of mental life. This idea of separation, he claims, has been haunting the history of Western philosophy from Plato and Aristotle to St. Augustine, Descartes and Kant.

For Descartes, the mental phenomenon has no place in the physical world. The philosopher had to assume the existence of a type of indivisible and intangible matter, the *res cogitans*, to support the existence of the mind, its thoughts, desires and volitions. cartesian dualism established a philosophical tradition in the West that crystallized the body as a solid object and the self, particularly the mind, ethereal in nature, mysteriously infused in the body.

Damásio (1994) will refer to cartesian dualism and the need to overcome it in the face of scientific evidence that showed where emotional thought takes place in the brain, and its significant influence on reason. In the 1950s, advances in linguistics, which depended on the use of computers and complex software, enabled the visualization of how cognitive processes can be generated from complex shapes using representations of things that are external to the computer. Stapp (1993, p. 21) wrote about this kind of analogy between the brain and a computer:

... an analogy can be seen between the brain and a computer, and with the help of mathematics developed in connection with artificial intelligence, some reductionist concepts about the brain were created. Some of those concepts were transferred to psychology, but they proved to be too elementary as they were based on the simplicity of the behaviourist approach that fails to account for all the complexities of human behaviour.

A superficial observation suggests that the impact of new scientific knowledge encounters fewer restrictions when dealing with the scientific field itself. However, when science and humanities are at stake, the difficulties seem to intensify. An emblematic example of these difficulties can be encountered when reading Gombrich's comment (2000, p. 17) on the article by Ramachandran and Hirstein (1999), *The science of art: A neurological theory of aesthetic experience*. In his remarks the art historian seeks to disqualify the scientists' opinion stating that:

To the historian of art, it is evident that the two authors' notion of 'art' is of very recent date, and not shared by everybody. They claim: 'The purpose of art, surely, is not merely to depict or represent reality – for that can be accomplished very easily with a camera – but to enhance, transcend, or even to distort reality' (Ramachandran & Hirstein, p. 16). They do not explain how one could photograph Paradise or Hell, the Creation of the World, the Passion of Christ, or the escapades of the ancient gods – all subjects that can be found represented in our museums. Nor is it more legitimate to generalize from certain Indian conventions of representing the female nude than it is for the academic tradition to take the Venus de Medici for the same purpose. Even a fleeting visit to one of the great museums might serve to convince the authors that few of the exhibits conform to the laws of art they postulate (Gombrich, 2000, p.17).

Massey (2009, p.18) reflects on the resistance between the fields of knowledge that he summarizes as a clash between the "how" and the "why". The author seeks to recognize the strength and weaknesses of each field, admitting how science can be more efficient in analytical reasoning, while the humanities are more so when it comes to synthesizing concepts. To that end, he exemplifies how neurology is apt at locating the components of the aesthetic experience in the brain, but is deficient in providing a better understanding of the production of *The Magic Flute* by Mozart.

Neuroscience is competent at connecting certain characteristics of aesthetic processes to specific events in the brain. However, artists and viewers are engaged in the entirety of the processes, being much more interested in the phenomenology of the artistic event, than in its scientific aspects (Massey 2009., p. 19). Nevertheless, artists are now using neuronal activity pattern sensors in the search to establish links between the narratives of their aesthetic experiences with the interagent. This fact leads to the realization of how interwoven the fields of knowledge are, and that scientific and technological advances reverberate throughout all areas of culture and knowledge.

The quest for knowledge, regardless of the methodological approach, encounters difficulties in any of the methods employed. Recognizing methodological possibilities and constraints inherent to science and humanities does not necessarily mean reducing them to a common denominator. Rather, this difference between points of view and methods should be celebrated as an important aspect of the diversity of alternatives in the search for knowledge. Therefore, their irreducibility is celebrated (Massey 2009 p. 22). It is of utmost importance, however, to recognize the need to establish conceptual bridges between the fields. This search should not be thought of as a mere epistemological exercise, but as an imperative that emerges from the demands presented by the scientific advances and the new technologies that confront the researchers with concrete and conceptual issues that are increasingly complex and multimodal. The fact that these issues are coming up is an important symptom that paradigmatic changes are on the horizon of the events.

Scientific advances, especially those from the neurosciences, and now from neuroesthetics, are at the heart of what Bryson (2003, p.19) considers the new arena of cultural development, "the neural interface". The new perspectives provided by scientific advances allow the products of art to be considered as being potentially able to directly access the internal brain activity, creating settings of image, space and time, forging new paths in the mind. The experience of reality stems from inter-subjective social constructs and can be considered from the point of view that it emerges from the collective triggering of neuronal synapses, sensory systems and consciousness, placing at the centre of our existence not the significant, but rather neural patterns and man's action in the environment.

LANGUAGE AND THOUGHT

For Pinker (1995, p.18) language is much more than a cultural artefact used to share information or say how certain systems work. Language is a distinct piece of our brain's biological machinery. Complex in nature, this specialized skill develops in children spontaneously, without effort or formal instruction. Among its many definitions, as for instance mental organ, neural system, or computer module, the one adopted herein is the one that defines language as an "instinct".

Donald (1991, p. 44) states that the instinct for language is the emergence of a process that occurred during the evolution of the human species since its inception. A history of this evolution, and that is based on Darwin's ideas, can be told through three stages. The first describes a pre-linguistic scenario in which the use of symbols depended on improvements in the ability to think. The second is determined by the use of vocalizations, although intelligence, gestures and facial expressions are also important elements in this evolutionary phase of language. The third step, according to the author, is related to the emergence of cognitive abilities dependent on language. This statement implies that thought in modern humans has at least two aspects: one that does not depend on language and is therefore archaic; and one that does depend. This condition establishes the need for linkage between non-linguistic representations with those representations immersed in language. This finding suggests the importance of language as a tool to supporting thought.

In this sense, Dennett (1994) wonders what happens to the brain of a human or hominid when it becomes equipped with words. To answer this question he considers the importance of language, analysing the foresight with which these beings are endowed. This kind of thinking is processed through the operation of mental models that allow them to predict future consequences of their actions without having to bear the consequences of these actions. Language is a tool that enables the articulation of long streams of thought, such as those required for anticipatory thinking. These aspects underlie other thought processes related to decision-making. The sophistication of these processes is fundamentally linked to language and provides a great evolutionary advantage.

Another important consequence of the emergence of language, and perhaps the most easily recognized, is the cultural one. A common language connects community members in an information-sharing network, providing individuals with formidable powers (Pinker, 1995). Language permeates successive refinement processes and discoveries of new standards in all areas of knowledge, in an accelerated and highly efficient manner. This aspect offers culture a characteristic of innovation and diversity. According to Dennett (1994, pp. 161-178):

Our human brains, and only human brains, have been armed by habits and methods, mind-tools and information, drawn from millions of other brains to which we are not genetically related. This, amplified by the deliberate use of generate-and-test in science, puts our minds on a different plane from the minds of our nearest relatives among the animals. This species-specific process of enhancement has become so swift and powerful that a single generation of its design improvements can now dwarf the R-and-D efforts of millions of years of evolution by natural selection.

Language is certainly the driving element of human culture and, in this context, alters biology's main role in the evolutionary processes by installing culture alongside it. Culture and evolution are inextricable, and language is the element of cohesion between these two areas of our existence. This statement is quite reasonable if one considers that, from a biological point of view, human beings barely differ from their ancestors of fifty thousand years ago. The difference between lifestyles, artefacts, tools and the complexity of the concepts that guide people's decisions is due, fundamentally, to the emergence of the culture that pervades the environment around them. For Gibbs (2007) man's body is not an object separated from culture. All aspects of his experience in the world are shaped by cultural processes, since conceptual theories and systems are inherently cultural.

The body functions as mediator between human beings and the spaces that surround them. And culture is transported in the semblance that the body projects in relation to the space, considered here in a broad sense, because the body that makes culture produces meanings and symbolism, materializes abstract ideas, materializes experiences. The holder of a language is the one who found the "minimum conditions to name their actions" (Almeida, 2001, p. 32). The issue involving language and experience is vital to building culture as seen throughout this text. For the moment it is important to reinforce the idea that language and cognitive processes are inseparable from culture.

A key aspect in the psychology of Vygotsky (Kaptelini & Nardi, 2006, p. 39) is the relationship between the mind, on the one hand, and culture and society on the other. These latter elements are not mere external agents, influencing the mind. On the contrary, they are made up of driving forces directly involved in the mind's production. Consciousness emerges in the body's interaction with culture through language in all its diversity. After the "invention" of natural language, culture enabled and continues enabling the emergence of a variety of new languages. The complexity, variability and the growing number of these new languages are perhaps the most emblematic aspect of the inventiveness that characterizes human culture (Fogliano & Camargo, 2010).

LANGUAGE AND CONSCIOUS EXPERIENCE

Kaptelinin and Nardi (2006), in the quest to build an integrative theoretical field for studies on the man-machine relationship and the issue of interactivity, discuss the unity of consciousness and activity. The latter is at the heart of their activity theory, which is defined as an "intentional interaction of the subject in the world, a process in which mutual transformations between the 'subject-object' poles are produced" (Kaptelinin & Nardi 2006., p. 31). Building on the concepts from the Russian school of psychology, notably on Vygotsky's ideas, the authors will define the concept of human mind as being:

Intrinsically related to the whole context of interaction between human beings and the world, that it is an organ of a special kind, emerging and developing in order to make interaction with the world more successful (Kaptelinin & Nardi 2006 ., p. 37).

From these considerations it can be concluded that consciousness has a very strong bond with the interactive experiences that are taking place in the environment. The idea that body and mind are inextricable is not only at the heart of contemporary philosophy, but also in the theories of language. Feldman (2006), when considering the neural basis for language, pursues in the subjective experience the idea that language originates in concrete experience. For the author "thought is structured neural activity. Language is inextricable from thought and experience" (Feldman, p.1).

In the neural theory context, language evolves from the concreteness of the individual's physical experience in his or her environment to the abstraction, built metaphorically (Lakoff & Johnson, 1999). The subjective experience is the basis from which abstract and conceptual words arise. Neurons and body are central to this process: people, as neural systems, understand abstract ideas because these concepts are mapped and activated in brain circuits involved in the experience. The metaphor is therefore not merely a linguistic trick or cultural figuration. Ramachandran and Hirstein (1999) also consider the metaphor to be a fundamental cognitive mechanism for the production of language, when defined as a mental "tunnel" between two concepts and percepts which seem dissimilar:

When Shakespeare says 'Juliet is the sun,' he is appealing to the fact that they are both warm and nurturing (not the fact that they both reside in our solar system!) (Ramachandran & Hirstein 1999., p. 31).

Through images taken using positron emission tomography, Ramachandran and Hubbard (2003) describe the neural mechanisms and brain structures related to metaphor production. The discourse on metaphor and culture shaped a paradigm shift in what concerns our understanding of creativity and knowledge acquisition. Based on these studies it can be argued that the most important developmental mechanisms of language and culture are supported in the neuronal machinery.

Experience, language and consciousness emerge from the bustling activity of neurons and shape reality, produce our individuality, social groups and all concrete and conceptual manifestations of culture. Understanding the nature of subjective experience, or *qualia* as the term was coined in the context of philosophy, is a very complex task and lies beyond the scope of this discussion. However, Aleksander's (2005) ideas about the subjective experience will suffice as a guide to the reflection proposed herein. According to the author, consciousness can be considered from five basic axioms (Aleksander 2005., p. 35):

1. It is part of, but separate from an "out there" world.
2. The perception of the world mingles with feelings of past experience.
3. The experience of the world is selective and purposeful.
4. Consciousness thinks ahead all the time in trying to decide what to do next.
5. Feelings, emotions and moods that determine what you do.

The axioms above provide knowledge about the main aspects of consciousness and can be applied both in the development of conscious machines and in the assessment of which aspects are present in the experiences in which one takes part in the world.

It is important to expand the concept of language beyond spoken language, which is the way it is usually thought of when considering the issue. For Johnson (2007) Western culture has over-appreciated the value of the word, and art has never been seriously considered as an essential mode of engagement with the underlying world. For Lakoff and Johnson (1999) the study of language can constitute a unified field of study of the languages of words, sounds, movements and all other forms of human expression such as photography, film, music, theatre, and design in all its fields. In the visual arts, images and their patterns, qualities, colours and rhythms are carriers of meaning (Fogliano & Camargo, 2010).

From this perspective, art can be considered a type of behaviour, a strategic game designed to engage human attention through its appeal to the preference for rich information patterns (Boyd, 2009, p. 85). It is important to stress that attention is one of the aspects of consciousness and that in this sense, we can consider that the game referred to by Boyd is a strategy to provoke conscious experiences.

This game takes place in a complex environment to allow socially developed minds, especially human minds, to access larger networks of abstract or concrete knowledge modules. This access enables the confrontation of new contexts, the evaluation of information and the production of inferences and scenarios for decision-making. This process occurs boosted by emotional systems, as described by Damásio (1994).

As previously mentioned, these systems constitute consciousness, which also has its evolutionary history, in which the emergence of language plays a decisive role. In this evolutionary process, the more complex does not supplant the more simple, but incorporates it in new ways creating new contexts or levels of complexity, which favour the development of new functions (Damásio 1994.). This may well serve as a definition for the emergency phenomena and applies similarly to the definition of metaphor.

LANGUAGE AND ATTENTION

Art as a mechanism for capturing the group's attention offers an interesting aspect in the understanding of art itself and as a producer of social cohesion.

To explain art we need to attend to attention. Art dies without attention, as people since Aristotle have noted, both within and outside evolutionary explanation. Art alters our minds because it engages and reengages our attention from nursery rhymes to rest-home singalongs. But art has rarely been considered in terms of the special role that engaging attention has evolved to play in human lives (Boyd, 2009, p.100).

Ramachandran and Hirstein (1999), when describing the function of exaggeration in the artist's quest to capture the essence of things and produce in the viewer an emotional or psychological state, wrote:

There may be neurons in the brain that represent sensuous, rotund feminine form as opposed to angular masculine form and the artist has chosen to amplify the 'very essence' (the rasa) of being feminine by moving the image even further along toward the feminine end of the female/male spectrum (Plate 4). The result of these amplifications is a 'super stimulus' in the domain of male/female differences (Ramachandran & Hirstein., p. 18).

A link between the ideas of Boyd and those of Ramachandran and Hirstein can be established. One of the best ways to co-opt the viewer's attention is to offer them a "super stimulus" capable of providing them a rich sensory experience. If so, the artist's role is to invent linguistic resources to build narratives capable of producing sensory stimuli, and drawing the group's attention to abstract or concrete aspects, present in the clashes with the environment. Most of the time, this activity, which makes the artist a sort of inventor, is related to techniques and technologies. The narratives produced in the field of art, as a result of what has been considered so far, can be seen as a cognitive strategy in the search for alternatives in the construction of reality.

SUBJECTIVITY

When considering the field of neuroesthetics, Changeux assigned art the role of establishing "intersubjective communication" that "implies motivations and emotions in harmony with reason, but with no apparent progress and constant renewal" (2013, p. 97). This activity is directly related to the neurocognitive processes that promote a greater development of consciousness and brain organization, leading to the expansion of the cerebral cortex, and "more particularly of the prefrontal and parietotemporal association and cingulate cortices, in close collaboration with the limbic system" (Changeux 2013., p. 98). This perspective seems to offer a universal and necessary approach to art as a strategic activity in our quest for permanence, allowing it to be considered as an evolving cultural trait of the human species.

Taking into account the impossibility of objective access to art and its production, it is appropriate to point out that although this issue arose in the context of reflection on art, with the case of the artistic avant-gardes (Fabbrini, 2006), it is not new in other fields of knowledge. The early twentieth century witnessed developments arising from the fields of physics and mathematics that dismantled the "myth of objectivity". Goldstein (2005) lists Gödel's incompleteness theorems, Einstein's theory of relativity and Heisenberg's uncertainty principle among these scientific developments. The impact of the ideas behind these immense contributions to culture cannot be considered in depth here, thus only a few of their consequences, particularly regarding the incompleteness theorem, will be considered.

For Goldstein the consequences of the incompleteness theorems have great impact:

The necessary incompleteness of even our formal systems of thought demonstrates that there is no nonshifting foundation on which any system rests. All truths – even those that had seemed so certain as to be immune to the very possibility of revision—are essentially manufactured. Indeed the very notion of the objectively true is a socially constructed myth. Our knowing minds are not embedded in truth. Rather the entire notion of truth is embedded in our minds, which are themselves the unwitting lackeys of organizational forms of influence. Epistemology is nothing more than the sociology of power. So goes, more or less, the postmodern version of Godel (Goldstein 2005., p. 25).

The inexorability of subjectivity places the construction of reality under the responsibility of collaborative, inter-subjective and essentially contradictory collective processes, as Gödel's paradoxes dismantled the law of excluded middle. The potentially conflicting readings referred to above are part of this scenario that herein is called postmodern. The universe that human beings are able to describe is precarious and contaminated by contradictions, being hybrid in its essence. The notion of purity has already came to an end in the world of biology and genetics. This same idea, associated with the law of excluded middle, also lost support in the field of logic, enabling the emergence of paraconsistent logic that perceives the universe taking contradiction into account.

AESTHETIC EXPERIENCE AND EMOTION

This topic intends to present a summary of Brown and Dissanayake's ideas (2009, pp. 48-50) in order to establish how art's narratives can modulate emotion in different experiential contexts.

According to those authors, an understanding of aesthetics should be grounded in a theory of human emotion, which includes the dimensions of *valence*, *intensity*, and *focus*. *Valence* refers to the fact that the vast majority of emotions fall binarily into the categories of positive and negative, good or bad. In contrast to this division, emotions may vary in a continuum of *intensity* along a graded scale from weak to strong. *Focus* describes emotion categories related to the narrative context: *outcomes*, *objects*, *agency*, and *social interaction.*

- The first *category of focus*, *outcomes*, deals with valenced reactions to the consequences of outcomes. They are the motivational emotions, ranging from happiness to sadness.
- The second category, that of *objects*, features valenced reactions to the aspects of objects and events. It is this category that comprises the aesthetic emotions, encompassing the range from liking/attraction to disliking/disgust.
- The agency, the third category of emotion, belongs to valenced reactions to the actions of agents and relates to the spectrum from approval to disapproval. The agency category comprehends what people think of as moral emotions, including assessments of praiseworthiness or shamefulness.
- The fourth *category of focus*, *social interaction*, involves valenced reactions to social interactions with others; and although similar to the moral emotions discussed in the previous category, it can be better understood by the comfort/discomfort emotional spectrum. When a subject feels threatened by someone else, and sees his opponent as being better or more competent than himself, or that the opponent can contribute to or frustrate his goals, it is this fourth category of emotion that is experienced. On the plus side, this involves emotions like love, trust, and affiliation, the feeling of belonging or wanting to belong to a group, that are at the heart of the experience of the arts.

This paper suggests that one of the most significant emotions that drives the arts is social affiliation, an emotion of strong reward value. Based on this idea, the arts create and reinforce a sense of social unity so as to promote cooperation and cohesion within social groups.

On the negative side of social interaction is a basic emotion that, like disgust, has complex and multifocus connotations, namely fear. Brown and Dissanayake say that fear "is an 'outcome' emotion related to predictions of negative consequences for future events" (Brown & Dissanayake 2009., p. 50). However, when considering the context of social interactions, fear is also a strong emotion related to the perception of people's intent to thwart the goals of others or hurt them.

Attentively reading emotions in different contexts can explain the assertiveness and ambiguity of the narratives present in portraits, as well as the socio-cultural source and importance of this narrative form (Fogliano, 2014).

NEUROESTHETICS: NEW DIRECTIONS IN THE DEBATE ABOUT THE ROLE OF ART

In recent years, the reflection on art has received important contributions that have rapidly and significantly transformed its conceptual bases. Perhaps the most striking are those from the scientific field. The use of functional magnetic resonance imaging (fMRI) has enabled the visualization of the dynamics of brain processing, making way for new hypotheses and considerations to be tested and established. Said contributions seem to present alternatives that shed light on issues that have been accompanying the evolution of Western thought about art and the aesthetic experience throughout many centuries.

Goguen (2000) discusses the evolution of the concept of art in the field of philosophy and points out that beauty is always taken as the measure of quality for art. This simple definition hides, however, a number of conceptual problems that have been the subject of debate in this field of knowledge during centuries. In that text, the author puts forward the best known approaches to this topic, all of them accompanied by pros and cons, and how these inconsistencies are part of the thought about art.

At one point, Goguen is forced to admit his difficulty or even impossibility to define art and beauty as well as to suitably classify the relationship between art and science. This would stem from the fact that "precisely what art is or what its role is in our lives" (Goguen 2000., p. 13) is unknown. Furthermore, that all brain activity that corresponds to the aesthetic experience is unconscious and thus inaccessible, since the aesthetic experience occurs beyond consciousness, in a fusion of objectivity and subjectivity of the individual taken to a state of "ecstatic epiphany" (Goguen 2000., p. 14).

Bringing an evolutionary perspective to the reflection on artistic activity, Thornhill (2003) uses scientific tools originating from darwinian thinking. Darwin's theory about brain design, human or nonhuman, relies on specific psychological adaptations. As with blood pressure regulation and habitat selection, aesthetic considerations are guided by different functionally specific psychological adaptations. In short, the topics regarding the aesthetic experience arise fundamentally from psychological adaptations (Thornhill 2003., p. 9). In this regard, it can be considered that the experience of beauty occurs unconsciously through adaptation processes internalized throughout evolutionary history.

Thus, the author deems that Darwin's theory of evolution can provide support to the idea that aesthetic values in humans derive from the promise of functionality in the environments in which they evolved. Consequently, aesthetic judgments are related to the probability of survival and perpetuation of the species. In this context, ugliness is related to the low likelihood of survival and reproduction. With this pair of opposites, Thornhill considers that aesthetic value lies on a scale between reproductive success and failure. This scale was engraved in human brains during the last few million years.

The theory of evolution has had a great impact on the reflection on art and the experience of beauty in art, as well as in areas besides the artistic phenomenon. Voland and Grammer (2003, p.1) underline the importance of evolution in the judgements developed in a wide range of human activities that involve their social and non-social environments. Basically, these judgments can be regarded in two domains: the things people like and those they do not like.

People's aesthetic preferences shape their behaviour, and this is reflected in the decisions whether to pursue or avoid things that are in the surrounding environment. The authors establish a retrospective, listing the different aspects involved in Evolutionary Aesthetics' successful contribution to the understanding of aesthetic experiences. After reading Konrad Lorenz they emphasize that the author considered that a human's cognitive structures are in fact adaptations arising from troubleshooting established in the human phylogenetic past and that:

In our past, those brains (or cognitive algorithms) that were able to process environmental information more efficiently were those who were selected for, generation after generation. However, Lorenz went further, he argued that in the course of evolution certain environmental stimuli which promoted reproductive success were connected to positive emotional reactions (as cited in Grammar & Voland, 2003, p. 4).

Scholars have found evidence to support the idea that human beings have specific adaptations in behavioural domains that are founded in the aesthetic field. An emblematic example of that evidence comes from studies showing that photos of attractive individuals activate specific brain regions. Emotion, an important achievement of the evolutionary process, as shall be seen later, is an important aspect of the conscious experience, and particularly of the aesthetic experience. Recent studies have shown that the creation of memories resulting from the conscious experience produced in emotional contexts occurs more efficiently (Phelps, 2004). This relationship between memory and emotion was internalized throughout the evolution of the human species, as described by Tooby and Cosmides in *Evolutionary Psychology and the Emotions* (2008).

The result of the research carried out by Ishizu and Zeki (2011), is an emblematic example of a scientific contribution that has provoked major changes in the scenario involving the Art and Science relationship in the debate on Aesthetics. In this work the authors present a theory of beauty and propose a definition of beauty based on neurocerebral dynamics that allow a better understanding of many aspects of the aesthetic experience, answering questions and solving secular conceptual impasses.

In this study, the cortical activity of several individuals was analysed while they were being exposed to auditory and visual sensory stimuli produced by music and images. Individuals were asked to evaluate the stimuli in terms of their beauty or ugliness. These subjective assessments were related to the measurements of the corresponding cortical activities. A strong correlation was noticed between the activity in a region located in the medial orbitofrontal cortex (mOFC) and the positive evaluations, related to the conscious experience of beauty. The results of this scientific work will still be greatly discussed both in the field of neuroesthetics and in the fields that classically study the subject. However, it is possible, although preliminarily, to list some interesting consequences for the discussion intended herein.

The first, and most important, it is that there must be a connection in the cortical processing that connects emotions, valuation processes, desire and beauty to positive judgments such as reward and pleasure (Ishizu & Zeki, 2011, p. 7). There is an important correlation between mOFC activation and the activation of other cortical areas (caudate nucleus) that in previous studies, were correlated with the cortical response to romantic love. Such findings indicate that there may be specializations in the brain for the creation of the experience of beauty and ugliness. It was found that ugliness is treated differently, in a more confined form in the amygdala and somatic motor cortex, while beauty, as was pointed out previously, recruits a greater number of cerebral cortices. On contemplating Damásio's contribution to neural dynamics on the evocation of the memories formed in experiences, it is possible to see how significant this fact is:

I suspect that explicit recalled mental images arise from the transient synchronous activation of neural firing patterns largely in the same early sensory cortices where the firing patterns corresponding to perceptual representations once occurred (Damásio, 1994, p. 101).

The aesthetic experience boosts expansion, creative thinking, and innovation. This is because new experiences will be judged in a broader and more diverse mnemonic context when they are confronted by memories formed fromevents stored in an emotionally positive context. This fact is even more significant when one remembers that most of the judgement about what is experienced depends considerably on memories of previous experiences (Solso, 1994, p. 27-49). It is for this reason that Brown and Dissanayake's (2009) definition of artification is so timely when they say that it is a human tendency to "make the ordinary extraordinary", a behaviour. In this disarmingly simple definition, one can find the engine of culture itself where positive emotions, present in the various activities of daily life in the context of artification, promote a continued expansion of meanings and complexification of patterns of action (decision making) and models of reality.

Another important consequence is the finding that the experience of beauty is absolutely subjective, although mOFC activation occurs in any individual regardless of ethnicity, culture or race. However, it is very important to consider that culture is the performance arena of individuals in their environment and is a powerful filter in the construction of the aesthetic experience and their consciousness. In other words, the emotional response of the individual depends on his ontogenesis. This fact seems to be on the authors' minds when they mention individuals who think rock and roll is more "rewarding" than the work of Richard Wagner (Ishizu & Zeki, 2011, p. 9).

ARTIFICATION AND POSTMODERNITY

The concept of artification presented by Brown and Dissanayake (2009) seems in line with the proposals present in the postmodernist paradigm shift described by Griffin (as cited in Best & Kellner, 1997). It inspires the need to establish new universal values, underlying the postmodern view, capable of replacing some "pernicious ideologies" such as: "individualism, anthropocentrism, nationalism, patriarchy, economism, consumerism, and militarism, in order to provide support 'for the ecology, peace, feminist, and other emancipatory movements of our time, while stressing that the inclusive emancipation must be from modernity itself" (Brown & Dissanayake 2009., p. 242).

The quest to establish a creative synthesis of premodern, modern and postmodern ideas points unequivocally to this inclusive process. Examining art in human culture must involve key aspects in any period of history. Understanding the issue of aesthetics therefore implies considering a perspective capable of reaching far beyond those views that characterize the Eurocentric perspective of the eighteenth century. This approach is in line with the blurring of the boundaries between scientific knowledge and unscientific "knowledge", so as to incorporate other fields of knowledge and values in science, involving a new unity among the scientific, ethical, aesthetic and religious institutions proposed by Best and Kellner (Best & Kellner 1997., p. 242).

The issue of subjectivity shifting the objective view to one more in connection with the world has been touched upon previously. It is worth noting now that this subjectivity includes the unconscious, that is, the understanding of the world lies beyond reason. The importance of the action of the unconscious in the construction of the conscious experience, which depends upon the ability to interact with the surrounding environment, depends fundamentally on the mechanisms underlying the perceptions of unconscious emotional states, which Damásio (2010) called feelings. These emerge in conscious thought and through the influence of emotions, constitute the foundations of human consciousness.

Damásio presents some very interesting considerations about consciousness and subjectivity, pointing out the importance of this last one to assign special features to humans, differentiating them from all other living beings:

Without consciousness, that is, a mind endowed with subjectivity, there would be no way of knowing that you exist, let alone who you are and what you think. Had subjectivity not begun, even if very modestly at first, in far simpler creatures, memory and reasoning probably would not have expanded in the prodigious manner they did, and the evolutionary road for language and the elaborate version of consciousness that human beings possess, would not have been paved. Had creativity not flourished, there would have been no song, no painting, or literature. Love would never have been love, just sex.

Friendship would have been mere cooperative convenience. Pain would never have become suffering – not a bad thing – but an equivocal advantage since pleasure would not have become bliss either. Had subjectivity not made its radical appearance, there would have been no knowledge and no one to take notice, and consequently there would have been no history of what the creatures did through the ages, no culture at all (Damásio 2010., loc. 149).

Until recently, emotion was not typically taken into account in the efforts to understand cognitive mechanisms such as memory. However, it became increasingly clear that it can no longer be neglected, on the contrary, it must be acknowledged that in fact emotion is rarely absent from daily tasks in the construction of episodic memories (Phelps, 2004). It is vital to relate experience, emotion and memory to understand how the repertoire of actions, identity and reality itself are built.

However, it is the memories built in emotionally positive scenarios, arising from aesthetic experiences, which provide greater diversity in the judgement of new experiences, providing a wider range of possibilities for evaluation and decision-making. For this reason, the aesthetics present in the production of the art field, and more broadly in artification processes, constitute the foundation from which to consider the subject's expansion, and as shall be seen below, of the society and culture into which he is inserted.

When one considers language, the aesthetic experience can be understood as a strategy that allows one kind of experience to be understood in terms of others, creating coherences structured by the natural dimensions of the experience. New metaphors are able to produce new understandings and thus new realities. For Lakoff and Johnson (p. 235) the metaphor involves both language, and all the sensory dimensions of the experience or other forms of language, and is not limited to just the art world, but can occur in any aspect of man's daily life, whenever he notices or creates for himself new coherences that do not belong to his mode of perception or thought.

Moreover, language serves another aspect of human life, beyond the individual one. For Damásio (2010, loc. 4476) language serves the mechanisms of homeostatic production at two levels: that of the individual, the biological level; and that of the social level, which he calls sociocultural homeostasis. Armed with conscious reflection, organisms whose evolutionary design was centred around regulation and tendency toward homeostatic balance invented forms of intersubjective articulation involving actions capable of organizing the action of the agents in a group in the most varied contexts. These actions involve, for example:

...consolation for those in suffering, rewards for those who helped the sufferers, injunctions for those who caused harm, norms of behaviour aimed at preventing harm and promoting good, and a mixture of punishments and preventions, of penalties and praise.

The problem of how to make this knowledge understandable, transmittable, persuasive, and mandatory was resolved with the narratives.

... and it should be no surprise that it pervades the entire fabric of human societies and cultures. It also should be no surprise that the sociocultural narratives borrowed their authority from mythical beings presumed to have more power and more knowledge than humans, beings whose existence explained all manner of predicaments and whose activity had the ability to offer succor and modify the future [...] Individuals and groups whose brains made them capable of inventing or using such narratives to improve

themselves and the societies they lived in became successful enough for the architectural traits of those brains to be selected, individually and groupwise, and for their frequency to increase over generations (Damásio 2010., loc. 4490).

The narratives that recruit diverse levels of experience implemented with the assistance of the different languages invented over more than thirty thousand years, constitute the strategy from which art was created to encourage cooperation in collective actions like hunting, food production, resisting the enemies, building infrastructure, etc. These activities are best understood if placed in a class of behaviours that include the production of art, education, design, that herein was treated as artification (Brown & Dissanayake, 2009). Through artification, man is extremely efficient at organizing efforts of collective action, being able to build consensus and diverse and complex realities that characterize human society.

Artified behaviours, like ritualized behaviours, attract attention, hold interest, and give shape to emotion. In fact, even when they are not art, the preferences investigated by evolutionary aesthetics and neuroesthetics are often artified or use artifications, such as when attractive colours and shapes emphasize or reinforce an object, event or message.

Based on the considerations made so far, one could add another concept to the list of post-modern universal concepts mentioned previously: artification. The human species depends fundamentally on the behaviour of artification to build its societies and relationships with nature. It must be remembered that the aesthetic experience is at the heart of this behaviour that builds consensus, unites collectivity and allows the creative expansion of consciousness and intelligence, integrates rationality and emotion, integrates man to himself, his fellow men and all forms of life, expanding his reality.

REFERENCES

Aleksander, I. (2005). *The world in my mind, my mind in the world: key mechanisms of consciousness in people, animals and machines*. Exeter: Imprint Academic.

Almeida, D. Di M. de. (2001). Corpo, Tecnologia, Cultura. In Corpo e cultura. São Paulo, SP: Xamã.

Best, S., & Kellner, D. (1997). *The post modern turn*. New York, NY: The Guilford Press.

Boyd, B. (2009). *On the origin of stories: Evolution, cognition and fiction*. Cambridge, MA: The Belknap Press of Harvard University Press.

Brown, S., & Dissanayake, E. (2009). The arts are more than aesthetics: Neuroaesthetics as narrow aesthetics. In S. Martin & V. Oshin (Eds.), *Neuroaesthetics*. New York, NY: Baywood Publishing Company Inc.

Bryson, N. (2003). Introduction. In W. Neidich (Ed.), *Blow-Up: Photography, cinema and the brain*. New York: D.A.P., Distributed Art Publishers.

Changeux, J. P. (2013). *O verdadeiro, o belo e o bem: Uma nova abordagem neuronal*. Rio de Janeiro: Civilização Brasileira.

Damásio, A. (1994). *Descartes' error: emotion, reason, and the human brain*. New York, NY: The Hearst Corporation.

Damásio, A. (2010). *Self comes to mind: constructing the conscious brain*. New York, NY: Pantheon Books.

Dennett, D. C. (1994). The role of language in intelligence. In J. Khalfa (Ed.), *What is Intelligence? The Darwin College Lectures*. Cambridge, UK: Cambridge University Press.

Dewey, J. (2010). *Arte como experiência*. São Paulo: Martins Fontes.

Donald, M. (1991). *Origins of the modern mind: Three stages in the evolution of culture and cognition*. Cambridge, MA: Harvard University Press.

Fabbrini, R. (2006). O fim das vanguardas. In *Cadernos de Pós-graduação da UNICAMP* (vol. 8). Retrieved November 18, 2015, from: http://www.iar.unicamp.br/dap/vanguarda/artigos_pdf/ricardo_fabrini.pdf

Feldmann, J. A. (2006). *From molecule to metaphor: a neural theory of language*. Cambridge, MA: MIT Press.

Fogliano, F. (2002). Fotografia e Complexidade. In L. Leão (Ed.), *Interlab: Labirintos do pensamento contemporâneo* (pp. 157–162). São Paulo: Editora Iluminuras.

Fogliano, F. (2012). A experiência estética: consciência, linguagem e narrativa. In C. Rocha, M. B. de Medeiros, & S. Venturelli (Eds.), *Art - Arte e Tecnologia // Modus Operandi Universal* (pp. 72–80). Brasília: Programa de Pós-Graduação em Arte da Universidade de Brasília.

Fogliano, F. (2013). Novos Paradigmas para a relação entre Imagem e Interação. In *Proceedings of #12. Art, Encontro Internacional de Arte e Tecnologia*. Brasília: Universidade Federal de Brasília.

Fogliano, F. (2014). Identidade e interação: Considerações sobre o retrato e suas funções. In *Proceedings of Simpósio Internacional de Inovação em Mídias Interativas - SIIMI* (pp. 55–66). Goiânia: Universidade Federal de Goiás.

Fogliano, F., & de Carmargo, D. C. F. (2010). Linguagem e materialidade na experiência fotográfica. In *Proceedings of Intercom - Sociedade Brasileira de Estudos Interdisciplinares da Comunicação XXXIII Congresso Brasileiro de Ciências da Comunicação* (pp. 1-14). Caxias do Sul, RS.

Gibbs, R. W. Jr. (2007). *Embodiment and cognitive science*. Cambridge, UK: Cambridge University Press.

Goguen, J. A. (2000). What is Art? Journal of Consciousness Studies, 7(8-9), 7-15.

Goldstein, R. (2005). *Incompleteness: the proof and paradox of Kurt Godel*. New York, NY: W. W. Norton & Company, Inc.

Gombrich, E. H. (2000). Concerning "The Science of Art": Commentary on Ramachandran and Hirstein. Journal of Consciousness Studies, 7(8-9), 17-20.

Ishizu, T., & Zeki, S. (2011). Toward a brain-based theory of beauty. PLoS ONE, 6(7). doi:10.1371/journal.pone.0021852

Johnson, M. (2007). *The meaning of the body: Aesthetics of human understanding*. Chicago: The University of Chicago Press. doi:10.7208/chicago/9780226026992.001.0001

Kaptelinin, V., & Nardi, B. A. (2006). *Acting with technology: Activity theory and interaction design*. Cambridge, MA: MIT Press.

Lakoff, G., & Johnson, M. (1999). *Philosophy in the flesh: The embodied mind and its challenge to western thought*. New York, NY: Basic Books.

Massey, I. (2009). *The neural imagination: Aesthetic and neuroscientific approaches to the arts*. Austin, TX: University of Texas Press.

Onians, J. (2007). *Neuroarthistory: From Aristotle and Pliny to Baxandall and Zeki*. New Haven, CT: Yale University Press.

Phelps, E. (2004). A Human emotion and memory: Interactions of the amygdala and hippocampal complex. *Current Opinion in Neurobiology, 14*(2), 198–202. doi:10.1016/j.conb.2004.03.015 PMID:15082325

Pinker, S. (1995). *The language instinct: The new science of language and mind*. London: Peguin Books.

Ramachandran, V. S. (2014). *O que o cérebro tem para contar: desvendando os mistérios da natureza humana*. Rio de Janeiro: Zahar.

Ramachandran, V. S., & Hirstein, W. (1999). The science of art: A neurological theory of aesthetic experience. *Journal of Consciouness Studies, 6,* 15–51.

Ramachandran, V. S., & Hubbard, E. M. (2003). The phenomenology of synaesthesia. *Journal of Consciousness Studies, 10*(8), 49–57.

Seeley, W. P. (2011). What is the cognitive neuroscience of art … and why should we care? *ASA Newsletter, 21*(2). Retrieved November 18, 2015, from: https://c.ymcdn.com/sites/aesthetics-online.org/resource/resmgr/Newsletters/31.2.pdf

Solso, R. L. (1994). *Cognition and the Visual Arts*. Cambridge, MA: MIT Press.

Stapp, H. O. (1993). *Mind, matter, and quantum mechanics*. Berlin: Springer-Verlag. doi:10.1007/978-3-662-08765-7

Thornhill, R. (2003). Darwin aesthetics inform traditional aesthetics. In E. Voland & K. Grammer (Eds.), *Evolutionary Aesthetics* (pp. 9–35). Berlin: Springer-Verlag. doi:10.1007/978-3-662-07142-7_2

Tooby, J., & Cosmides, L. (2008). The evolutionary Psychology of the emotions and their relationship to internal regulatory variables. In M. Lewis, J. M. Haviland-Jones, & L. F. Barret (Eds.), *Handbook of Emotions* (3rd ed.; pp. 114–137). New York, NY: Guilford Press.

Voland, E., & Grammer, K. (2003). Introduction. In E. Voland & K. Grammer (Eds.), *Evolutionary Aesthetics* (pp. 1–5). Berlin: Springer-Verlag. doi:10.1007/978-3-662-07142-7

Wilson, E. O. (1999). *A Unidade do Conhecimento*. Rio de Janeiro: Editora Campus.

KEY TERMS AND DEFINITIONS

Aesthetic Experience: According to Dewey (2010) refers to the experience as appreciation, perception and enjoyment by the subject who is in contact with the artistic object. What Ishizu and Zeki (2011) realized is that the aesthetic experience is a cognitive phenomenon of a subjective nature, independent of the specific properties of the objects and includes attributes constituted within and outside the formal standards of beauty. It is known from these studies that there is an objective way to understand and objectively measure the conscious and aesthetic experience by observing the state of excitation of neurons situated in a cortical structure of the brain called the medial orbitofrontal cortex (mOFC).

Artification: According to Brown and Dissanayake (2009) is a behaviour. It is what people do when they create all kinds of aesthetic objects or ritual practices that require a series of cognitive, neural and cultural skills, where emotion plays an important role. Artification can be considered a superclass of activities in which social, religious and political activities are included. The inclusion of art and design as categories belonging to artification is very important to discuss these fields under a new approach, the darwinian perspective.

Attention: Attention is included in the list of mechanisms of human consciousness like memory, anticipation, emotion, and intentionality. Attention must be taken into consideration to explain art. Art dies without it, as people since Aristotle have noted, both inside and outside the evolutionary explanation. Art alters the mind because it engages and re-engages the attention from the nursery songs to the distracted hum (Boyd, 2009).

Conscious Experience: For António Damásio (1994, 2010), the conscious experience can be understood as an aspect of the conscious mind that allows the subject to know that the experiences lived belong to him and can be arranged in his memory to be intentionally used in communicative, emotional processes, etc., without the conscious experience there is no language, no culture.

Culture: Dewey (2010, p. 97) claims that "just as the evolutionary growth of the individual, from the embryo to maturity, results from the organism's interaction with its surroundings, culture is the product not of efforts undertaken by men in the void, or only with themselves, but of the prolonged and cumulative interaction with the environment". Here it is important to consider the need to expand the context of this definition beyond art, and by doing so artification would be the mechanism underlying the idea of interactive accumulation that Dewey, so appropriately, used to define culture.

Emotion: Is a cognitive mechanism that affects thinking and behaviour. For Damásio (1994, 2010), emotions are unconscious patterns of action. Such patterns can be of genetic origin, or internalized throughout life. This internalization is associated to mechanisms related to emotion and memory that transfer part of the processes under conscious control to the unconscious. This mechanism gives the conscious mind more time for analysis and planning.

Language: For Pinker (1995) language is much more than a cultural artifact that we use to provide information or say how certain systems work. Language is a distinct piece of the biological machinery of our brains. Complex in nature, this specialized skill develops in children spontaneously, without effort or formal instruction. Among its many definitions, among which are mental organ, neural system, or computer module, the author prefers to consider language as an "instinct". There is a close relationship between language and conscious experience. For Feldman (2006) language is intimately related to what scholars label as personal experience, subjective experience, or phenomenology. Philosophers have coined a technical term to refer to conscious experience: "qualia".

Neuroesthetics: Emerging field of knowledge establishing conceptual bridges between the advances made in neurosciences and the reflection on art. Neuroesthetics is particularly interesting when its object of study is contemporary art. This is because the classic conceptual tools are ineffective in the development of a critical analysis of the products resulting from this production.

Postmodernity: Even if there is not a strict definition for postmodernism, the production gathered under this category of thought leads to the depletion of classical principles, unable to cope with many of the central themes of the theory and practice of contemporary culture. The repositioning of the issue of aesthetic experience in the context of neuroesthetics is an emblematic example of scientific contribution triggering great change in the scenario that involves the relationship between art and science in the debate on aesthetics. Recently, the emergence of a theory of beauty based on neurological studies has permitted a definition of beauty grounded on neurocerebral dynamics. The results of which have paved the way to clarify secular conceptual impasses. Neuroesthetics can be listed in the realm of post-modern thought among various scientific advances, such as the study of complexity and chaos, the self-organizing systems, theories of relativity and quantum mechanics, for example.

Subjectivity Subjectivity: In recent years, the construction of reality has been framed as the result of intersubjective cultural effort, of the construction of collective narratives able to express the stories of the various groups that make up a society. When considering the field of neuroesthetics, Changeux (2013) assigns to art the role of establishing "intersubjective communication" which "implies motivations and emotions in harmony with reason, but with no apparent progress and in constant renewal" (Changeux 2013, p. 97). This activity is directly related to neuro-cognitive processes leading to the further development of the consciousness and organization of the brain, but also to the expansion of the cerebral cortex and "more particularly the cortices of prefrontal association, parietotemporal and cingulate, closely linked to the limbic system" (Changeux 2013, p. 98).

Chapter 5
A Study on the Interface between Arts and Sciences:
Neuroesthetics and Cognitive Neuroscience of Art

Alexandre Siqueira de Freitas
Universidade Federal do Sul da Bahia, Brazil

ABSTRACT

This chapter discusses issues related to two fields of knowledge: neuroesthetics and cognitive neuroscience of art. These two fields represent areas that link historically dichotomic instances: nature and culture. In the first section, the author introduces a brief discussion on this dichotomy, reified here as science and art/aesthetics. Based on a preliminary analysis of these fields, as well as potential interfaces and articulations, the author then situates neuroesthetics and cognitive science of art. In both cases, the main definitions, usual criticisms, and comments on potential expectations regarding the future of these two areas will be presented.

INTRODUCTION

Arts and sciences are, in their most basic senses, "two great engines of culture", as coined by Stephen Wilson (2002, p. 5). According to Wilson, they consist of *stimuli*, sources of creativity and identity marks. However, the ways through which these engines converge, diverge, and sometimes merge present substantial and complex issues. Even though it's possible to trace analogies (this is our intent here), relating the creative processes from both terrains (arts and sciences) or the visual aspects of its products, some qualities attributed to each field often seem to remain distinct.

In order to propose a brief illustration, the author would like to recuperate the term "engine", used by Wilson. On one hand, we have an art object whose movement of axes, gear wheels and clock hands are the main attraction to its public. On the other hand, we have an engine that presents the exact same components, but, in this case, the public's main interest is focused on the movement of clock hands, that indicate time. These two pictures are, at once, similar and distinct. Regarding their semantic values

DOI: 10.4018/978-1-5225-0510-5.ch005

and cultural symbolism, they are quantitatively identical and qualitatively different. The first object is primarily aesthetical and ideally gratuitous, if we think from a Kantian perspective. The second one, by its turn, belongs to a primarily cognitive sphere, from all the scientific fields. However, both objects present, to varying degrees, aesthetic and cognitive elements. This example unveils some of the potential tensions and ways through which one might notice the divergences and convergences between art and science products.

In order to raise epistemological questions related to fields of knowledge focused on art and sciences – neuroesthetics and cognitive neuroscience of art –, the author believes it's important to analyze, even if briefly, the duality between arts and sciences, including thougts on aesthetics, a central issue to the subject discussed here. These "two cultures" divided by an abyss of mutual incomprehension – as C. P. Snow (1990) presented the dichotomy between sciences and humanities in 1954 –, unveil, at the present day, signs of change in this landscape. These signs consist of small cracks in humanistic and scientific epistemologic strucutures, in their modern conception. Cracks that may indicate major ruptures, future transformations. And they tend to get more and more intense in our current historic moment, which some would call postmodernity. Therefore, this work aims at discussing these transformations, which are present in certain researches and new theoretical fields. The author will firstly describe some of the elements that were (and still are) used in the construction of fields we call science, arts and aesthetics – all of them embedded with signification and history. Secondly, the author will address the neuroesthetics and the cognitive neuroscience of art separately, with a brief history of each, critiques and expectations.

BACKGROUND: SCIENCE, ART AND AESTHETICS

Science, as we know it today, bears great expectations about "how" and "why" general *phenomena* occur. It's primarily focused on the cognition of the "natural world" and reflects the necessity of extracting order structures from a vast, chaotic background from time to time. According to Júlio Plaza (2003, p. 38), influenced by Ortega's ideas, knowledge is the mental effort that extracts information, or language, from disorder. Plaza also argues that in order to actually know something we must reach its very "being". This "being" is not achieved through perception, but through an intellectual model. The production of knowledge is thus the construction of intellectual models that results from a work proccess based on complex information (Srour *apud*Plaza, 2003, p. 38). Scientific knowledge is grounded on the faith in empirical information, with which scientists usually create massive theoretical structures to link observation and reality (Wilson, 2001, p. 19). Science pursues ways of effectively acting upon reality and, therefore, "builds conceptual models or representations that reflect, with some isomorphy, aspects of the world's objective organization" (Vieira, 2006, p. 48).

We are facing, now, one of the great pillars of science: objectivity. In its core resides a cumulative expansion movement driven by the specificities of the operations that guide observation. Concepts, principles, and codes are combined in the quest for "objective and universal truths", as defined by neuroscientist Jean-Pierre Changeux (2013, p. 97). Even though the history of science proves these truths are transitory, the driving forces behind scientific knowledge are not as dynamic as common sense believes – maybe due to the troubled distinction between science and technology (which is not the purpose of this work). According to Thomas Kuhn's classic and essential *The Structure of Scientific Revolutions* (1962), even truths, or better yet, scientific paradigms, are constantly changing. Such paradigms, models created to address the biggest number of observations in inside a coherent landscape, suffer the tension of what Kuhn called anomalies. That goes on until a new conceptual landmark – a potential rival to the preceding one – is created. Therefore, objectivity (in its mobility) is linked to two inseparable instances: man's inventiveness and the enhancement of instruments, the latter frequently associated with mathematics. As for subjectivity, its influence is usually avoided by the scientific method through control mecahnisms and, posteriorly, by the critique and comments on the results published. This movement reveals an inescapable intersubjectivity, which is also inherent to the science experiment. In such a wide scope, empirists underscore the importance of observation, rationalists focus on the logical process of construction and derivation, and critical theorists, among other branches of knowledge, see science as the modern disillusionment (Wilson, 2001, p. 12). Theorists in this last branch point out to underestimated social forces and metanarratives, though they structure questions and paradigms in the scope of scientific investigations. Bruno Latour, author of *Science in Action* (1987) is one of these theorists. The persuasive influence of gender, class, national identity, and history are also underscored. Besides these three groups, Wilson mentions the constructivists, who question our hability to find truths that are applicable to all times and cultures. Objectivity is always questioned and conditioned in different ways, but, despite suspicion, it endures in the core of Western civilization.

The relation between science and the natural world (usually in opposition to the cultural) is another essential issue for its characterization. Science emphasizes an observation of the natural world as reality, implying an essentially ordered character of nature. The author says "essentially" because, when analyzed through perception, nature is usually seen as chaotic. As argued above, scientists are responsbile for the objective organization of the world. And science fiercely believes in the potential discovery/organization of the world.

The scientific thinking, as Plaza (2003, p. 38) puts it, is a productive process that aspires universality. Still according to Plaza and following Bense's precepts, knowledge lies on the intervention of intelligent beings who identify things in the world in order to be cognizant of it. Such identification happens in the natural world (according to the principle of exteriority) and functions according to different rules, in order to establish and understand things and *phenomena*.

Ideally, science continually tests and refines its hypotheses, enhancing its instruments, defining its concepts and intellectually interfering in symbolic objects, such as representations, intuitions, and observations of all kinds.

Art and Aesthetics

It's natural and maybe fundamental that a concept as polissemic as "art", carrying so many different meanings through history, especially in the "posthistoric" time in which we live[1]. "There really is no such thing as Art. There are only artists" – wrote Gombrich in his *The Story of Art* (1951, p. 12). Taking art as a kind of social combination, the author of this celebrated work chooses to present art history as a story about artists. Gombrich's choice is certainly appropriate in terms of historic dimension, since art objects results from the encounter between the artists and their desires, conceptions, cultural and social circumstances, selected materials and media. However, for the purposes of an epistemologic study, one must look at art as a subject, inspite of its inseparability from environment. After all, every product ever made by men is also the result of social encounters and combinations.

In order to clarify a few ideas on the concept of art, the author turns to Italian thinker Luigi Pareyson, whose conception is structured on three historically recurrent definitions: art as action, as knowledge and as expression (2001, p. 21). All along Western history, these definitions opposed, excluded and dialogued with each other in different ways. In Acient history, action was art's predominant attribute, embedded in its etymology: *ars*, from the Latin, and *techné*, from the Greek. In the Renaissance, the cognoscitive element was highlighted, and art became a way of knowing and interpreting the world. Finally, Romanticism underscored art as expression, and the canon shifted from beauty as a model to the beauty of expression.

Regarding the executive aspect of art, Pareyson argues that:

(...) art consists of intensive, eminent, absolute production and realization, to the point that it was actually called creation. It produces, though, not only organisms that are autonomous and independent (as in nature), but also radically new objects, which are reality's very own evolution, an onthological innovation. (Pareyson, 2001, p. 25).

Through knownledge, embodied in congnoscitive elements, art takes on "functions from other human activities, i.e., from science, philosophy, religion, or morality" (*Pareyson, 2001*, p. 24). But the relation between "art-as-knowledge" and "art-as-action" is not necessarily peaceful.

Art doesn't have, per se, a revealing or cogniscitive role, and it's not even limited to knowledge, especially if we think of a contemplative kind of knowledge. The emphasis on art's cognoscitive, visible, contemplative, and theoretical attributes weakened its most essential and fundamental aspects – its executive and accomplishing dimension – and did a disservice to art theory and practice. (Pareyson, 2001).

Art-as-expression highlights, unsurprisingly, art's expressive aspect, which, by its turn, allows us to think of art as language, and even to establish semantic theories. However, while art is certainly expression, this is not its essence.

(...) art also has, alongside other attributes, an expressive character. At most, one could add that, without it [the expressive character], it could not even be art, once it would lack that sense of finished humanity, which is the condition to any human production. (Pareyson, p. 22).

Therefore, while such conceptions (art-as-action, art-as-knowledge, art-as-expression) bear essencial artistic marks, they should not be essentialized or isolated, because all three of them are based upon vigorous historic engines.

Another interesting way of understanding the art phenomenon is presented by semiotician Max Bense (1971) and summed up by Júlio Plaza (2003, p. 39): the definition of art as a view of the world as "something made", as part of the creative structure. This is different from seeing the world as "meaning and signification", a semantic and communicative perspective, or "something given" (the physical dimension), according to the causal model. While the latter is highly determined, the semantic and communicative model is defined in a conventional way, and the model of the world as a "given entity" – the aesthetic state one – is singularly and ambiguously determined. Science aims at determining and establishing ideas, by identifying, transposing, translating and ordering physical states. In the case of aesthetic states, such determination happens in a slight way, as in the distribution of colors seen in a painting by Ticiano (*Plaza, 2003*, p. 39).

Even though art's expressive dimension approximates it to language in a certain way, aesthetical information transcends semantic information, once it is open to much higher unpredictability and improbability of signs, due to its slight determination and low degree of codification (*Plaza, 2003*). This is a game of possibilities of perception, more than a game with rules, which is the case of one of the most important dimensions of science.

As for the art-as-knowledge modality presented above, Plaza places it according to a perspective that diverges from science. While science addresses knowledge from an abstract perspective and applies it universally, at any time and place, art addresses it in a concrete plan made of individual, irreplaceable, and unique objects. Even though semantic information might subsist, art is presented to the world primarily as aesthetic information.

It's also important to mention that the idea of art bears certain contradictory elements such as the two definitions presented by Étienne Souriau in his *Vocabulary of Aesthetics* (2010, p. 178). On one hand, art is defined as a set of precepts that ensure a good artistic realization – possibly associated with the notion of *beaux-arts* and the rise and consolidation of academies in the 17th and 18th centuries, a movement in line with the Enlightenment ideals. On the other hand, art is considered as knowledge that escapes rules and guidelines – the *savoir-faire* goes beyond information and procedures acquired. Still according to Souriau, some claim that a work of art reaches its higher aesthetic quality when rules are applied in a way that the technical *savoir-faire* is no longer apparent (*Souriau, 2010*).

After exploring the artistic terrain and the potentiality of its multiple definitions, we must address the concept of aesthetics (even if briefly) in order to rectify some usual misconceptions. These come not only from the common sense, but are also closely related to issues to be discussed in this work, especially those rising from neuroesthetics.

Aesthetics, as most of the disciplines, is unique and multiple. It might be possible to offer a relatively clear definition of it, but hardly an exact one (Freitas, 2015, p. 48). Seeing it as art contemplation might be a correct, yet vague assumption. However, the more precise we get, the more subject to contradictions and errors we will be, once Aesthetic's meaning lies on its uses, which usually are very diverse. The problem of considering it a simple contemplation of art resides in the fact that other instances or disciplines, such as Poetics, Criticism, and Art Theory, might interfere and create conceptual misunderstandings and a few methodology problems.

Aesthetics, as we call it today and as a field pursuing a space as discipline, was introduced by the German phylosopher Alexander Gottlieb Baumgarten (1714-1762), as the "science of beauty" in 1735 (Souriau, 2010, p. 725). Opposing to *noêta*, facts of intelligence, and to *aistheta*, facts of sensibility, Baumgarten took the sensible as a central subject and thus made the discipline's autonomy possible. Aesthetics was only possible thanks to an epistemological turn, ocurred in the 16th and 17th centuries, that changed the way we conceive the world and how it presents itself to our sensations. The emancipation of the visual arts and its attempt to join the liberal arts during the Renaissance contributed to the rise of the modern concept of art, or at least initially, of the arts.

In the 20th century Aesthetics faced countless obstacles, and many of them were related to changes in the way art was seen. Nothing seems classifiable and the art object, as a sensible and clearly delimited totality, is frequently set aside. In the beggining of the last century, Aesthetics influenced the work of philosophers as Benjamin and Adorno. Benjamin introduces the "loss of the aura" in the work of art in "the age of its reproduction", while Adorno claims that the art objects don't have an intrinsic essence[2] (Talon-Hugon, 2010, p. 89-92).

In the second half of the 20th century, Étienne Souriau (2010, p. 728-729) distinguished five variations of aesthetics, which we summarize as follows:

1. **Philosophical Aesthetics:** The earliest one, introduced by Plato and consisting of a reflexive and analytic method.
2. **Psychological Aesthetics:** Created in the second half of the 19th century and focused on understanding how art and ideals of beauty influence thought. It includes psychoanalytical aesthetics.
3. **Sociological Aesthetics:** Its main theorists are Marx, Taine, and Guayau, and it's focused on establishing a place for art in society.
4. **Comparative Aesthetics:** Remains exclusively focused on strictly aesthetic issues. If focuses on similarities, transpositions and mutual influences of different kinds of art.
5. **Morphological Aesthetics:** Uses the analytic and even the mathematic method. It's prospective and aims at unveiling inventive principles and ways of building structures in the interior of works of art.

Even though all these variations are, to a certain degree, imbricated, psychological aesthetics (addressing aspects related to cognitive dimensions and to "art-as-knowledge") and morphological aesthetics (focused on understanding form, in its broad and specific senses) seem to be the closest to issues regarding the interfaces between sciences (neurosciences) and artistic activities.

To conclude our thoughts on aesthetics as discipline, the author reproduces an introduction by Carole Talon-Hugon, whose definition is based on similarities, convergences and filiations of different concepts of aesthetics: "Aesthetics is the analysis of a certain field of objects where the terms 'beautiful', 'sensible' and 'art' are dominant." (2010, p. 4). Even though it sounds slightly vague, this assertion is interesting because it allows us to historically analyze Aesthetics from this triple base and unveils an interchange of these poles, as in Pareyson's definition of art. Thus, the core shifts from one concept to another: art, the sensible, and beauty. These three instances are always present, but never in balance. For example, the idea of beauty, as we'll see further in this work, hasn't been central since the 20th century, when ideas of art and the sensible became more important.

After these relatively long yet necessary considerations on issues regarding science, art, and aesthetics, we turn to relatively recent fields situated in this disciplinary "tension zone". We will attempt to establish connections between disciplines and to approach the expansion of the taxonomy of fields that were fairly stable until recently.

MAIN FOCUS: NEUROESTHETICS AND COGNITIVE NEUROSCIENCE OF ART

The name "neuroesthetics" reveals an attempt to include different theoretical fields which, since the decline of the Renaissance, became autonomous and tended to diverge due to epistemological issues. In the present day, these interdisciplinary branches are included in a dynamic center of thought focused on renewing traditional views and highlighting new understandings and perspectives about all kinds of *phenomena*.

The knowledges related to the application of scientific and biologic methods to the investigation of artistic works and processes are relatively old, even though not always under the name of "neuroesthetics". Neuroesthetics is the confirmation of this history and founded a few specific guidelines and nuances. It's hard to define its historic origins, due to the coexistence, in the last centuries, of scientific currents and puctual observations simultaneously focused on the analysis of aesthetic perspectives and biologic facts.

Neuroesthetic's recent origins were located by Jean-Pierre Changeux (2008, p. 97) in his own investigations, as well as in Alexander Luria's. Luria, a Russian neuropsychologist who was very active during the mid-20th century, conducted a series of investigations attempting to reveal details of the interaction between mind and brain in cinema, articulating film theory, psychology, biology and philosophy (Valisseva, 2013). Luria contributed to the rise of studies focused on the interfaces between arts and sciences and influenced important scientists and researchers, such as neurologist and writer Oliver Sacks.

We could also mention I. A. Richards, an important English literary critic, as one of the main precursors of neuroesthetics or even of neurosciences applied to arts. Richards advocates, in his 1926 essay "Science and Poetry", the use of psychology of instincts and emotions as an effective yet underdeveloped tool for textual interpretation (Ashton, 2011). The psychology of instincts is strongly related to evolutive biology and, therefore, to ideas that originated the neurosciences applied to arts.

Changeux defines neuroesthetics as a terrain focused on "raising questions on the neural bases of the contemplation and the creation of the work of art" (2013, p. 97). The author believes that congnitive activities, specially conscience and artistic practices, are related to a better development of the brain organization. Changeux also reflects on "neurohistory", including investigations on color, beauty, empathy, sinesthesia, among others themes. The field called neuroesthetics includes all of these ideas.

However, it's not possible to discuss neuroesthetics without mentioning the British neurobiologist Semir Zeki, who coined the concept. Zeki's concept of "visual brain", defined as a specific region of the frontal cortex where the aesthetic experience primarily takes place, was established in the last decades with his particular interest on visual mechanisms. In the 1970s, Zeki conducted a pioneer study on specialized and overlapped "visual channels", located in multiple areas, in different regions of the cerebral cortex (Changeux, 2013, p. 118). The neurobiologist thus proposes two major laws of the visual brain: *constancy* and *abstraction*. The first one is related to the brain's ability to retain the object's constant and essential elements, while abstraction liberates it from memory limitations (Zeki, 2001, p. 51-52).

Besides the mentioned general definitions, which are related to an understanding of biologic and neurobiologic bases of the aesthetic experience, Zeki places art as the extension of an essential brain function: knowledge acquisition. In the description of neuroesthetics provided in his website[3], the scientist does not refuse subjectivity, but expresses his wish to articulate it with a common neural organization that allows us to communicate through art[4]. His investigations are guided by questions on how ordinary visual activations can provoke emotions. Besides that, Zeki also argues, in his website, that celebrated artists could be, to a certain extent, successful neuroscientists who instinctively understand common emotinal and visual organizations, as well as how the brain works. Zeki's investigations often refer to themes as beauty, neutrality and ugliness. Based on the registry of brain activity through advanced methods, he analyzes responses, their intensity, and the regions in the cerebral cortex where they take place. Different painting categories are analyzed and reactions (high and low frequencies of brain activity) to those considered "beautiful" or "ugly" are registered. High brain activity in response to beautiful images would represent a reward reaction. The scientist also explains, in his introduction to his website, the great expectations usually attributed to neurosciences, even more than to arts. Zeki believes that the neurosciences are the appropriate field for the achievement of a proper understanding of human nature. He also believes that, by understanding neural laws we can also understand all kinds of human activity (morality, religion, law, etc.).

Regarding the idea of beauty, critics usually claim that neuroesthetics is related to the ways by which scientists handle this concept. This theme is also related to a need for defining certain parameters and, for that reason, specialists from this field are constatly looking for the neural bases of beauty. In an attempt to dialogue with aesthetic and philosophical terrain, they focus on a discourse about beauty that is no longer popular among art and philosophy thinkers. Semir Zeki, for example, tries to find in Kant a perspective that would be appropriate to his own scientific investigation: a more stable idea of beauty[5]. Beauty as discussed by Kant – something that pleases universally – can easily fit into neuroesthetic purposes. However, a brief look at the history of this concept reveals otherwise.

Aesthetic ideals vary, alternate, change, and coexist from time to time. Western art history clearly shows us the diversity in this concept. Beauty has been associated with simetry, reason, light, color, romantic ideals, techonologic ideals, consumism, etc. And all these associations are still partially true. The pursue of neural bases of beauty, in an attempt to universality, usually tend to become a failed investigation, if we consider the wide range of types of "beauty" created by historic and geographic situations. The search for universal principles in the work of great masters of the past (like some neuroesthetics theorists do) seem to be the search for an univesal ideal or some static entity in the interior of an extremely slippery, fluid concept. Therefore, it might be probematic to address established beauty as empirical experiment. Early in the 19th century Gustav Fechner – somehow precursor of neuroesthetics with his experimental psychology – already knew that the appreciation of beauty was inherently subjective and that such analysis should serve as background to all other investigations (Conway; Redhing, 2013). A "discovery" of universal beauty, by means of the intensive use of neuroimaging technologies, could become a normative and fruitless idea to artistic appreciation itself. According to Conway and Rehding, scanners that reveal areas of higher activation in a certain region of the cerebral cortex (especially the medial orbitofrontal cortex, mOFC) are actually measuring preferences, and not the beauty of the objects observed (*idem*). Futhermore, studies reveal the existence of highly creative people who suffered damages in this region, often regarded as essential to artistic perception (*ibidem*). The technology applied to the comprehension of neural mechanisms would, then, bear certain analytic pitfalls. Even aesthetic components in color images of brain processes can, many times, suggest particular interpretations that, under the influence of the scientific method, could succumb to the desire for regulation.

Still regarding beauty, science usually tends to relate it to the idea of art. As we already argued, this idea has long been rejected by the art field and is almost inexistent in the current artistic discourse. "Beauty" is definetly not an issue for the aesthetics developed in the last decades inasmuch phylosophical analysis. For example, neuroesthetics theorists erroneously identify artistic appreciation as beauty appreciation. This common association tends to corroborate the idea (usually revealing some prejudice from the arts and humanities) that scientists are naive and conservative. A brief look at the choices made by some of the most representative theorists from neuroesthetics, or neuroscientists focused on art, in terms of the type of art analyzed would show otherwise. Ramachandran often focuses on Indian art, Zeki on European modernists and Kandel on the study of Vienna Secession, an art movement from his hometown. While it might be natural that people are guided by familiar patterns, subjective convictions might also influence the results of their investigations. This could lead to a universalization of particular value attributes, according to Conway and Rehding (2013).

As we analyze the history of definitions from neuroesthetics (in the work of its precursors or not), we realize that this terrain is more hybrid than its name suggests. It combines principles from perpection psychology, evolutive biology, phenomenology, anatomy and brain function (in all its branches), radiology applied to neuroimaging, and obviously philosophy, aesthetics, and art criticism and theory. Likewise, we can notice that even though some researchers working in the interfaces between art and science could call filiate themselves to neuroesthetics, they don't. Such posture is probably due to their opposition to epistemological concepts from neuroesthetics that are often criticized.

In order to discredit neuroscientific studies applied to art (or neuroesthetic studies), John Carey, a famous literary critic, compared neuroscientist Vilaynur Ramachandran and his colleague, philosopher of mind William Hirstein, to comedians Laurel and Hardy (Hirstein, 2012). Carey believes that neurosciences could offer nothing but a limited perspective of art[6]. One might question the radical character of the British professor's criticism regarding scienfific studies focused on art (in this case, poetry). However, we can't deny that it reveals a certain discomfort in the pretense encounter of neurosciences and art practices. The reductionist perspective mentioned by Carey results from the fact that science does not admit the coexistence of given truths, while art can (and should) allow the existence of different truths in terms of values, practices and appreciation.

The belief in science, with a limiting perspective of the artistic phenomenon, can also be explained from other perspectives. Science generally reflects a certain obsession humans have for mastering and controlling objects. We must standardize and, therefore, reduce. A mountain turns into a cone; the moon turns into a sphere. Pleasure, wether in appreciation of art or drug use, turns into a reward in the limbic system. But the degrees of complexity of human beings, of things and their processes tend to keep scientists at a certain distance from the object analyzed. "Science manipulates things and refuses to habit them" is the first sentence in Marleau-Ponty's *Eye and Mind* (2004, p. 13). The phenomenologist questions the fact that science sees all beings as "general objects", "as if they meant nothing for us and were, thus, subject to our schemes".

Scientific reductionism can also turn into the common scientific practice of applying similar research models to different situations. "When a model is successful in a set of questions, science applies it to all other situations" (Merleau-Ponty, *idem*). That explains modern science's tendency to follow trends and the proliferation of "neurodisciplines", such as neuroethics, neurosemiotics, neurophenomenology, neuroesthetics, and neuromarketing.

But how can we practice science without establishing standards, without pointing what's right and what's wrong? The refusal of judgement might imply subjection and, therefore, the risk of discreditation. It's hard, in science, to avoid the right-and-wrong pitfall. It's easy to succumb to the desire for standards. Since this desire is almost a rule, standardization tends to create some tension when it approaches fields like art, which doesn't even have clear limits.

Even though artists usually work according to a wide range of rules and techniques, the regulation of practices and artistic values and goals can disturb them immensely. Phillip Ball (2013), writer and former editor for the *Nature* magazine, recalls, for example, Paul Klee and his interpretation of a study on the use of color, written by Wilhelm Ostwald, amateur paintor and Nobel prize winner in Chemistry. Ostwald praises the possibility of creating harmony with balanced tones and criticizes the "incorrect use" of blue in a Ticiano's painting. For Klee, Bauhaus professor and famous colorist, the scientist's normative thoughts represent the absolut abandon of the artist's rich and complex soulfulness.

Another important issue that might represent a critique to neuroesthetics, or to neurosciences applied to art, is the misunderstandings regarding causes, effects and meanings of the work of art. These scientific investigations fail to distinguish the effects of art from its meaning. There is a tendency to deny the interpretative dimension of these researches, because they believe they have a better understanding of brain processes, evolution mechanisms and the ways through which environment and body format these processes (Ashton, 2011). Neurosciences are relatively successful in their analysis of causes and effects of the work of art. However, the interpretations, the broader understandigns, depend on responses that are more closely related to meanings and senses. In other words, most of the neuroscientists believe, equivocally, that the discovery of causes and effects equals the discovery of what the artwork means.

Furthermore, according to Jennifer Ashton (*idem*), neuroscientists also tend to mistake meaning for feelings inspired by the artwork, during its conception or its appreciation. For example, different paintings can have similar effects and very distinct meanings. The identification of how and where in our brains the aesthetic responses take place does not tell us what they actually are or mean. This is, hence, a background, epistemological question that neurosciences address with difficulty. Thus, the field gets subject to fierce and justifiable criticism.

The initial arguments regarding neuroscience's limited perspective towards art, especially the neuroesthetic's, is appropriately validated by arguments related to epistemologic and conceptual questions that rise from the tension between scientific and artistic/humanistic fields. The practice of neuroesthetics, at least as conducted by the most frequently mentioned theorists, can easily provide us, almost naturally, with arguments that question the field's epistemologic foundation. By reading the texts that founded the field one can quickly identify an apparently exaggerated expectation regarding discoveries and the future of the discipline. For example, Semir Zeki highlights, in the introduction to his definition of neuroesthetics in his website, the belief in the possibility of defining art. According to him, that would happen if we could incorporate references to our brains (which is responsible for art creation, execution and appreciation). However, the polissemy inherent to the concept of art is the condition to its very existence. And we also must highlight the fact that when neuroesthetics theorists discuss art, they are actually limited to visual arts. This is another example of how exaggerated science's expectations probably are: In order to respond to the criticism aimed at neuroesthetics, William Hirstein claimed that it's possible to understand the processes of artistic creation due to the fact that the brain is an entirely physical system[7]. Even though we can reveal with higher precision the plasticity of neural networks and complex brain processes through sofisticated technology and sagacious methods, we are still not able to ensure the real apprehension of what the work of art might mean.

Therefore, we notice that the research field focused here can be subject to endless criticism, which is justified by certain authors' perspectives which we tried to include in this text. Most of these comments are basically an understanding of the simplification with which scientists approach the art field. While, on one hand, the demonstrations and understandings of brain processes related to art are hugely complex, on the other hand they tend to a reductionist perspective of the background concerning the field's own epistemologic basis.

However, it's important to notice that this epistemologic tension and the uncomfortable situation and conceptual problems caused do not annul the importance of results and discoveries produced by neurosciences (especially neuroesthetics) in its dialogue with artistic facts.

Cognitive Neuroscience of Art

In the beginning of this millennium, the neurosciences have become broader and established dialogues with different knowledge instances, wether hard sciences or humanities. Not surprisingly, the art field faces a series of tensions and problems (as those mentioned here and related to neuroesthetics) in its dialogue with sciences. The criticism that denounces neurosciences' reductionist perspective regarding art is not limited to neuroesthetics. The same happens with the mutual mistrust between, on one hand, artists, philosophers, and art theorists and, on the other hand, neuroscientists. This criticism aimed by philosophers and art theorists at reductionist perspectives frequently rises from methodological questions proposed by the neurosciences and its attempts to explain aesthetic dimensions of art. These neuroscientific explanations about art are rarely sufficient to address all the aesthetic properties of an object and the experiences related to it (Seeley, 2013). However, we must remember that any approach to an artwork (or to most objects and events) in a specific disciplinary field is necessarily limited to a mere perspective of a multifaceted object.

Now the author would like to turn to some definitions and discussions on the cognitive neuroscience of art, a field that, despite a few similarities, represents an alternative to neuroesthetics.

According to William P. Seeley's insightful definition (2011), cognitive neuroscience of art is a branch from empiric aesthetics that is focused on applying neuroscientific methods in order to study our relations with artworks. These methods come from a broad research scope that approaches all of our psychologic relations with the world. Through its contact with humanities (especially philosophy), neurosciences can, among other things, dialogue with philosophical theories that approach the entirety of our relations with artworks and artistic practices.

Empirical aesthetics, which, acccording to Seeley (*2011*), created the cognitive neuroscience of art, was originated in the 19th century with Gustav Fechner, who, in 1871, wrote "On experimental aesthetics"[8]. Fechner was a rigorous scientist, filiated to the monist philosophical current. He also wrote poems and commedy plays, such as "*Vergleichende Anatomie der Engel*" (Comparative anatomy of angels). In this play from 1825, Fechner gives us a detailed description of men and then ellaborates a series of analogies relating them to angels' spiritual states. Fechner's creativity, empiric perspective, methods, and experiments later made him the father of future cognitive neurosciences of art.

Even though art is not limited to rules and guidelines, artistis develop glossaries related to specificities and general subjects from their backgrounds, along with composition strategies that produce behavioral responses in their public (or consumers, according to cognitive neuroscientists)[9]. Artworks are not limited to guidelines, but most of them follow internal rules which are inherent to the singularity of the object or to certain styles and genres. These are rhytmic patterns in a broad sense, internal narrative agreements, ways of handling space. Ultimately, they are elements which are inherent to the poetics of a given artwork or a certain group of artworks.

Cognitive sciences consist of a group of scientific disciplines that focus on the description, explanation and, in some cases, simulation of mechanisms of though as well as the whole system of information processing, regarding the acquisition, retention, use, and transmission of knowledge (Miller, 2003, p. 144). According to Miller, cognitive sciences include six disciplines: philosophy, neurosciences, informatics, psychology, linguistics, and anthropology.

Hence, cognitive neuroscience of art can be defined as the study of how art consumers acquire, represent and manipulate the information that is present in the structure of an artwork. It is the set of *stimuli* intentionally projected in order to produce emotional, perceptive, and cognitive responses (Seeley, 2011). In this context, the work of art is seen as a form. This idea is related to the work of philosopher Luigi Pareyson, mentioned earlier in this work. According to Pareyson, the concept of form is appropriate to the work of art, and "formativity" qualifies the artistic activity[10]. Therefore, artworks consist of forms. However, they are live forms, "organisms that are autonomous and independent, like those from nature" (as previously quoted in this text).

Thus, cognitive neuroscience of art is seen as a tool for modeling the artwork consumer's processes and behaviors (Leote; Oliveira; Baraúna, 2015; Seeley, 2011). So, when we understand the analysis of the work of art as an attempt to explain how it works, the definition of this neurosicence as a tool is very appropriate. This modality of cognitive science is, then, presented as a potential "key" for understanding the nature of art and the aesthetic experience. And the author believes this key is consciously situated with others that already exist and have long been applied. Such is the case of those implicitly aggregated to art theories, music theories, film analysis, discourse analysis, phenomenology, etc.

In comparison to neuroesthetics, cognitive neuroscience of art tends to include a wider range of possibilities, due to its desire to comprehend the art phenomenon and its multiple expressions. Meanwhile, neuroesthetics, following the paths created by precursor Semir Seki, is usually limited to visual perception. Besides that, the epistemologic positioning of the cognitive neuroscience of art solves an important problem of neuroesthetics. It's specifically focused on art and not on aesthetics, while the latter "does not encompass non-aesthetic semantic or ontological questions about the nature of art or our engagement with artworks" (Seeley, 2011). We get, then, to the distinction between art philosophy and aesthetics. Issues related to the nature and onthology of art belong to art philosophy, while aesthetics, as the science of sensible knowledge, includes other questions than those related to art. Since philosophy is considered a cognitive science, it must naturally dialogue (in order to approach the art field) with neurosciences and other disciplines that might be envolved (psychology, linguistics, anthropology, among others). We should also notice that semantic elements of the artwork have regulating funcions in our neurophysiologic systems, and sometimes it defines the aesthetic character of our relation with specific artworks (*Seeley, 2011*). Even though the problem of signification is inherent to art (and that's why it's so difficult to define it as language), the semantic question plays a central role in the relation between the artwork and its public.

Mark Rollins suggests the notion of artworks as entities which are full with exogenous forces that are preponderant in the process of information transmission (2004, p. 101). Artworks must be seen as "atentional strategies" – an idea that, somehow, is in line with Pareyson's concept of artwork as form and of the artistic activity as formativity. At the same time, the expresive/communicative aspect of art is highlighted, with emphasis on the artwork, and not on the artist. Somehow, art becomes something that communicates and informs. In this context and with the rise and consolidation of new technologies, we can imagine, as does Wilson (2001, p. xxi), that the arts inform and enrich scientific investigations, and vice versa.

Once the basis of this new and specific modality of neuroscience is well established, we will obviously face methodological issues, due to the vastness and the complexity that are inherent to the relation between artwork and consumer. This problem becomes even more complex when we refer to the art definitions presented in this text and realize that we can only define what is sufficiently determined – which is not the case of the works of art, that consist of aesthetic states. However, isn't science the practice through which one extracts order structures from chaos? Then, we must find ways (μέθοδος, methods, *hodos* = way) that allow us to deal with the relations between semantic contents and aesthetic states presented in artworks. Besides that, it's necessary to achieve certain theoretical models or pertinent explanations of the nature and the onthology of art.

Other researchers besides Seeley see the cognitive neuroscience of art as a more appropriate alternative than neuroesthetics for the comprehension of perceptive processes in art (Leote et al., 2015). The first one seems to nurture more realistic expectations towards the future of this field of investigation. Cognitive neuroscience of art could, for example, confirm our best theories about artworks and enrich our philosophical practices (Seeley, 2011). It could also, maybe, help us understand the mechanisms for art perception and reveal onthological aspects of the artistic activity.

CONCLUSION

The analysis of the criticism aimed at the neurosciences with focus on art, presented in this text, especially in the section that discusses neuroesthetics, reveals a series of tensions in the contact between arts and sciences. These tensions rise, in most of the cases, from mutual ignorance regarding background issues that guide each field. Terminological issues and conceptual clashes represent a substantial problem in Zeki's neuroesthetics. In this discourse, two different fields with distinct ideologies clash, and one of them becomes subordinated to the other. Anyway, the comprehension of the creative processes and functions/functioning processes of our brains necessarily encompasses the compreehension of perception and reward processes, as well as neurologic and psychologic questions, among others. And these might be interesting questions for theoretical fields and art making.

Rosangella Leote argues that the art field needs to understand natural processes frequently investigated by science, "and use this situation in favor of actions focused on art, in any kind of experience" (2012, p. 164). The artistic production can be enriched by investigations developed by the neuroesthetics, the cognitive neuroscience of art and, more generally speaking, by the neurosciences focused on art. It can also contribute to these theoretical fields – not only through the attenuation of the borders between art and science, but also through a fruitful dialogue of mutual enrichment.

The consolidation of these new research fields as fertile terrains for knowledge and action requires the development of a comprehension of how art consumers generally acquire, represent, and manipulate the information produced by the formal structure of *stimuli*. That would allow a more comprehensive understanding of perception mechanisms and the modeling of processes and behaviors (if pertinent). That means to keep tracking the paths created by Gustav Fechner back in the 19th century, suggesting readings and models with scientific rigor, avoiding static categories and, besides that, remaining aware that clarifications about the artwork's signification processes do not represent the meaning itself. The comprehension of the ways through which mind works, of evolutive developments and of restrictions imposed by the environment and the human body on these processes allows us to ellaborate a rich panorama of our relation with art (Ashton, 2011).

The results of efforts made by researchers from the neuroesthetics and the cognitive neuroscience of art will probably be better understood and received by the academy once they offer serious investigation of our aesthetic relations with works of art. They also must offer effective instruments for the analysis of artworks and styles, so that it becomes possible to understand how different kinds of art work from the biologic perspective.

Despite the epistemological differences between sciences and humanities/arts, one might say, from a scientific point of view, that the discovery of perceptive processes and the comprehension of the structural fucntioning of works of art offer interesting possibilities. Several results of investigations on senses and arts are emerging and will continue to be published in the future. Consequently, several fields of knowledge will fructify. From the perspective of humanities – especially the aesthetics – and arts, the dialogue with science could offer incitement and new ways of seeing, hearing and experiencing art.

REFERENCES

Ashton, J. (2011). *Two problems with Neuroesthetic Theory of Interpretation.* Emory College of Arts and Sciences: nonsite.org. Available from http://nonsite.org/issues/issue-2/two-problems-with-a-neuro-aesthetic-theory-of-interpretation

Ball, P. (2013). *Neuroaesthetics is Killing Your Soul: Can Brain Scans Ever Tell Us Why We Like Art?* Nature Publishing Group. Available from http://www.nature.com/news/neuroaesthetics-is-killing-your-soul-1.12640

Bergeron, V. (2011). *What Should We Expect from the New Aesthetic Sciences?* Aesthetics: American Society for Aesthetics. Available from http://www.aesthetics-online.org/articles/index.php?articles_id=55

Carey, J. (2010). *What Good Are the Arts?* Oxford, UK: Oxford University Press.

Changeaux, J.-P. (2013). *O verdadeiro, o belo e o bem.* Rio de Janeiro: Civilização Brasileira.

Conway, B. R., & Rehding, A. (2013). Neuroaesthetics and the Trouble with Beauty. *PLoS Biology, 11*(3), e1001504. doi:10.1371/journal.pbio.1001504 PMID:23526878

Danto, A. C. (2014). *After the End of Art: Contemporary Art and the Pale of History.* Princeton, NJ: Princeton University Press.

Freitas, A. S. (2015). *Rencontre des arts: correspondances entre œuvres sonores et visuelles au xxe siècle.* Paris: Harmattan.

Gombrich, E. H. (1951). *The History of Art.* New York, NY: Phaidon Publishers Inc.

Hirstein, W. (2012). Neuroaesthetics: Responding to the Critics. *Psychology Today.* Available from https://www.psychologytoday.com/blog/mindmelding/201212/neuroaesthetics-responding-the-critics

Kuhn, T. S. (1996). *The Structure of Scientific Revolutions.* Chicago, DC: The University of Chicago Press. doi:10.7208/chicago/9780226458106.001.0001

Latour, B. (1987). *Science in Action.* Cambridge, MA: Harvard University Press.

Leote, R. (2012). Multisensorialidade e Sinestesia: Poéticas Possíveis?*Proceedings of 6th International Conference on Digital Arts, ARTECH.*

Leote, R., Oliveira, H. C., & Baraúna, D. (2014). *Teoria neurológica da experiência estética de Ramachandran e Hirstein: apresentação, comentários e proposta de aplicação.* Paper presented at the *5º* Encontro Internacional de Grupos de Pesquisa: Realidades Mistas & Convergências entre Arte, Ciência e Tecnologia. Escola de Artes e Comunicações ECA-USP, São Paulo.

Miller, G. A. (2003, March). The Cognitive Revolution: A Historical Perspective. *Trends in Cognitive Sciences, 7*(3), 141–144. doi:10.1016/S1364-6613(03)00029-9 PMID:12639696

Pareyson, L. (2001). *Problemas da Estética.* São Paulo: Martins Fontes.

Plaza, J. (2003). *Arte/ciência: uma consciência.* Revista Ars, Universidade de São Paulo.

Rollins, M. (2003). Pictorial Strategies and Perceptual Content. In H. Hecht, R. Schwartz, & M. Atherton (Eds.), *Looking Into Pictures: An Interdisciplinary Approach to Pictorial Space* (pp. 99–112). Cambridge, MA: MIT Press.

Seeley, W. P. (2011). *What is the Cognitive Neuroscience of Art ... and Why Should We Care?* American Society for Aesthetics – Aesthetics on-line. Available from http://www.aesthetics-online.org/newsletter/31.2.pdf

Snow, C. P. (2012). *The two cultures*. Cambridge, MA: University Press. doi:10.1017/CBO9781139196949

Souriau, E. (2010). *Vocabulaire d'esthétique*. Paris: Presses Universitaires de France.

Talon-Hugon, C. (2010). *L'Esthétique*. Paris: Presses Universitaires de France.

Vassilieva, J. (2013). Eisenstein/Vygotsky/Luria's project: Cinematic Thinking and the Integrative Science of Mind and Brain. *Screening the Past Publications, 38*. Available from http://www.screeningthepast.com/2013/12/eisenstein-vygotsky-luria%E2%80%99s-project-cinematic-thinking-and-the-integrative-science-of-mind-and brain/

Vieira, J. A. (2007). *Teoria do conhecimento e arte*. Fortaleza: Lab. Estudos da Oralidade.

Wilson, S. (2002). *Information Arts: Intersections of Art, Science and Technology*. Cambridge, MA: The MIT Press.

Zeki, S. (2001). Artistic Creativity and the Brain. *Science Magazine, Essays on Science and Society, 293*(5527). Available from http://www. sciencemag.org/content/293/5527/51.full

KEY TERMS AND DEFINITIONS

Art: The main definition of art used in this chapter is by Luigi Pareyson, whereby the concept of art is structured on three historically recurrent definitions: art as action, as knowledge and as expression.

Cognitive Sciences: A group of scientific disciplines that focus on the description, explanation and, in some cases, simulation of mechanisms of though as well as the whole system of information processing, regarding the acquisition, retention, use, and transmission of knowledge.

Critical Theorists: Theorists that see science as the modern disillusionment. They point out to underestimated social forces and metanarratives, though they structure questions and paradigms in the scope of scientific investigations.

Empirical Aesthetics: Empirical study of esthetical proprieties of art.

Epistemology: A branch of philosophy concerned with the nature and scope of knowledge.

Formativity: Term coined by Luigi Pareyson to qualifies the artistic activity.

Science: Science pursues ways of effectively acting upon reality and, therefore, builds conceptual models or representations that reflect, with some isomorphy, aspects of the world's objective organization.

Visual Brain: Defined by Semir Zeki as a specific region of the frontal cortex where the aesthetic experience primarily takes place.

ENDNOTES

1 "Posthistoric" is the term used by US phylosopher and critic Arthur Danto (2014) to discuss the "end of art", which is actually the end of a master narrative that guides Western art and embodies certain homogeneity.

2 This is a reference to Walter Benjamin's famous text: "The Work of Art in the Age of Mechanical Reproduction", published in 1936.

3 <http://neuroesthetics.org/statement-on-neuroesthetics.php>

4 No discipline can actually be clearly and staticaly delimited. However, art seems to question disciplinary borders even more fiercely, especially in the present day.

5 The reference to Kant in presented in his website and also in an interview conceded to Brazilian newspaper *O Globo*, in the section *Prosa e Verso* of the edition from 12/30/2006, available at: http://www.umsl.edu/~carrolljc/Documents%20linked%20to%20indiex/darwin4%20%281%29.pdf

6 John Carey's critique can be found in: Carey, John. *What Good Are the Arts?* Oxford: Oxford University Press, 2010.

7 See https://www.psychologytoday.com/blog/mindmelding/201212/neuroesthetics-responding-the-critics.

8 Vincent Bergeron also mentions Gustav Fechner, but associates him to neuroesthetics, as an activity that investigates where and how different components of our aesthetic responses are implemented in our brains (2011, p. 6). Bergeron's idea of neuroesthetics is very similar to the definitions of cognitive neuroscience of art mentioned in this text.

9 The author believes the term "consumer", often mentioned by William Seeley, can also be applied to artistic situations of interactivity. The author prefers the terms "spectator" and "interator".

10 "Formativity", according to Pareyson, asserts the process, the dialog between the artist and raw materials, as well as the work's driving force over the artist himself/herself.

Chapter 6
Neuroaesthetics:
Insights into the Aesthetic Experience of Visual Art

Ana Teresa Contier
Universidade Federal de São Paulo (Unifesp), Brazil

Laila Torres
Hospital Israelita Albert Einstein, Brazil

ABSTRACT

The aesthetic experience has been discussed throughout the history of mankind by philosophers and art historians, becoming a universal part of human experience, which leads us to some great interdisciplinary questions. It has been the subject of study by neuroscientists and neuro-psychologists since the 2000s. This recent evolution of neurology studies in the field of art, is due to in vivo brain imaging techniques, especially functional neuroimaging. Furthermore, recent research has provided evidence of cognitive interaction during the perception of an artwork indicating that the perceptual experience of art is not merely a passive one. This article reviews important studies in neuroaesthectics of visual art that point out that the aesthetic experience is related to the distribution in the neural architecture, suggesting the involvement of sensory-motor areas, emotional centers, reward system, memory and language.

INTRODUCTION

A painter is revealing something which no one has ever seen before and translates it into the absolute concepts of painting. That is, into something other than reality. (Paul Cézanne)

The relationship between perception and scientific knowledge is one of appearance to reality. It befits our human dignity to entrust ourselves to the intellect, which alone can reveal to us the reality of the world...(Descartes)

DOI: 10.4018/978-1-5225-0510-5.ch006

Humans are the only animals that creat art. The evolutionary role that this ability plays is not yet fully understood. After all, why do we express ourselves by creating art? And why do we appreciate art? Considering the quoted statements by Cézanne and Descartes at the opening of this chapter, we can certainly come to the realization that one same object observed by us and by an artist takes on new meaning when painted with specific shadowing, light volume and coloring techniques, and not only that, this new form becomes an additional stimulus to our senses and as it is perceived by our mind it can potentially contribute to changing our worldview. Perception modulates our reality? This and the previous questions were the main underlying theme for studies in philosophy, history and aesthetics for many years. For example, the French phenomenological philosopher Merleau-Ponty argues that perception is the key component in the production of the artistic object and in its consideration by the viewer:

Suffice it to say that even when painters are working with real objects, their aim is never to evoke the object itself, but to create on the canvas a spectacle, which is sufficient unto it. The distinction, which is often made between the subject of the painting and the manner of the painter, is untenable because, as far as aesthetic experience is concerned, the subject consists entirely in the manner in which the painter on the canvas constitutes the grape, pipe or pouch of tobacco. Does this mean that, in art, form alone matters and not what is said? Not in the slightest. I mean that form and content – art and the world of perception what is said and the way in which it is said – cannot exist separately from one another. (p.45)

These issues were also addressed by experimental aesthetics in the nineteenth century in the works of Gustav Fechner (as cited by Shimamura,2012). One can definitely argue that art and aesthetic experience have been objects of those research fields and that there is a clear interest in the influence of perception on the relationship between the art object, the artist and the viewer.

In the late 90s, with the advancement of scientific research on the brain and mind, a new scientific field starts to emerge: neuroaesthetics. Neuroaesthetics was a term created by Semir Zeki in 1999, in his article "Art and the brain" and it refers to the study of the neural basis of aesthetics. Zeki's approach to art is based on understanding how the "visual brain" works, in order to obtain a true knowledge about the world. Neuroaesthetics is a relatively young field within cognitive neuroscience and is concerned with identifying the neural basis of aesthetic experience in various works of art, such as painting, music, photography and sculpture. Since the late twentieth century and early twenty-first century, many studies focus on our response to beauty to understanding the neural basis of perception. Chatterjee (2010) pointed out the study of beauty in art has helped the advancement of this research field and it surely remains incontrovertible that art is not always concerned with beauty, and visual artists do not always aim to create beautiful artwork. Another question in this field is the investigation of meaning and emotional representation in art. This article will explore the neuroaesthetic approach to the experience of art.

As we shall see, researchers have not reached a consensus regarding the definition of aesthetic experience, but it is clearly a concept tied to art behavior. Shimamura (2012) devised a model for aesthetic experience known as I-SKE, an acronym where I stands for Intetion, S for Sensation, K for Knowledge and E for Emotion. The model deals with all those four elements that are implicated in the aesthetic experience: the starting point in this framework is the artist's intention when producing an artwork that will be aesthetically appraised. The outcome of this intention is the artwork that will in turn elicit mental effects that are involved in the artwork's appreciation and fall in three categories: sensation, knowledge and emotion. According to Shimamura (2012) I-SKE framework's main contributions are taking the artist's intention and beholder's knowledge (culture, world view and personal experience) into account, these are the key points that set this model apart from previous research. It follows from the model that a greater aesthetic appreciation will rely on sensation, knowledge and emotion to be effectively balanced and well conjugated in the beholder.

This model presents the major components that make part of the aesthetic experience but it still not sufficient in order to capture the complexity of this experience in neurobiological terms. Scientific investigations by Zeki (1999); Gabrielle Starr (2013); Chatterjee (2009, 2010); Leder, Belke, Oeberst and Augustin (2004) presented different approaches about neuroaesthetic but also common foundations such as focusing on understanding the brain mechanisms of the subject ´s aesthetic experiences. In order to guide us throughout this discussion, we will need a working definition for this concept and it shall by the one given by Cinzia and Gallese (2009):

aesthetic experience is one that allows the beholder 'to perceive-feel-sense' an artwork (from the Greek aisthese-aisthanomai), which in turn implies the activation of sensorimotor, emotional and cognitive mechanisms (p. 682).

We find their approach thorough and it synthesizes the recent research in the field of neuroaesthetics. In order to accomplish this, they relied on neurophysiological methods such as electroencephalography (EEG) and functional magnetic resonance imaging (fMRI).

Today, the use of technics like neuroimaging, has been contributing in advancing our understanding of the cognitive basis of aesthetic appreciation, since it can provide a new analytical level of evidence in the understanding of cognitive processes (Nadal & Skov, 2013). Researchers undertook human brain imaging experiment, using fMRI, which is a noninvasive technique for measuring and mapping brain activity. According to the department of radiology of University of California, San Diego,

it is being used in many studies to better understand how the healthy brain works, and in a growing number of studies it is being applied to understand how that normal brain function is disrupted in disease (n.d, para 1).

Many researchers have used paintings as stimuli within their experiments and have been able to establish which areas across the brain are integrated during the aesthetic experience. Vartaniana and Skov (2014) discuss those include areas related to memory, emotion, visual and motor systems. We shall go through the main concepts and leading researchers in the field of neuroaesthetics, particularly those related to the aesthetic experience. Our goal is to show how this relatively new research field can contribute to the understanding of art behavior without failing to explore the challenges it faces. In order to illustrate some of the most interesting findings in this field of research, we will discuss some of the works of the world famous Brazilian modernist painter, Candido Portinari, (1903-1962): the drawing "Crib" (1931), "Denise with the cat" (1961) and "Portrait of Maria Jordão Arruda and her daughter" (1926).

AESTHETIC EXPERIENCE

According to Cinzia and Vittorio (2009),

Neuroaesthetics has been, so far, mainly concerned with visual perception, with a particular focus on how the properties of artworks are visually processed (p.686).

However, the evidence reviewed consistently suggests that a visual description of art works, just like the perception of any visual object, is only the beginning of the aesthetic experience. Besides the perceptual processing of the object itself, the aesthetic experience includes the emotional response to the object and, when relevant, an explicit judgment about the object's beauty. In fact, research results indicate that emotion and sensorimotor processes are related to the aesthetic experience. Gabrielle Starr (2013) substantiates the notion that aesthetic perception involves the projection of the *self* and the reintegration of sensations, in a process that relies on both the visual and motor systems. Throughout the assessment of what is seen, the subject adjusts its own perception of the world, assigning new meanings to elements that make up its world view. It works as if it were a learned vision in which the reward is found by seeking the pleasant and beautiful, being able to distinguish it from the disgusting and ugly. In some way, the experience and concept of beauty are related to the brain centers for reward and pleasure. The integration of sensory systems and their relationship to other brain areas suggests a perceptual network involving emotion, but also attention, values, memory and learning.

Usage of neuroimaging methods has been instrumental to several discoveries in the field of neuroaesthetics. The following studies we deal with one aspect that has been the subject of much research: the question of beauty. Results indicate an integration of several areas in the brain when subject to different stimuli: paintings, music and mathematical formulae.

In the simplest fMRI experiment a subject alternates between periods of doing a particular task and a control state, for example, subjects are asked to view pictures of paintings and listen to brief musical excerpts, during a short period of time, and then rate them as beautiful, neutral or ugly, while the activity of their brains are imaged. With this type of method it is possible to identify a "single area or set of areas whose activity would correlate with the experience of beauty or ugly, regardless of whether it was derived from an auditory or visual source" (Ishizu and Zeki, 2011, p. 1).

Another example of fMRI usage can be found in a study published by Jacobsen, Schubotz, Hofel, and Cramon (2006). In this study subjects were asked to judge on the symmetry and beauty of geometrical shapes while being imaged. Parietal and premotor areas were activated when subjects were making symmetry judgements. Differences in activations were found when subjects were making aesthetic judgements, in which case the activated areas included the frontomedian cortex, bilateral prefrontal cortex, posterior cingulate, the left temporal pole and the temporoparietal junction.

Aesthetic experience is therefore the product of the integration of various systems: sensory system, emotional system and cognitive system all take part in the processes that make up aesthetic experience. In the case of visual art, a painting reflects light which, as it continually reaches our eyes, is the stimulus that will trigger a series of processes in the visual system leading up to the decoding of the painting. The neurobiological properties of this system, which we will shortly present, make it possible for us to identify shadows and contours, recognize patterns and infer shapes. Kandel (2012) and Starr (2013) agree with important contributions from attention, memory, emotion and empathy, a mental image of the painting is created while the aesthetic experience (pleasant or not) takes hold.

We shall see next the most import features of these mechanisms and how they contribute to art behavior. As a means to illustrate the underlying concepts, we will present paintings from the Brazilian modernist painter Cândido Portinari. Portinari's work is known internationally and, as Fabris (1990) puts it, *what really characterizes the art of Portinari is the quest for a national imagery in which thematic and formal elements merge to express a Brazilian reality (…) (pp. 38-39).*

His relationship with the question of Brazilian national identity is well established in critical studies of art. We believe that choosing Portinari to illustrate the aesthetic concepts we have started to discuss is useful in demonstrating that while many of his works are linked to Brazilian themes, they are, at the same time, fine examples of how graphical elements and choice of themes end up stimulating the nervous system and creating the aesthetic experience as previously defined.

PERCEPTUAL ANALYSIS

As our main focus throughout this discussion lies on the visual arts, we shall discuss the neurobiological properties of our brain's image processing. This brief introduction to the neurobiology of vision is not meant to be a detailed and thorough review, since for our current purposes it is sufficient to provide an understanding of the basic mechanisms and their relation to cognition and emotion in the creation of an aesthetic experience. As Kandel (2012) put it, we live between two worlds at once: the external world that enters through the eye's fovea and the internal world of perception, cognition and emotion. Research of neurological content allowed for the observation that the subject cognitively interacts with the artwork, therefore implying an active and projective relationship and not a passive and receptive one.

Research has clearly established how our eyes behavior fundamentally differs from that of a camera. Kandel and Schwartz (2000) show that a camera registers light intensity on a single plane of a vision field on a point by point basis. The brain on the other hand breaks up scenes in multiple components, identifying background and foreground, so that it can determine which light stimulus belongs to a certain object.

The visual processing that is carried out when we look at an artwork starts from the moment the first light particles reach our eyes until we make out a picture in our head can be hierarchically divided in two main phases. The first phase happens in the photoreceptor cells in our eyes' retina. They are exposed to light particles (photons) of different energy levels. Biochemical processes in those cells are responsible for converting that sensory input into nervous signals that convey information about color and contours. The second phase happens when those basic nervous signals issued by neurons in the retina are processed by the primary visual cortex and other visual areas in the cerebral cortex. Movement detection, identification of depth, and of foreground and background are made possible by those cortical areas. This processing of visual information by the nervous system is parallel and hierarchical in nature and allows for the distinction between form and content, color and space localization (Chatterjee, 2010).

Under Zeki's (1999) perspective, the function of art is a brain function, and as such, the physiological properties of the brain are involved in the creation of a specific perception of an artwork. The same author argues that constancy is a fundamental propriety of vision that allows aesthetic appraisal. The fact that our visual system is capable of establishing such constancies makes it possible for us to recognize faces as belonging to same person, even when seeing them from different angles, wearing different facial expressions or after aging.

Visual processing will assign a single color to a surface even when there are small variations in light wavelengths coming from an object's surface. Faces and other objects will be recognized as one same entity regardless of varying angles in which they face the beholder. The perception of contours in visual art means that artists are not required to portray all the wealth of details that real world object have. The brain of the beholder has the ability to infer borders and to make complex guesses, so that even very simple drawings without color or shadows will evoke very specific meaning associated to real world objects. A simple drawing with just a few contour lines for eyes, nose and mouth will be enough to trigger the recognition of a face. The retina extracts limited visual information form the external world, leaving the brain with the continuous task of making creative guesses that lead to visual flexibility that artists have investigated throughout the centuries. Out of all variations and distortions that artists were able to explore given the nature of our brains' elaborate guesses, the most significant feat is undoubtedly the ability to create a sense of perspective in a two dimensional image. The process of ascribing meaning to an image also depends on memory, because associations with past experiences can be relevant in that process, as well as the identification of cultural symbols also retrieved from memory (Kandel, 2012). Such memories also carry an emotional content, which also plays a role in the relationship that we will establish with an artwork. For instance, when seeing an painting that reminds the beholder of something unpleasant or of a previous traumatic experience, it is likely that the ensuing association between the depicted scene and the beholder's own past experience will interfere with the its aesthetic appraisal.

Attention is a modulator of visual stimuli, as we see the world, we perceive first the parts that demand more attention. Guided by attention, our first gaze starts with a quick scanning of the whole scene. These are called saccades (literally meaning jerks, in French) and they allow the eye fovea to explore the environment. After and between subsequent saccades, our gaze becomes fixed at certain points and this is what makes it possible for us to consciously see (Kandel, 2012).

Let us ponder those neurobiological properties of vision, while we consider the following drawing: The Crib (Figure 1).

The Crib is a biblical scene known by many, namely the birth of Jesus Christ. The scene is portrayed with a very small third dimension for most objects (there is almost no depth), the bull being the main exception, but our brain is able to recognize the famous scene with simple sketches. This is achieved not just by contemplating the actual sketches, but by tapping into the wealth of world knowledge we

Figure 1. "Crib" (1931) by Cândido Portinari

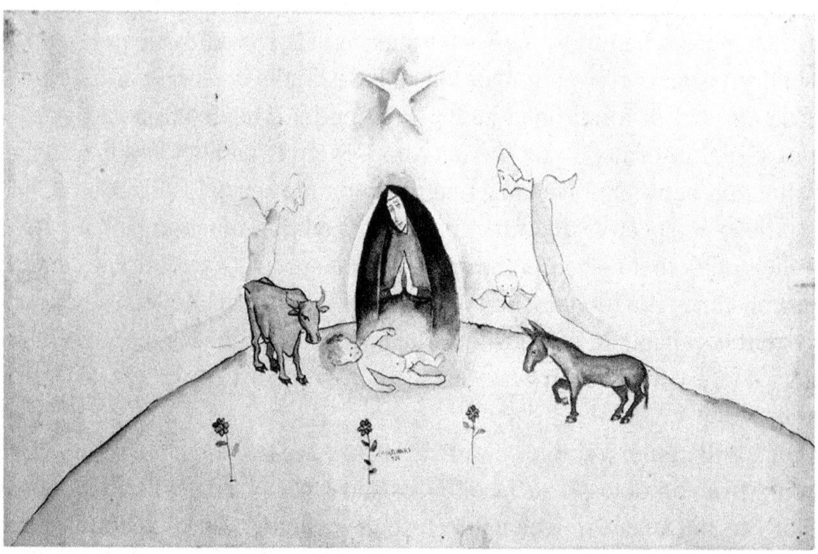

have stored through life. The light colors used to paint the scenery and the fact that brighter colors are used on central objects direct our attention to the drawing's center. Our attention will be drawn to Mary wearing a dark veil with Jesus glowing underneath. This is made easy by the presence of additional objects working as boundaries for our attention: the star above, a brightly colored donkey to one side and a bull to the other. According to Kandel (2012), color also plays an important role in the emotional response of the beholder.

We perceive colors as associated to different emotional contents and we also know that this identification can vary across subjects and even in a same person at different moments. This could also be one of the reasons why a painting may cause different responses in a same person at different times and also across different beholders. In this drawing, we can establish a link between yellow and sanctity and we also note that it is used to bring attention to some of the scene's elements. Another important observation is our ability to infer a more complete image relying upon basic sketches. Consider the angel at the right: it does not have a body, its wings and face are enough to allow us to recognize it as an angel.

Research on the perception of faces has indicated some of the cortical areas that are thought to be associated with recognition and processing of face information. While the fusiform face area is believed to be implicated in face recognition (Kanwisher, McDermott, & Chun, 1997), there is research that indicates that the superior temporal sulcus plays a role in the detection of eye gaze direction (Campbell, R., Heywood, C.A., Cowey, A., Regard, M., & Landis, T., 1990) and in processing of facial expressions (Tseng, Tseng, Liang, Hung, Tzeng, Muggleton & Juan, 2014). The reward system has also been connected with processing of face emotions, activity on bothg right and left amygdala has been tied to early emotional face processing (Liu, Chen, Su, Hsieh, & Chen, 2014).

A testament to the great attraction that faces exert on mankind is the staggering number of portraits and self-portraits that were painted throughout the history of art. For example, Cândido Portinari painted more than 40 portraits throughout his career (Portinari Project). The following oil painting is called "Denise with the cat" (1961) (Figure 2).

The eyes, facial expression and body posture represent the depicted character's emotion. In this picture Denise's face reveals tenderness through her eyes and the way that she holds her cat. Looking at face is not only judging whether it is attractive or not. When looking at a face we normally quickly scan eyes first and then the mouth, though attention focus can be guided by various factors such as intention, interest, previous knowledge freshly recruited from memory. In this case, our infancy's memory is also probably recruited as a sense of empathy is created.

Embodiment

The embodiment approach of aesthetic experience is related to two different issues, according to Torabi Nami and Ashayeri (2011):

First, the relationship between embodied simulation-driven empathic feelings in the observer and the representational content (here, the visual art), next, the relationship between embodied simulation-driven empathic feelings in the observer and the visible traces of the artist's creative gestures p. 9.

The most direct implication of these relationships is our capacity to simulate actions, emotions and sensations of others. This comes about through the activation of our internal representation of bodily states that are associated with social external stimuli. In these circumstances, the same brain areas are

Figure 2. "Denise with the cat" (1961) by Cândido Portinari

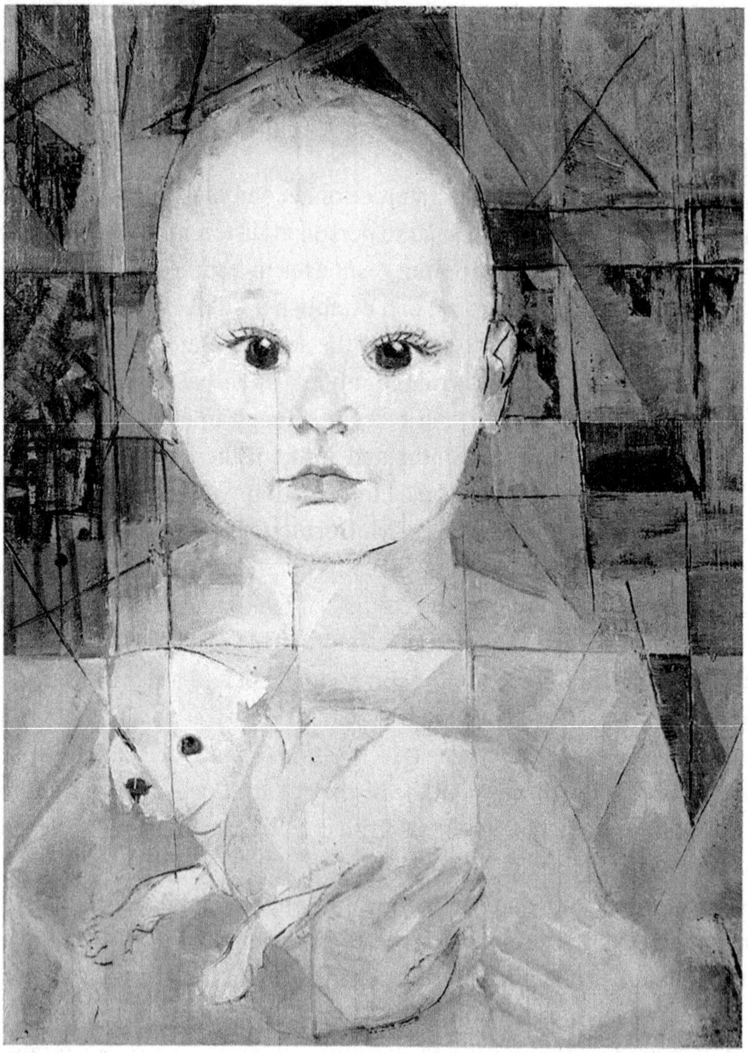

activated when dealing with first hand or third person sensations, or even experiences of action and emotion. This suggests, according to Freedberg and Gallese (2007), that in evolutionary terms there probably is a very old mechanism that allows us to have a direct understanding of objects through experience. A mechanism that also applies to the empathic understanding of the internal world of others.

For example, Gallese, Keysers and Rizzolatti (2004), suggest that self-initiated actions and the visual perception of an object receiving identical actions carried out by someone else, evoke the same neural response. Although an important aspect of this observation is that the visual perception of objects is not activated in the mirror neurons, as these neurons are activated only when observing a specific action. These observations lead to the hypothesis that the brain is inherently relational. We use our bodies as a model that allows us to find the way to the experience of the other. This supports the theory that the body plays a significant role in the development of empathy and that the process occurs unconsciously.

The fact of being able to spatially perceive reality represented in two-dimensional images as well as the representation of actions and movements, combined with the psycho-analytically proven phenomenon of projective identification and the current discovery of mirror neurons, suggests a basis for our mysterious ability to experience an intense emotional and affective relationship with artistic images. Research has shown that there is more activation in the motor cortex during the observation of attractive faces (Kawabata & Zeki, 2004) and it has also been shown that movements made by a face that is being observed will activate mirror neurons in the person watching those movements (Knoblich & Sebanz, 2006). These results raise questions about possible similar effects that could develop when we are observing an artwork. In fact, Keysers; *et al* (2004) investigated brain activity unfolded by the sight of a touch action as depicted in a movie. Their research shows that similar patterns of activation in the somatosensory cortex are found when test subjects see a touch action or when they are actually touched.

In Portinari's painting "Portrait of Maria Jordão Arruda and her daughter" (Figure 3) we can observe the mother's care and effort while holding her daughter, and the child's joy as expressed by her smile. Both faces are instrumental in the presentation of an emotional content related to maternal affection. Additionally, as it has been discussed in previous research on mirror neurons, looking at the action of

Figure 3. "Portrait of Maria Jordão de Arruda and her daughter" (1926) by Cândido Portinari

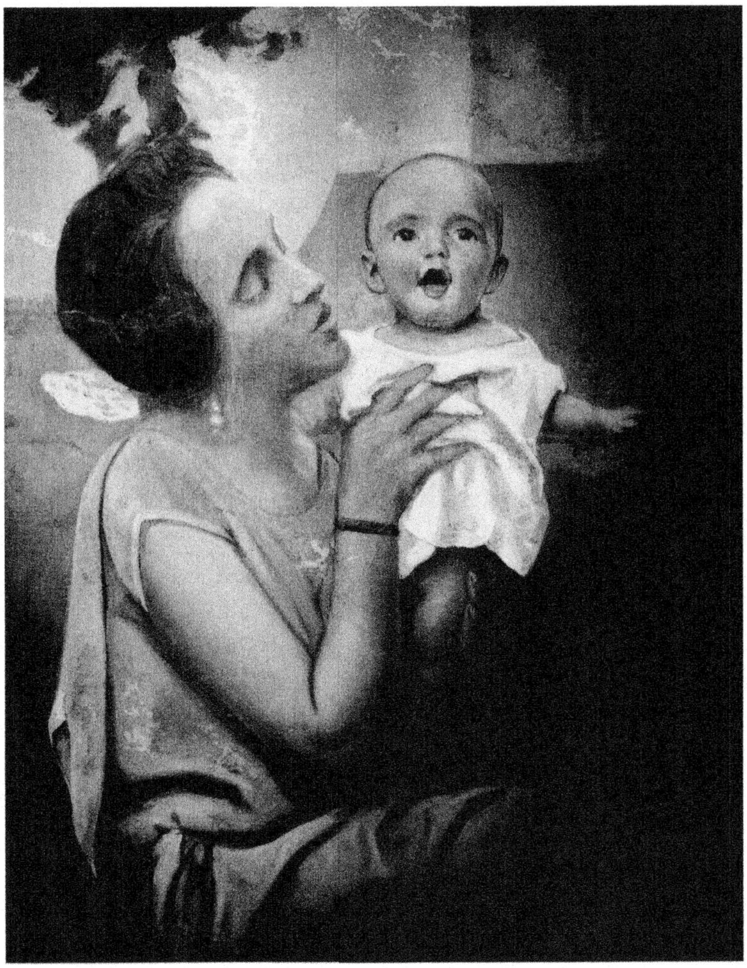

someone carrying her daughter, as can be seen in this painting, will probably activate the motor cortex of the beholder. The close affective relationship between mother and daughter that one can easily perceive in this painting will activate the semantic and emotional memory of beholders. This example illustrates current discussions in studies of this field, as research increasingly points to the integration of different systems as driving the creation of aesthetic experience.

CHALLENGES

Recent research (Chatterjee, 2010; Brown, 2009; Skov, 2009) points out some challenges that need to be addressed within the field of neuroaesthetics. Authors of these studies agree that, because this research field is relatively recent, protocols and methods need to evolve in order to avoid common pitfalls. One first challenge that needs addressing is the creation of a common framework that could be used by researchers. There are no equivalent standardized models for art to the language models that define main linguistic elements and their relationships (Skov, 2009). Fundamental definitions for the concepts of style and aesthetic appraisal have not been completely established. Furthermore, there is currently only a small volume of aesthetic data associated with sensory perception that is structured and available for research and the same holds for the lack of established criteria to differentiate art from non-art (Brown, 2009). As Skov (2009) points out, there is a need for the creation of an art object model, as well as model for the aesthetic experience. Would the aesthetic experience be linked to beauty and ugliness or would it reach beyond and encompass all experience evoked by art?

The definition of such a framework would also aid in telling apart research about brain properties from neuroaesthetics research. According to Chatterjee (2010), abstract paintings can be used to investigate how the brain deals with non determined visual stimuli, but that in itself is not an investigation on aesthetics. The same author reminds us that researchers need to question how their inquiries on the involved brain physiological properties and systems lead to a better understanding of the relevant aesthetic concerns. Questioning if a research hypothesis would be more appropriately approached by another research field is also important.

A second challenge is the risk of reductionism. This happens if experiments are not based on empirically verifiable hypothesis, which ends up leading to insufficient results (Chatterjee, 2010).

Another important challenge is related to available techniques and to data interpretation. Cinzia and Gallese (2009) point out that more objective results could be obtained by putting subjects in an appropriate state of mind while being imaged or under EEG, but that there are certain difficulties in achieving that. This kind of challenge is also present when researching aesthetic appreciation, i.e. the judgments that a subject attributes to an art object. When interpreting results of such studies, researchers need to look for contrasts and similarities between activations and patterns occurring in different situations and test settings, in order to avoid mistakenly associating aesthetic value with common or baseline features.

One last significant challenge faced by neuroaesthetics is integrating diverse research fields such as neuroscience, art and philosophy. Hyman (2010) criticizes previous work by Zeki and Ramachandran for failing to address this challenge. The critique of these studies is that they draw heavily from neurology but fail to fully incorporate the relevant developments of other knowledge areas. According to his vision, studies by those authors make generalizations on the concept of art, but Hyman still believes neuroscience can contribute to the understanding of visual art, provided that the developments of other research fields are taken into account. Skov (2009) sums it up with his remark,

Just as a model of what constitutes art is a prerequisite to investigations of how the brain computes art, studying the neurocognitive mechanisms that cause art behavior is of vital importance to the mission of traditional aesthetics: figuring out what art is. (p.11)

CONCLUSION

Sight is not isolated, sight is rooted in its corporality, as a sensibility and as a motricity. (Bosi, 2006, p.66)

The field of neuroaesthetics within cognitive neuroscience is still a relatively young research area. Studies indicate that perceptual systems, memory and emotional centers are activated in face of aesthetic experience". These areas are recruited after the visual system receives and processes a painting's visual stimuli, identifying patterns, detecting form and content, as well as shadow and lighting. Bearing no resemblance to the processing of a camera, the beholder creates the image. The active process of attention will guide the way we look at a painting, capturing at first a glimpse of the artwork in its entirety, and proceeding to select what is most interesting. It has been observed that when we are looking at a face, attention is generally concentrated in the eyes and mouth because they represent emotion and create empathy. But memory and personal interests also inform and direct our gaze. The fact that we categorize objects and recognize them as cultural symbols tapping into our semantic memory supports the idea of an active visual perception.

Research on mirror neurons brought evidence that the beholder establishes an empathic relation towards artworks. When observing an artwork, the beholder will actively place itself in the portrayed context, putting itself in the shoes of portrayed characters. Embodiment in aesthetics has also provided evidence that pictorially represented action does activate motor regions of the beholder's brain.

The complexity of this research field is self-evident and amongst the challenges it faces are the creation of a common framework and the availability of more data which will both help standardizing methods and getting more relevant results. As Chatterjee, Thomas, Smith and Aguirre (2009) point out, the study of the neural basis of aesthetic experience has so far, focused on the processing of aesthetic judgment, experience and the characterization of aesthetic reward.

Neuroaesthetics proves itself as a very promising area, considering that it provides for a deeper understanding of the neural processes that underlie an aesthetic experience, having repercussions in various areas of society: education, art, philosophy and other segments.

ACKNOWLEDGMENT

The authors thank the JOÃO CÂNDIDO PORTINARI (PROJETO PORTINARI- http://www.portinari.org.br).

REFERENCES

Bosi, I. A. (2006). Fenomenologia do olhar. In A. Novaes (Ed.), *O olhar* (p. 66). São Paulo: Companhia das Letras.

Brown, S., & Dissanayake, E. (2009). The arts are more than aesthetics: neuroaesthetics as narrow aesthetics. In M. Skov & O. Vartanian (Eds.), *Neuroaesthetics* (pp. 43–57). Amityville, NY: Baywood.

Campbell, R., Heywood, C. A., Cowey, A., Regard, M., & Landis, T. (1990). Sensitivity to eye gaze in prosopagnosic patients and monkeys with superior temporal sulcus ablation. *Neuropsychology*, *28*(11), 1123–1142. doi:10.1016/0028-3932(90)90050-X PMID:2290489

Cézanne, P. (n.d.). *Paul Cézanne: French Draftsman and Painter*. Retrieved from http://www.theartstory.org/artist-cezanne-paul.htm

Chatterjee, A. (2010). Neuroaesthetics: A coming of age story. *Journal of Cognitive Neuroscience*, *23*(1), 53–62. doi:10.1162/jocn.2010.21457 PMID:20175677

Chatterjee, A., Thomas, A., Smith, S. E., & Aguirre, G. K. (2009). The neural response to facial attractiveness. *Neuropsychology*, *23*(2), 135–143. doi:10.1037/a0014430 PMID:19254086

Cinzia, D. D., & Vittorio, G. (2009). Neuroaesthetics: A review. *Current Opinion in Neurobiology*, *19*(6), 682–687. doi:10.1016/j.conb.2009.09.001 PMID:19828312

Descartes. (1984). The Principles of Philosophy. In *The Philosophical Writings of Descartes*. Cambridge, UK: Cambridge University Press.

Fabris, A. (1990). *Candido Portinari*. São Paulo: Perspectiva.

Freedberg, D., & Gallese, V. (2007). Motion, Emotion and Empathy in Aesthetic Experience. *Trends in Cognitive Sciences*, *11*(5), 197–203. doi:10.1016/j.tics.2007.02.003 PMID:17347026

Gallese, V., Keysers, C., & Rizzolatti, G. (2004). A unifying view of the basis of social cognition. *Trends in Cognitive Sciences*, *8*(9), 396–403. doi:10.1016/j.tics.2004.07.002 PMID:15350240

Hyman, J. (2010). Art and Neuroscience. In *Beyond Mimesis and Convention*. Dordrecht: Springer Netherlands. doi:10.1007/978-90-481-3851-7_11

Ishizu, T., & Zeki, S. (2011). Toward a brain-based theory of beauty. *PLoS ONE*, *6*(7), e21852. doi:10.1371/journal.pone.0021852 PMID:21755004

Jacobsen, T., Schubotz, R. I., Hofel, L., & Cramon, D. Y. (2006). Brain correlates of aesthetic judgment of beauty. *NeuroImage*, *29*(1), 276–285. doi:10.1016/j.neuroimage.2005.07.010 PMID:16087351

Kandel, E. (2012). *The Age of Insight. The quest to understand the unconscious in art, mind, and brain, from Vienna 1900 to the present*. Published by Random House.

Kandel, E. R., Schwartz, J. H., & Jessell, T. M. (Eds.). (2000). *The constructive nature of visual processing. In Principles of neural science* (Vol. 4, pp. 556–576). New York: McGraw-Hill.

Kanwisher, N., McDermott, J., & Chun. (1997). The Fusiform Face Area: A Module in Human Extrastriate Cortex Specialized for Face Perception. *The Journal of Neuroscience*, *17*(11), 4302-4311.

Kawabata, H. & Zeki, S. (2004). Neural correlates of beauty. *Journal of Neurophysiology*, *91*(4), 1699-705.

Keysers, C., Wicker, B., Gazzola, V., Anton, J. L., Fogassi, L., & Gallese, V. (2004). A touching sight: SII/PV activation during the observation and experi- ence of touch. *Neuron*, *42*(2), 335–346. doi:10.1016/S0896-6273(04)00156-4 PMID:15091347

Knoblich, G., & Sebanz, N. (2006). The social nature of perception and action. *Current Directions in Psychological Science*, *15*(3), 99–10. doi:10.1111/j.0963-7214.2006.00415.x

Leder, H., Belke, B., Oeberst, A., & Augustin, D. (2004). A model of aesthetic appreciation and aesthetic judgments. *British Journal of Psychology*, *95*(4), 489–508. doi:10.1348/0007126042369811 PMID:15527534

Liu, T. Y., Chen, Y. S., Su, T. P., Hsieh, J. C., & Chen, L. F. (2014). *Abnormal Early Gamma Responses to Emotional Faces Differentiate Unipolar from Bipolar Disorder Patients*. BioMed Research Internacional.

Merleau-Ponty, M. (2004). Conversas, 1948- Maurice Merleau-Ponty. São Paulo: Martins Fontes.

Nadal, M., & Skov, M. (2013). Introduction to the special issue: Toward an interdisciplinary neuroaesthetics. *Psychology of Aesthetics, Creativity, and the Arts*, *7*(1), 1–12. doi:10.1037/a0031842

Portinari, J. C. (2000). Projeto Portinari. *Estudos. Avançados*, *14*(38). Retrieved from http://fmri.ucsd.edu/Research/whatisfmri.html

Shimamura, A. (2012). *Toward a Science of Aesthetics: Issues and Ideias. In Aesthetic Science connecting minds, brains, and experience* (pp. 4–26). Oxford University Press, Inc.

Skov, M. (2009). Neuroaesthetic problems: a framework for neuroaesthetic research. In M. Skov & O. Vartanian (Eds.), *Neuroaesthetics* (pp. 9–26). Amityville, NY: Baywood.

Starr, G. (2013). *Feeling Beauty: The Neuroscience of Aesthetic Experience*. MIT Press. doi:10.7551/mitpress/9780262019316.001.0001

Torabi Nami, M., & Ashayeri, H. (2011). Where Neuroscience and Art Embrace: The Neuroaesthetics. *Basic and Clinical Neuroscience*, *2*(2), 6–11.

Tseng, L. Y., Tseng, P., Liang, W. K., Hung, D. L., Tzeng, O. J. L., Muggleton, N. G., & Juan, C.-H. (2014). The role of superior temporal sulcus in the control of irrelevant emotional face processing: A transcranial direct current stimulation study. *Neuropsychology*, *64*, 124–133. doi:10.1016/j.neuropsychologia.2014.09.015 PMID:25261612

Vartaniana, O., & Skov, M. (2014). Neural correlates of viewing paintings: Evidence from a quantitativemeta-analysis of functional magnetic resonance imaging data. *Brain and Cognition*, *87*, 52–56. doi:10.1016/j.bandc.2014.03.004 PMID:24704947

Zeki, S. (1999). *Inner vision: An exploration of art and the brain*. New York: Oxford University Press.

KEY TERMS AND DEFINITIONS

Crib: A drawing by Cândido Portinari made in 1931 using India ink and watercolor on paper. It measures 36,4 cm by 56,7 cm. Portinari drew it to illustrate his poem Natal (Christmas) which was published in the Brazilian magazine Bazar.

Cândido Portinari (1903-1962): Brazilian modernist painter who was born on the 30th of December 1903 in Brodowski, a small town in the Brazilian state of São Paulo. His first known work was a drawing of poet Carlos Gomes, made when he was 10 years old. At 15 he was invited to paint stars on the dome of a church on his hometown. In 1919, he decides to go to Rio de Janeiro to study painting in the National School of Fine Arts (Escola Nacional de Belas Artes). In 1958 the Palais des Beaux Arts (Brussels) exhibits 50 Years of Modern Art and Candido Portinari was the only Brazilian artist to participate. In 1962 he was preparing to exhibit 200 works in Milan City Hall (Italy), when he passed away victim of intoxication caused by the paints.

Denise and the Cat: An oil on canvas painted by Candido Portinari in 1960 with a height of about 37 cm and a width of 27 cm. Denise is the granddaughter of the artist and has inspired more than 10 works in the years of 1960 and 1961. Amongst those are: "Denise and horse" (1960), "Denise with the little dog" (1960), "Portrait of denise" (1961) and "Denise standing by the table" (1961).

Electroencephalography: (EEG): An electroencephalogram (EEG) is a test that measures and records the electrical activity of your brain. Special sensors (electrodes) are attached to your head and hooked by wires to a computer. The computer records your brain's electrical activity on the screen or on paper as wavy lines. Certain conditions, such as seizures, can be seen by the changes in the normal pattern of the brain's electrical activity.

Emotional Centers: The biological basis for emotions are perhaps first investigated by researchers William James and Carl Lange at the end of the nineteenth century. Working independently in 1884 (James) and 1885 (Lange) they posit a similar theory for emotion. It became known as the James-Lange theory and it basically identifies emotions with physiological changes caused by external stimuli. Walter Cannon and his doctoral student Philip Bard rule out James-Lange theory by interpreting several experimental results that lead to a different theory for the neural basis of emotion. The Cannon-Bard theory links emotions with the activity of thalamus and hypothalamus. One important distinction from the James-Lange theory is that within the Cannon-Bard framework, physiological changes related to emotions are independent of the emotional feeling. Hypothalamic and thalamic discharges are responsible for bodily emotional responses as they reach muscles and viscera and simultaneously for our feelings as as they reach our brain's cortex. In 1937, James Papez described a series of elements forming a cyclic neural pathway that according to his research were responsible for the brain's emotional processing. That cyclic pathway became known as the Papez circuit. Confirming Cannon and Bard's previous findings, Papez circuit includes the thalamus. Important research on the Papez circuit was carried out by Paul MacLean who extended the circuit to include other important emotional areas and called it the limbic system in 1949. Among the extensions introduced by MacLean is the amygdala (in 1952), now considered to play a role in emotional processing.

fMRI: A functional neuroimaging procedure using MRI technology (medical imaging technique used in radiology to image the anatomy and the physiological processes of the body in both health and disease), that measures brain activity by detecting changes associated with blood flow.This technique relies on the fact that cerebral blood flow and neuronal activation are coupled. When an area of the brain is in use, blood flow to that region also increases. MRI scanners use magnetic fields and radio waves to form images of the body. The technique is widely used in hospitals for medical diagnosis, staging of disease and follow-up without exposure of the body to ionizing radiation)

Frontal Lobe: A part of the cerebral cortex and is made up by the prefrontal cortex, orbital cortex, motor and premotor cortices, and Broca's area. It is involved in higher cognitive functions such as attention, voluntary movement, decision making and language.

Gustav Fechner (1801-1887): German physicist, natural philosopher and experimental psychologist that published Aesthetics' Kindergarten (Vorschule der Aesthetik) in 1876. This work was one of the first to deal with aesthetics as a science and as such it proposed the use of scientific methods and the analysis of empirical data in the pursuit of aesthetics studies.

Maurice Merleau-Ponty (1908-1961): French phenomenological philosopher who posited that the human mind phenomenon was dynamically brought about by the interactions between brain, body and the external world. His first published work in 1942 was *La Structure du comportement* (The Structure of Behavior) was followed by the *Phénoménologie de la perception* (Phenomenology of Perception) *in* 1945 and several others before the incomplete The Visible and the Invisible which was posthumously published in 1964.

Motor Cortex: Located in the cerebral cortex and made up of primary motor cortex, premotor cortex and supplementary motor area. The primary motor cortex generates impulses that pass through the spinal cord and control movement execution. The premotor cortex is responsible for the preparation of movement, the sensory guidance of movement, and direct control of some movements focusing on body muscles. The supplementary motor area contributes to the planning of sequences of movements and coordination of the two sides of the body.

Portinari Project: Created to preserve Cândido Portinari's cultural memory through the organization and systematization and dissemination of the work and life of this Brazilian painter. It is an endeavor in surveying, organizing and digitizing images, that has accomplished: the cataloging of 5300 paintings, drawings and engravings of Portinari's works accompanied by 25000 documents about his life. Digitized images are made available to the public on the project's web site http://www.portinari.org.br. The project was established within the Pontifical Catholic University of Rio de Janeiro (PUC-RJ) and fosters cooperation amongst researchers from different research fields, focusing on science and technology in order to develop methods that could be of use to other institutions faced with similar challenges. The Center for Art Education and Social Inclusion is one of the arms of the project that plays a social and educational role by integrating the artist's work with Brazilian culture in an accessible way to children.

Portrait of Maria Jordão de Arruda and Her Daughter: An oil on canvas painted by Cândido Portinari in 1926. It measures 69.5 cm by 54 cm and belongs to a private collection. The artist painted it three years after the depicted Maria Jordão de Arruda had died, using a photography.

Reward System: The mesolimbic dopamine system controls the responses of individuals to natural rewards like sex, social interaction and appetite, becoming a determining factor in motivation. Activation of these pathways leads to encouraging the individual to repeat the action in order to be rewarded. That activation is also related to effects in memory centers where motivating experiences are recorded for future repetition, while the opposite also happens to unpleasant experiences which are remembered so that they can be avoided.

Sensory Areas: The primary sensory cortex areas and the multimodal association areas. The primary sensory cortex areas of the brain are related to major senses like: visual, auditory, taste, smell, somato-sensory, vestibular and visceral sensations. Multimodal (or multisensory) neurons in multimodal areas are responsible for handling different senses in an integrated fashion, enabling a coherent representation of all sensory information received from the primary sensory cortex areas.

Chapter 7
Understanding the Interdisciplinary Meaning of Beauty to Neuroscience:
Designing Beauty to Neuroscience

Bruno H.S. Araujo
Universidade Federal de São Paulo, Brazil

Ibrahim Elias Nasseh
Universidade de São Paulo (USP), Brazil

ABSTRACT

The search for the source of beauty has exercised the speculation of philosophers and writers throughout the ages. The classical conception is that beauty consists of an arrangement of integral parts into a coherent whole, according to proportion, harmony, symmetry, and similar notions. The opposing view is that art is a fully subjective enterprise and our preferences are shaped by our values and experiences. To study beauty, neuroscience evokes the juxtaposition of concepts such as symmetry, geometry, proportion, judgment, affection, grotesque, erotic, sensitive with certain loci in the human brain. However, to explain and understand the aesthetic experiences at the network level, neuroscience have to incorporate the ambiguity in their research field. And with the continuous increasing of knowledge and comprehension of how beauty is perceived by the human brain, the world in its wholeness will be highly impacted, since we will start to live in pleasant societies.

INTRODUCTION

To deliberate about beauty, we first need to contemplate different types of considerations and approaches that, in history, were indorsed on this subject. Therefore, in an epistemic census, the key is to establish the possible heuristics endeavor and to elect the "epistemic figures"[1] of sense and possibilities for a new understanding.

DOI: 10.4018/978-1-5225-0510-5.ch007

The theory of beauty with universal principles was attempted at the first time by Plato (Panofsky, 2000), whose established separation between art and beauty, in opposition rules. To Plato, the beauty, has a completely existence, independent and higher than the physical reality, however it also has a universal true value, what make it undeniable. In this way, beauty cannot be defined or held in any touchable manifestation. On the other hand, for Plato, art was *mimesis*, imitation and therefore unchangeable. It had, consequently, the same calling for misunderstanding and confusion that the sensations have on us. The true beauty could only be contemplated by the intellect. It is the Plato's theory of Ideas, a conception of the universe in its wholeness, which was mobilized to give logical basis to the concept of beauty.

Some reconciliation between beauty and art only came up from Cicero who was the first to allow the artist the ability to create superiors forms more than simple copying or *mimesis* the nature. The true artist's model is inside, by the contemplation of his own intellect. He can thus correct the "mistakes" of nature and create a perfect model of beauty. We need to keep in mind that such reconciliation takes place, however, through a distortion of the platonic concept of art and idea (Panofisky, 2000). In Cicero and over the antiquity, the transcendental Plato's doctrine of idea, as harmonic and changeless principle of the universe, and the art as a simple sensitive and objective figuration (in the sense that its reality come true in physical objects like paints or sculptures, so transients and mortal) gave raises to opposite paradigms which art releases itself from his eminently sensible externality to became the achievement of a mental interiority. In this way, the idea falls from its pedestal of transcendental essence to be a thought or internal concept in human immanent consciousness.

In Panofky (2000), the Plato's concept of idea and its development over the time is the main topic to explain the split of the historiography of beauty into certain ages. But although we can see the history of art and beauty through this prism as like evolutionary, which could invoke an Apollonian ideal, harmonic or positivist - whether we would take a most appropriate term to some philosophy of science - we also can see the history of beauty in an opposite sense. This sense is expressed as Dionisiac, chaotic and open.

The categorizing effort, which historiography is a lessee, perhaps can be understood through the key of the critique of historical culture on which Nietzsche weaves fierce criticism.

In this sense the historiographical anachronism does not recognize the becoming[2] as a stream in which all meanings transit. In this sense, the historiography demand to find the same or the identity under the otherness. This goal in such classificatory approach which in the same time excludes and imprisons the hallmarks of a unique kind of art.

Thus,

... the historiography is bogged down in the furrows that it knocks: the scholar is surprised when comes across the reference to " medieval´s " authors in Renaissance texts; so the reviews Ghiberti, dealing with optics, brings authors such as Al-Hazen ... (Novaes,1994,p.10).

Meanwhile, from this concept of genealogy of Nietzsche - which term he uses to conceptualize the authentic "historical spirit" – what happens is the reverse of historiography offers.

(genealogy) will pretend, in first place, to determine the differences instead of to forge identities, it will be attentive to mutations of meanings and suspicious of concepts supposedly unambiguous. Thus, it will investigate the history with no claim to find there the realization of any eternal ideal, thus it will

not put our present there at the origin, as if there was a predestination to be performed in a prior sense to be deployed. It shows that history is a succession of disparate meanings, without any predetermined unification, a succession of interpretations, dominations that alternate. (Moura, 2005,p 144).

Thereby it is the dispute of power by itself that moves the historical development and there is no supra-historical morality that determine it, least of all retain it in a certain context that eternize in some closed periods, being much more like a coming and going of non-perennial and dialectical realities. This concept of game was precisely a main subject to Heraclitus - appropriated by Nietzsche - which decouples the reason as a necessity in history and starts to think on your own contests and your own disparity.

Placed this view of concepts of art and beauty in a problematizing outlook, the new conceptual field that neuroscience brings can invoke the juxtaposition of concepts such as symmetry, geometry, proportion, judgement, affection, grotesque, erotic, sensitive - and many other significant concepts to the genesis and development of what is meant as beauty - with certain *loci* in the human brain, or to certain mathematical relationships, or even to certain universality. Neuroscience not only does, but also settles the method and the necessary conditions, without which the science would never reach the necessary rigor of the study of its object nor the success of your enterprise. We shall see how this happened in neuroscience in the forward paragraphs.

It is part of the modus operandi of science reduce and abstract objects, decomposing them - just like recommending Descartes in his Discourse on the Method (Descartes, 1973). Thus, topics such as art and beauty will be more resisted to this kind of approach. This resistance may benefit science if it remains opens to the knowledge from other disciplines such as philosophy and conceptual ideas created by the artists themselves. This resistance increases progressively throughout history and set requirements for the universal concept of beauty becomes progressively difficult. This may occurs most acutely after modernity and avant-garde art where what is considered beautiful seems to collapse and the very term beauty seems to be emptied within the philosophical thought itself as seen in Dewey, Collungwood and Lyotar (Jonhson, 2010).

But, what in art and beauty may seem a particular case of knowing, in several other scopes it has been a rule. In the core of the interdisciplinarity, we will testimony the increasing emerge of this new-old way of knowing, as a sort of comprehensive and genealogical epistemology.

How many different areas and cognitive resources, like geographic, economic, political, diplomatic, physical, engineering, were vital to mobilize for the construction of the European Council for Nuclear Research (CERN), the largest particle collider in the world?

Substantiating in the whole idea of interdisciplinarity and even transdisciplinarity, many of the concepts about the beauty that science employ, were provided by philosophy and even the art itself. In this way, the art as it was understood in its classical period provided to science epistemic figures of which beauty was comprehended as proportion and harmony, that furnish the basis of the prime researches on the beauty issues (Umberto, 2004).

THE CLASSIC CONCEPTION OF BEAUTY

During an aesthetic appreciation, in general, what is analyzed is the physical properties of the object. However, as we know, appreciation cannot only be reduced to this quantitative criterion, since holistic contemplation attributed to the object is also important (Walker, 2006). A biologist should instead suggest that the objective[3] side of beauty stems from built-in predispositions for certain features, colours, shapes or proportions. The opposing view is that art is a fully subjective enterprise and our preferences are shaped by our values and experiences. The real answer is likely to lie somewhere in the middle – after all, new art apprentices learn basic common skills such as proportion, perspective and symmetry before represent their overall vision of the world (Bayles & Orland, 2007). In this account, beauty is at least in part objective[3].

The classical conception is that beauty consists of an arrangement of integral parts into a coherent whole, according to proportion, harmony, symmetry, and similar notions. This is a primordial Western conception of beauty, and is embodied in classical and neo-classical architecture, sculpture, literature, and music wherever they appear. Aristotle says in the Poetics that

to be beautiful, a living creature, and every whole made up of parts, must ... present a certain order in its arrangement of parts" (Aristotle, volume 2, 2322 [1450b34]) and in the Metaphysics: "The chief forms of beauty are order and symmetry and definiteness, which the mathematical sciences demonstrate in a special degree (Aristotle, volume 2 1705 [1078a36]), (Sartwell, 2014).

This view, as Aristotle implies, is sometimes boiled down to a mathematical formula, such as the golden ratio.

It is important to note that the idea of 'symmetry' in ancient texts is richer than its current implication of bilateral similarity, though it incorporates that as well. It also refers to the sorts of harmonious proportions characteristic of objects that are beautiful in a classical sense.

Vitruvius, the ancient Roman architect, gives as good a characterization of the classical conception as any, both in its complexities and in its underlying unity: Architecture consists of Order, which in Greek is called taxis, and arrangement, which the Greeks name diathesis, and of Proportion and Symmetry and Decor and Distribution, which in the Greeks is called oeconomia. (Vitruvius, 1970).

Order is the balanced adjustment of the details of the work separately, and, as to the whole, the arrangement of the proportion with a view to a symmetrical result.

Proportion implies a graceful semblance: the suitable display of details in their context. This is attained when the details of the work are of a height suitable to their breadth, of a breadth suitable to their length; in a word, when everything has a symmetrical correspondence.

Symmetry also is the appropriate harmony arising out of the details of the work itself: the correspondence of each given detail to the form of the design as a whole. As in the human body, from cubit, foot, palm, inch and other small parts come the symmetric quality of eurhythmy. (Vitruvius, 1970)

The golden ratio (1.618) has been studied widely and has attracted interest like no other number in mathematics. It is believed by many to be the basis of aesthetic quality in the visual and auditory arts and has been employed in architecture, painting, music, and industrial design (Livio, 2003).

In one of the oldest empirical and most classic studies in psychology, Fechner (1876) presented a range of rectangles whose width-to-height ratios varied from 1:1 to 2:5. and he found that observers preferred those near the golden ratio. On the order hand, Davis and Jahnke (1991) used squares, rectangles, circles, ellipses, trapezoids, and parallelograms and they found no preference for the golden ratio. Instead, they observed a strong preference for shapes whose sides were in a 1:1 ratio, even when these were pitted directly against golden ratio shapes. They conclude that symmetry is the driving factor behind preference of geometric shape. Other studies support this conclusion. Höge (1995) had participants draw "beautiful" quadrangles and found more of them corresponded to square-like proportions. Boselie (1984) created a variety of different geometric patterns, including kite shapes and star-like shapes. In one version of these patterns there were multiple 1:1 ratios of side lengths. These ratios were absent in corresponding shapes. In the majority of cases, the shapes with more unity ratios among their side lengths were preferred.

In addition to the golden ratio, several other numbers have been proposed to have special status with regard to aesthetic judgments. One of these is the plastic number (1.321). This number emerges from the Padovan series created by three ones and then generating each additional digit by adding those that come two or thrce digits back (1, 1, 1, 2, 2, 3, 4, 5, 7, 9, 12, 16, 21, ctc). The ratios of adjacent number pairs in this sequence converge on the plastic number. Stewart (1996) argues that the plastic number should be used as a design principle. Another ratio very close in value is the Pythagorean number (1.333) formed by the ratio 4:3, sometimes known as the Pythagorean triple. Shortess et al (1997) took measurements of art paintings, art supplies, scientific figures, television and movie screens, custom frames, and other objects, all of which had median ratios close to 1.3. The prevalence of this number among these materials suggests that it may also be considered attractive.

It has been shown that groups of neurons in the brain are able to respond to different characteristics, such as proportion and when the signals from these individual processing centers are interpreted, it generates a positive feeling – a sense of beauty (Di, Macaluso, & Rizzolatti, 2007). Briefly, Di et al. (2007), showed images of 15 masterpiece sculptures from the Classical and Renaissance periods, to volunteers who had no experience in art theory. As they rated the pictures, the scientists scanned their brains. Some of the images were subtly altered. In their original form, the sculptures were proportioned according to the golden ratio. In the altered versions, the images had different proportions but were the same in every other respect. The viewers' evaluation of the stimuli, as expressed in the aesthetic judgment condition, showed that the original images were mostly evaluated positively (76%), whereas the modified images were generally scored with a negative rating (63%). In summary, their findings shown that the presence of a parameter, like the golden ratio, can elicit a specific neural pattern underlying the sense of beauty in the observer.

The Other Side of the Coin

When we start to passing through the modern and contemporary art and their conceptual fields, what seems uniform and unambiguous seems to undo in mirages. When reference thinkers like Merleau-Ponty, Derrida, Husserl, Heidegger, Adorno and Benjamin are faced with the representatives of a new conceptual field in the art such as Picasso, Braque, Matisse, Cezanne, Mallarmé, Stravinsky, Duchamp and Schöenberg, we find ourselves dealing with an increasingly complexity and astonished ambiguous in the semantic field. This does not happen by a simple increase of knowledge as a whole, but through the recognition of the very ambiguity inherent in the cognitive objects of art and beauty and endurance that the very historical time with cultural barriers, economic, social, scientific and linguistic. In summary, the subject of art demands a continuous updating to make sense (Figure 1). Baudelaire recognizes this resistance in his notes:

This is in fact an excellent opportunity to establish a rational and historical theory of beauty, in contrast to the academic theory of an unique and absolute beauty; to show that beauty is always and inevitably of a double composition, although the impression that it produces is single – for the fact that it is difficult to discern the variable elements of beauty within the unity of the impression invalidates in no way the necessity of variety in its composition. Beauty is made up of an eternal, invariable element, whose

Figure 1. The emphasys on symmetry, proportion and geometry is put into practice by Vitruvius and many architects and architectural authors in the Renaissance. They went out of their way to build according to rules and methods that were considered necessary for beautiful and good architecture during the classical period. Proportion, as golden ratio, which is commonly incorporated in classical architecture, might be universally pleasing to the mind, however, these proportions are completely absent from successful modern buildings such as Frank Gehry's Guggenheim museum in Bilbao, Spain (B). This is an expressive example of how is beauty evolving over time with a sequence of variation of its interpretations throughout history. A: Example of Doric project with notes of geometry and proportion in classical architecture Adapted from Tavola IX from Vignola (1562). B: Frank Gehry's Guggenheim museum in Bilbao, Spain.

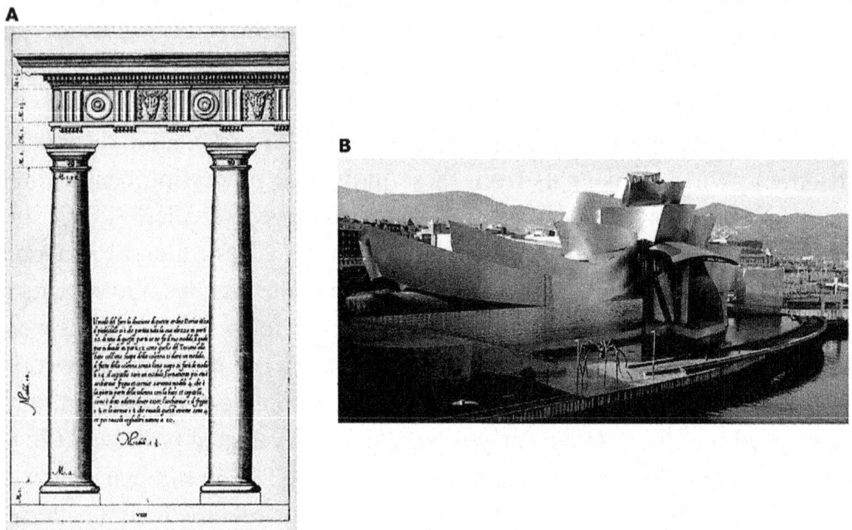

quantity it is excessively difficult to determine, and of a relative, circumstantial element, which will be, if you like, whether severally or all at once, the age, its fashions, its morals, its emotions. Without this second element, which might be described as the amusing, enticing, appetizing icing on the divine cake, the first element would be beyond our powers of digestion or appreciation, neither adapted nor suitable to human nature. (Baudelaire, 1995, p.3)

A golden thinker who denounces the incompleteness and ambiguity that lies to the conceptual universe of modern and contemporary was Merleau-Ponty.

In his aesthetic work and within a general theory of knowledge as is evident in the Phenomenology of Perception (Merleau-Ponty, 2012), the author, through a dialogue with modern art, science and psychology, sees aesthetics as a perfect sensitive knowledge in a certain way. The object priority of Merleau-Ponty, is not the field of art or beauty, but the sensation as aisthesis and, in a comprehensive way, all forms of appearance, at the same time, complementary and distinct of a logical framework. (Abbagnano, 2007)

Merleau-Ponty criticizes science not to draw its boundaries or describe its objects but to point its incompleteness. For the author, modern art is thought in a phenomenological perspective. In phenomenology the objects have an existence only as related and co-existing with a particular consciousness.

In this way the objects are captured in a global manner in which cultural determinants, social, poetic, sentimental and sensorial aspects are a unit that is not perceived only conceptually as the example of lemon and molasses (Merleau-Ponty, 2002), but with real world objects that emergence as unitary and indivisible in its physical embodiment in its all-round relationship to human consciousness (Figure 1).

Despite millions of nuances that can weave about the phenomenal object it appears to us as a complete informational complex of consciousness. I feel the object, I realize its temperature, odor and etc... Groping their recesses, we immediately connect memories, awake feelings before even think about it. Everything about it tells me in every thing, all that is in it for me. This perception of reality implies the obvious ambiguity of values that can exist in complex object as art. So we see the realization of the modern artist more like a reality experimentation field and its many missteps by the intricate universe of human consciousness than just the enjoyment of the beauty's site. This impression can be generalized to other areas of knowledge. Thus, sciences also reviews its concepts and relativize its limits, relaxes the margins of what you can see. To deal with new observations of the empirical world, creates mathematical models that imply a reinterpretation of reality in its radical nature. Creates concepts such as non-spatiality (since the same object can be in two different places) and not causality (as the simultaneity of events does not establish a causal link even in a basic temporal time-line), like in Quantum Theory.

Returning to the initial proposal pointed in the beginning of this chapter, neuroscience can also benefit itself from the same procedures achieved in others areas of knowledge. Rising questions and not resolving some dubiousness as if they do that exist, and not necessarily doing that seeking the extinction of ambiguity. Perhaps neuroscience should incorporate the ambiguity in its conceptual horizon, and not extinguishing it. To achieve this goal, neuroscience has philosophy, art and science itself (in its most paradigmatic representative – Physics) as a conceptual body of knowledge or epistemic pictures, pioneers in the treatment of this paradox. Although this objective appears extremely hard to achieve, we cannot doubt about this possibility, in the future horizon of events. Citing Gaston Bachelar;

Missing, rectify, diversify are types of dynamic thinking fleeing the confidence of the unity, and they finding in homogeneous systems more obstacles than stimulus. In short, the man moved by the scientific spirit wonders, but for once, the better questioning.......The Epistemologist must take the facts as if they were ideas by entering them into a system of thought. A fact misunderstood by a time remains for the historian, a fact. For the epistemologist, it is an obstacle, a counter-thought. (Bachelar, 1996).

Even the empirical world in which neuroscience is based, points to this ambiguity and multiplicity of possibility, as we shall see.

THE MULTIPLICITY AND AMBIGUITY OF BEAUTY IN NEUROSCIENCE

Neuroesthetics received its formal definition in 2002 as the scientific study of the neural bases for the contemplation and creation of a work of art (Nalbantian, 2008). This field has emerged at the intersection of psychology, neuroscience and aesthetics, and its diverse aims and methods reflect its multidisciplinary nature. (Nadal & Pearce, 2011) Since our minds are part of the world, and how we think and experience and act has been molded by the world over years of evolution and the experience of beauty comes from the interactions between our minds and the world (Chatterjee, 2013).

In spite of our brain my be evolved to find some objects universally beautiful and even if experiences of beauty maybe shared universally, cultural influences and context affect these experiences (Chatterjee, 2013)

The majority bibliographic data of neuroscience suggests that we do not have specific brain network for beauty (Muñoz & Martín-Loeches, 2015). In other words, it is proposed that beauty emerges from cognitive integration and processing of different information by the brain of the subject who perceive it, like: visceral reaction, perceptual analysis, imagery, implicit memory, reward and judgment. Such cognitive process might have at least three components. The perceptual processing of the object itself, the emotional response to the object and, when relevant, an explicit judgment about the object's beauty. Today, the use of technics like neuroimaging, has been contributing in advancing our understanding of the cognitive and neural underpinnings of aesthetic appreciation, since it can provide a new analytical level of evidence in the understanding of cognitive processes (Nadal, & Skov, 2013). Researchers undertook human brain imaging experiment, using functional magnetic resonance imaging (fMRI). fMRI is a technique for measuring and mapping brain activity that is noninvasive. It is being used in many studies to better understand how the healthy brain works, and in a growing number of studies it is being applied to understand how that normal brain function is disrupted in disease.

In the simplest fMRI experiment a subject alternate between periods of doing a particular task and a control state, for example subjects are asked to view pictures of paintings and listen to brief musical excerpts, during a short period of time, and then rate as beautiful, neutral or ugly, while the activity of their brains are imaged. With this type of method is possible to identify a single area or set of areas whose activity would correlate with the experience of beauty or ugly, regardless of whether it was derived from an auditory or visual source.

A review of the neurobiology of beauty reveals that the experience of visual, musical, and moral beauty all correlate with activity in different brain areas (Figure 2). More recently, fMRI studies found that when consumers are exposed to a higher level of beautiful products, brain activations are similar to getting rewards; in contrast, exposure to a lower level of beauty results in similar brain activations as feelings of nausea or unfairness (Chatterjee, 2011). Studies have reported that attractiveness activate areas within the orbitofrontal cortex, the nucleus accumbens or the ventral striatum (Aharon et al., 2001; Ishai, 2007; Kampe, Frith, Dolan, & Frith, 2001; Kranz & Ishai, 2006; O'Doherty et al., 2003) and that the amygdala has a nonlinear relationship to it (Figure 2) (Winston, O'Doherty, Kilner, Perrett, & Dolan, 2007).

These regional activations, within neural circuitry dedicated to reward systems, are interpreted as reflecting the emotional valence attached to attractiveness (Senior, 2003). The particular emotional valences are those involved in the expectation of rewards and satisfaction. The judgment of beauty, as distinct from its emotional evocations, involves parts of the prefrontal córtex (Nakamura et al., 1998).

Many, if not all studies that have addressed the neural correlates of the experience of beauty using fMRI have found high activity in medial orbitofrontal cortex (mOFC) (Zeki, Romaya, Benincasa, & Atiyah, 2014). The orbitofrontal cortex occupies the ventral surface of the frontal part of the brain. In terms of its neuroanatomical connectivity, the orbitofrontal cortex is uniquely placed to integrate sensory and visceral motor information to modulate behavior through both visceral and motor systems. This has led to the proposal that the orbitofrontal cortex is an important part of the networks that are involved in emotional processing, reward, decision-making and hedonic experience (Kringelbach., 2005).

The results for many studies shown that the mOFC was the only common cortical area activated by stimuli that were judged to be beautiful, regardless of whether the origin of the stimulus (example: visual

Figure 2. Image showing different brain areas: A: Lateral view of the brain showing in blue the Frontal Cortex. B: Sagital view of the brain showing in pink the location of the medial orbitofrontal cortex. C: Limbic system – The structures of the limbic system are involved in motivation, emotion, learning, and memory. The two most studied areas in the limbic system are highlighted, in red the amygdala and in grey the hippocampus. D: Figure showing the functional regions of the cerebral cortex - the motor areas are highlighted.

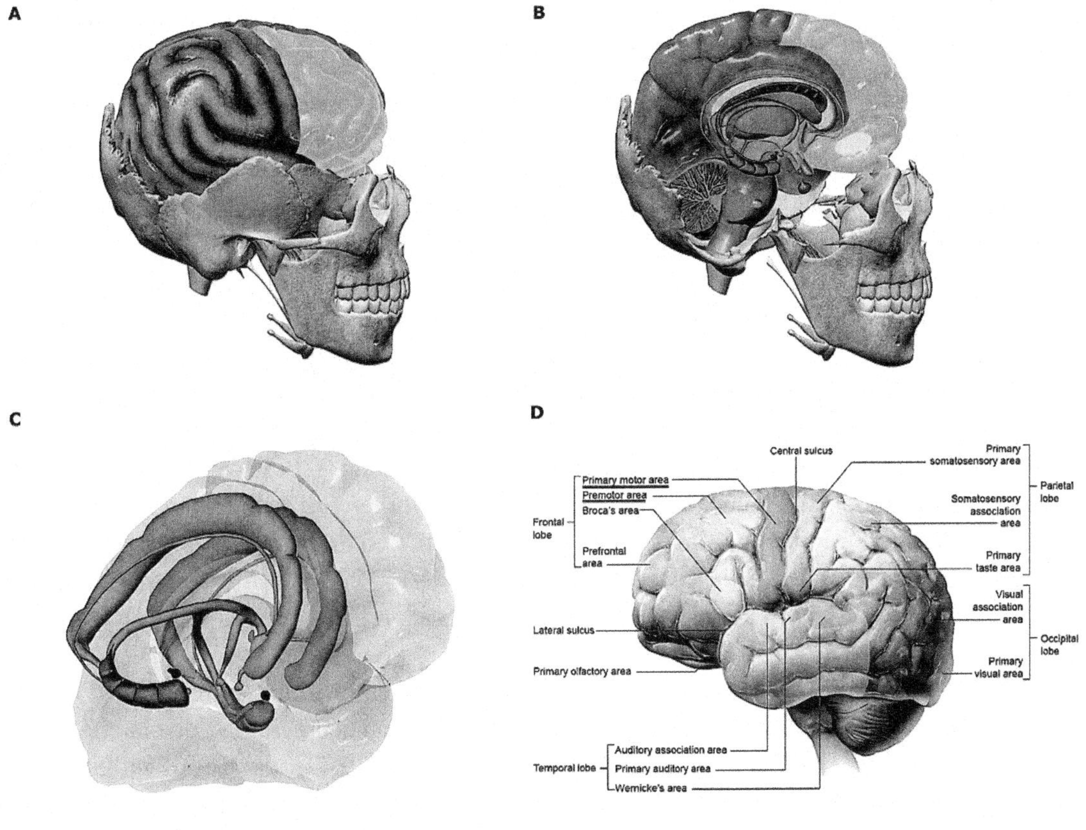

or auditory) (Ishizu & Zeki, 2011). For example, it is know that mOFC is activated in response to facial attractiveness (Tsukiura & Cabeza, 2011; Cloutier, Heatherton, Whalen, & Kelley, 2008; O'Doherty et al., 2003) and even in studies that involved judgments of beauty vs. symmetry of geometric shapes, the activation of mOFC was reported (Jacobsen, Schubotz, Höfel, & Cramon, 2006).

Probably, the most amazing experiment done on this field is from Zeki et al. (2014). What made this research so interesting and unusual was the different type of methodology employed to explore beauty. Instead of do analysis of shapes, art objects or human faces, Zekis et al. (2014) used mathematical formulae to try to understand the neurobiological basis of beauty. In this work, the authors used fMRI to image the brain activity of 15 mathematicians when they viewed mathematical formulae that they had previously rated as beautiful, neutral or ugly. The results showed that the experience of mathematical beauty also correlates with activity in the mOFC.

For the general people who are not mathematician the mathematical formulae appear dry and inaccessible but to a mathematician an equation can be beauty. What make this so interesting is that the even abstract source as mathematics could be correlated with something beauty, and moreover this work also shown that bases of beauty is not sustained only in the geometric shape of the object (on the object itself), like symmetry, it is more than these, it is something that is also related to your life history.

With this in mind, the integration of the processes of perceptual analysis, implicit memory, explicit classification, cognitive evaluation (Leder, 2013) and the induced emotion by the object in the individual (Brattico, Bogert, & Jacobsen, 2013) is the base of the evaluative process to judge if something is beautiful or not.

With the accumulative scientific progress on the understanding of neuroanatomy and the bibliography collection available on various experimental trials about the neural correlates of different physiological conditions and neuroimaging. Researchers start to design experiments to known if it is possible to modulate brain activity without the use pharmacological treatment or any invasive technique such as deep brain stimulation. Using non-invasive stimulation techniques Chib et al. (2013) produced behavioral changes in volunteers, making them to judge faces as more attractive than before their brains were stimulated, in other words they were able to manipulate the beauty judgment of the volunteers.

The two predominant means of non-invasively stimulating the brain are transcranial magnetic stimulation (TMS) and transcranial direct current stimulation (tDCS) (George & Aston-Jones, 2009). Both of these stimulation methods result in changes in brain function by causing neurons resting membrane potential to depolarize ("activating") or hyperpolarize ("deactivating"). Positive stimulation (as in the cases of TMS and anodal tDCS) increases the neuronal excitability and more spontaneous cell firing ("activating"). Negative stimulation (in the cases of cathodal tDCS) decreases the neuronal excitability and decreased spontaneous cell firing ("deactivating") (Chib et al., 2013). The only bottleneck of these stimulation techniques is the problem of accessing the deep brain regions such as medial areas in frontal cortex.

A number of studies have associated increases in ventromedial prefrontal cortex (VMPFC) activity (Au-Young, Shen, & Yang, 1999; Tzschentke, & Schmidt, 2000) and decreases in dorsolateral prefrontal cortex (DLPFC) activity (Bertolino et al., 2000; Pycock, Kerwin, & Carter, 1990), with increases in midbrain activity. With these relationships in mind, the researchers hypothesized that excitatory/anodal and inhibitory/cathodal tDCS electrode placement over VMPFC and DLPFC, respectively, would result in the remote activation of the ventral midbrain and that remote activation would manifest behaviorally as increases in participants' rewarding appraisals.

To test their hypothesis, the researchers asked volunteers to judge the attractiveness of groups of faces both before and after the volunteers' brains had been stimulated with tDCS. The study participants rated the faces while inside a fMRI scanner, which allowed the researchers to evaluate any changes in brain activity caused by the stimulation.

Those in the main group rated the faces presented after stimulation as more attractive than those they saw before stimulation. There were no differences in the ratings from the control groups. This change in ratings in the main group suggests that the researchers were able to activate the ventral midbrain, and that result in changes in brain activity in this deep-brain region associated with rewarding appraisals.

While it looks trivial in our daily lives, the complexity in the acquisition and interpretation of a certain stimulus to generate a response, whether beautiful or ugly, is huge and involves several brain areas. As incredible as it may seem, researchers have shown that both beautiful and ugly stimuli modulate activity in the same brain areas and it seems that the difference between judging opposite situations as beauty and ugliness, appears to be the activation intensity of these brain areas. The findings suggests that the orbitofrontal cortex is activated more strongly for things considered beautiful than those seen as ugly, while in the motor cortex the response appears to be opposite (strongly in ugly) (Kawabata & Zeki, 2004).

The activation of motor cortex is of special interest, since it is not expected activation of this region in tasks that are not related to motor activity, such as during aesthetic judgment. With this in mind, brain activity seems to support that beauty and ugliness are not independent esthetic categories, sharing most, if not all, of the same brain areas. The activation of the same areas, during the judgment of beauty and ugliness, may explain, at least in part, why some people are attractive to ugliness. Such considerable similarity might result as problematic for a plain elucidation of whether beauty and ugliness are actually related or independent in the human brain (Martín-Loeches, Hernández-Tamames, Martín, Urrutia, 2014).

CONCLUSION

The Learn'd Astronomer

When I heard the learn'd astronomer,/
When the proofs, the figures, were ranged in columns before me,/
When I was shown the charts and diagrams, to add, divide, and measure them,/
When I sitting heard the astronomer where he lectured with much applause in the lecture-room,/
How soon unaccountable I became tired and sick,/
Till rising and gliding out I wander'd off by myself,/
In the mystical moist night-air, and from time to time,/
Look'd up in perfect silence at the stars.
(Walt Whitman)

REFERENCE

Abbagnano, N. (2007). *Dicionário de filosofia.* Martins Fontes.

Aharon, I., Etcoff, N., Ariely, D., Chabris, C., O'Connor, E., & Breiter, H. (2001). Beautiful faces have variable reward value: FMRI and behavioral evidence. *Neuron, 32*(3), 537–551. doi:10.1016/S0896-6273(01)00491-3 PMID:11709163

Alberti, L. B. (1989). *Da Pintura.* Campinas, Brazil: Editora da UNICAMP.

Aristotle, . (1984). *The Complete Works of Aristotle, in two volumes* (J. Barnes, Ed.). Princeton, NJ: Princeton University Press.

Au-Young, S. M., Shen, H., & Yang, C. R. (1999). Medial prefrontal cortical output neurons to the ventral tegmental area (VTA) and their responses to burst-patterned stimulation of the VTA: Neuroanatomical and in vivo electrophysiological analyses. *Synapse (New York, N.Y.), 34*(4), 245–255. doi:10.1002/(SICI)1098-2396(19991215)34:4<245::AID-SYN1>3.0.CO;2-D PMID:10529719

Bachelar, G. (1996). *A Formação do Espirito Científico: Contribuição para uma psicanálise do conhecimento*. Rio de Janeiro, Brazil: Contraponto.

Baudelaire, C. (1995). The Painter of Modern Life and Other Essays. London: Phaidon Press.

Bayles, D., & Orland, T. (2007). *Art and Fear. Observations on the perils (and rewards) of artmaking*. Eugene, OR: Image Continuum Press Edition.

Bertolino, A., Breier, A., Callicott, J. H., Adler, C., Mattay, V. S., Shapiro, M., & Weinberger, D. R. et al. (2000). The relationship between dorsolateral prefrontal neuronal N-acetylaspartate and evoked release of striatal dopamine in schizophrenia. *Neuropsychopharmacology*, *22*(2), 125–132. doi:10.1016/S0893-133X(99)00096-2 PMID:10649825

Boselie, F. (1984). Complex and simple proportions and the aesthetic attractivity of visual patterns. *Perception*, *13*(2), 91–96. doi:10.1068/p130091 PMID:6504684

Brattico, E., Bogert, B., & Jacobsen, T. (2013). Toward a neural chronometry for the aesthetic experience of music. *Frontiers in Psychology*, *4*, 206. doi:10.3389/fpsyg.2013.00206 PMID:23641223

Chatterjee, A. (2011). Neuroaesthetics: A coming of age story. *Journal of Cognitive Neuroscience*, *23*(1), 53–62. doi:10.1162/jocn.2010.21457 PMID:20175677

Chatterjee, A. (2013). The illogic of beauty. In *The Aesthetic Brain: How We Evolved to Desire Beauty and Enjoy Art*. Oxford, UK: Oxford University Press. doi:10.1093/acprof:oso/9780199811809.003.0010

Chib, V. S., Yun, K., Takahashi, H., & Shimojo, S. (2013). Noninvasive remote activation of the ventral midbrain by transcranial direct current stimulation of prefrontal cortex. *Translational Psychiatry*, *11*(3), e268. doi:10.1038/tp.2013.44 PMID:23756377

Cloutier, J., Heatherton, T. F., Whalen, P. J., & Kelley, W. M. (2008). Are attractive people rewarding? Sex differences in the neural substrates of facial attractiveness. *Journal of Cognitive Neuroscience*, *20*(6), 941–951. doi:10.1162/jocn.2008.20062 PMID:18211242

Cunningham, M., Barbee, A., & Philhower, C. (2002). Dimensions of facial physical attractiveness: The intersection of biology and culture. In G. Rhodes & L. Zebrowitz (Eds.), *Facial attractiveness. Evolutionary, cognitive, and social perspectives* (pp. 193–238). Westport, CT: Ablex.

Davis, S. T., & Jahnke, J. C. (1991). Unity and the golden section: Rules for aesthetic choice? *The American Journal of Psychology*, *104*(2), 257–257. doi:10.2307/1423158

Descartes, R. (1973). Discurso do método. In *Coleção Os Pensadores* (J. Guisnburg & P. B. Júnior, Trans.). Abril Cultural.

Di Dio, C., Macaluso, E., & Rizzolatti, G. (2007). The golden beauty: Brain response to classical and renaissance sculptures. *PLoS ONE*, *2*(11), e1201. doi:10.1371/journal.pone.0001201 PMID:18030335

Dutton, D. (2010). *The Art Instinct: Beauty, Pleasure, and Human Evolution*. Oxford, UK: Oxford University Press.

Fechner, G. T. (1876). *Vorschule der Aesthetic*. Berlin, Germany: Breitkopf & Härtel.

George, M. S., & Aston-Jones, G. (2009). Noninvasive techniques for probing neurocircuitry and treating illness: Vagus nerve stimulation (VNS), transcranial magnetic stimulation (TMS) and transcranial direct current stimulation (tDCS). *Neuropsychopharmacology*, *35*(1), 301–316. doi:10.1038/npp.2009.87 PMID:19693003

Höge, H. (1995). Fechner's experimental aesthetics and the golden section hypothesis today. *Empirical Studies of the Arts*, *13*(2), 131–131. doi:10.2190/UHTQ-CFVD-CAU2-WY1C

Ishai, A. (2007). Sex, beauty and the orbitofrontal cortex. *International Journal of Psychophysiology*, *63*(2), 181–185. doi:10.1016/j.ijpsycho.2006.03.010 PMID:16759727

Ishizu, T., & Zeki, S. (2011). Toward a brain-based theory of beauty. *PLoS ONE*, *6*(7), 21852. doi:10.1371/journal.pone.0021852 PMID:21755004

Jacobsen, T., Schubotz, R. I., Hofel, L., & Cramon, D. Y. (2006). Brain correlates of aesthetic judgment of beauty. *NeuroImage*, *29*(1), 276–285. doi:10.1016/j.neuroimage.2005.07.010 PMID:16087351

Jonhson, G. A. (2010). The retrieval of beautiful: thinking though Merleau- Ponty's aesthetics. Evanston, IL: Northwestern University Press.

Kampe, K., Frith, C., Dolan, R., & Frith, U. (2001). Reward value of attractiveness and gaze. *Nature*, *413*(5856), 589. doi:10.1038/35098149 PMID:11595937

Kawabata, H., & Zeki, S. (2004). Neural correlates of beauty. *Journal of Neurophysiology*, *91*(4), 1699–1705. doi:10.1152/jn.00696.2003 PMID:15010496

Kossovith, L. (1996). Permanência e Renovação nas Artes. *Discurso*, *26*, 83–92.

Kossovith, L. (2006). *Tradição Clássica. Desígnio: revista de história de arquitetura e do urbanismo. nº5*. São Paulo, Brazil: Ed. Annablume/FAU-USP.

Kranz, F., & Ishai, A. (2006). Face perception is modulated by sexual preference. *Current Biology*, *16*(1), 63–68. doi:10.1016/j.cub.2005.10.070 PMID:16401423

Kringelbach, M. L. (2005). The human orbitofrontal cortex: Linking reward to hedonic experience. *Nature Reviews. Neuroscience*, *6*(9), 691–702. doi:10.1038/nrn1747 PMID:16136173

Kumar, M., & Garg, N. (2010). Aesthetic principles and cognitive emotion appraisals: How much of the beauty lies in the eye of the beholder? *Journal of Consumer Psychology*, *20*(4), 485–494. doi:10.1016/j.jcps.2010.06.015

Leder, H. (2013). Next steps in neuroaesthetics: Which processes and processing stages to study? *Psychology of Aesthetics, Creativity, and the Arts*, *7*(1), 27–37. doi:10.1037/a0031585

Livio, M. (2003). *The Golden Ratio: The Story of Phi, the World's Most Astonishing Number*. New York, NY: Broadway Publishing.

Martín-Loeches, M., Hernández-Tamames, J. A., Martín, A., & Urrutia, M. (2014). Beauty and ugliness in the bodies and faces of others: An fMRI study of person esthetic judgement. *Neuroscience*, *26*(277), 486–497. doi:10.1016/j.neuroscience.2014.07.040 PMID:25086316

Merleau-Ponty, M. (2002). *Causeries. 1948, établies et annotées par S. Ménasé*. Paris: Seuil.

Merleau-Ponty, M. (2012). *Phenomenology of Perception* (D. Landes, Trans.). London: Routledge.

Moura, C. A. R. (2005). *Nietzsche: civilização e cultura*. São Paulo, Brazil: Martins Fontes.

Muñoz, F., & Martín-Loeches, M. (2015). Electrophysiological brain dynamics during the esthetic judgment of human bodies and faces. *Brain Research*, *1594*, 154–164. doi:10.1016/j.brainres.2014.10.061 PMID:25451119

Nadal, M., & Pearce, M. T. (2011). The Copenhagen Neuroaesthetics conference: Prospects and pitfalls for an emerging field. *Brain and Cognition*, *76*(1), 172–183. doi:10.1016/j.bandc.2011.01.009 PMID:21334125

Nadal, M., & Skov, M. (2013). Introduction to the special issue: Toward an interdisciplinary neuroaesthetics. *Psychology of Aesthetics, Creativity, and the Arts*, *7*(1), 1–12. doi:10.1037/a0031842

Nakamura, K., Kawashima, R., Nagumo, S., Ito, K., Sugiura, M., Kato, T., & Kojima, S. et al. (1998). Neuroanatomical correlates of the assessment of facial attractiveness. *Neuroreport*, *9*(4), 753–757. doi:10.1097/00001756-199803090-00035 PMID:9559951

Nalbantian, S. (2008). Neuroaesthetics: Neuroscientific theory and illustration from the arts. *Interdisciplinary Science Reviews*, *33*(4), 357–368. doi:10.1179/174327908X392906

Novaes, A. (1994). *Artepensamento*. São Paulo, Brazil: Cia das Letras.

O'Doherty, J., Winston, J., Critchley, H., Perrett, D., Burt, D. M., & Dolan, R. J. (2003). Beauty in a smile: The role of medial orbitofrontal cortex in facial attractiveness. *Neuropsychologia*, *41*(2), 147–155. doi:10.1016/S0028-3932(02)00145-8 PMID:12459213

Panofsky, E. (2000). Ideia: a evolução do conceito de belo. *Nature*, *286*, 74–76.

Sartwell, C. (2014). *Beauty*. The Stanford Encyclopedia of Philosophy. Retrieved from http://plato.stanford.edu/archives/spr2014/entries/beauty/

Senior, C. (2003). Beauty in the brain of the beholder. *Neuron*, *38*(4), 525–528. doi:10.1016/S0896-6273(03)00293-9 PMID:12765605

Shortess, G. K., Clarke, J. C., & Shannon, I. (1997). The shape of things: But not the golden section. *Empirical Studies of the Arts*, *15*(2), 165–165. doi:10.2190/3UBV-2LJU-T2T9-D7XN

Stewart, I. (1996). Tales of a neglected number. *Scientific American*, *274*, 102–102.

Symons, D. (1979). *The Evolution of Human Sexuality*. Oxford University Press.

Tsukiura, T., & Cabeza, R. (2011). Shared brain activity for aesthetic and moral judgments: Implications for the Beauty-is-Good stereotype. *Social Cognitive and Affective Neuroscience*, *6*(1), 138–148. doi:10.1093/scan/nsq025 PMID:20231177

Tzschentke, T. M., & Schmidt, W. J. (2000). Functional relationship among medial prefrontal cortex, nucleus accumbens, and ventral tegmental area in locomotion and reward. *Critical Reviews in Neurobiology*, *14*(2), 131–142. doi:10.1615/CritRevNeurobiol.v14.i2.20 PMID:11513242

Umberto, E. (2004). *História da beleza* (E. Aguiar, Trans.). Rio de Janeiro, Brazil: Editora Record.

Vignola, J. (1562). Regola delli cinque ordini d'architettura. Rome: n.p. Facsimile.

Vitruvius, . (1970). *On Architecture, Frank Granger, trans.* Cambridge, MA: Harvard University Press.

Walker, S. (2006). Sustainable by Design: Explorations in Theory and Practice. London: Earthscan.

Winston, J., O'Doherty, J., Kilner, J., Perrett, D., & Dolan, R. (2007). Brain systems for assessing facial attractiveness. *Neuropsychologia*, *45*(1), 195–206. doi:10.1016/j.neuropsychologia.2006.05.009 PMID:16828125

Zeki, S., Romaya, J. P., Benincasa, D. M., & Atiyah, M. F. (2014). The experience of mathematical beauty and its neural correlates. *Frontiers in Human Neuroscience*, *8*(68), 1–12. PMID:24592230

KEY TERMS AND DEFINITIONS

Aesthetic: The branch of philosophy dealing with the nature of art, beauty, and taste, with the creation and appreciation of beauty.

Art: The expression or application of human creative skill and imagination, typically in a visual form such as painting or sculpture, producing works to be appreciated primarily for their beauty or emotional power.

Beauty: A combination of qualities that pleases the aesthetic senses.

Deep Brain Stimulation (DBS): DBS is a neurosurgical procedure, involving the implantation of a medical device called a neurostimulator, which sends electrical impulses, through implanted electrodes, to specific parts of the brain.

Functional Magnetic Resonance Imaging: Functional neuroimaging procedure using magnetic resonance imaging technology that measures brain activity by detecting changes associated with blood flow.

Genealogy: In philosophy is a historical technique in which one questions the commonly understood emergence of various philosophical and social beliefs.

Historiography: The study of the methodology of historians in developing history as an academic discipline, and by extension is any body of historical work on a particular subject.

Medial Orbitofrontal Cortex: Prefrontal cortex region situated in the ventral medial surface of the frontal lobes of the brain.

Neuroesthetic: Branch of the neuroscience that study the neural bases for the contemplation and creation of a work of art.

Prefrontal Cortex: In mammalian brain anatomy, the prefrontal cortex is the cerebral cortex which covers the front part of the frontal lobe. This brain region has been implicated in planning complex cognitive behavior, personality expression, decision making, and moderating social behavior.

Transcranial Magnetic Stimulation: Noninvasive procedure that uses magnetic fields to stimulate nerve cells in the brain.

ENDNOTES

1 Images that endures during history and have the capacity to change his concept during time.

2 Understood as Nietzsche organized.

3 In this context objective means features of the object.

Chapter 8
Design and Emotion:
Contributions to the Emotional Design

Paula da Cruz Landim
Universidade Estadual Paulista (UNESP), Brazil

ABSTRACT

This chapter has the purpose to contribute to the scientific organization and assist in the process of building a theory related to the interrelation of design and emotion. It is intended that the issues addressed here raise discussions and lead to deepening these relations. In this sense this text presents aspects and situations surrounding the design in contemporaneity from, the cultural and economic relations, the meanings of the object and design of products in material culture, the aspects of industrialization and globalization, the senses of excitement and consumption and through the establishment of discourses concerning the design and the role of designers in contemporary.

INTRODUCTION

One of the most recurrent questions, under different aspects, concerns the design definition. Fact that it is understandable, after all these questions have to be constantly observed and updated as it varies in time, historical and socio-cultural context. Thereby, when studying design as a professional performance space, and also on acting within the spectrum of the field of knowledge in design, it is clear that there is a myriad of definitions and the situation becomes more forceful when it comes to the definition of emotional design.

In this sense and in the theoretical and critical systematization, it is understood that it is essential to establish a focus on the relation between design and emotion. Even if it means being in mutant and flexible land due to various issues, among them especially the aspects of cultural dynamics.

DESIGN AS A DYNAMIC FIELD

The design is constituted by social, economic and cultural transformations. For most people the feeling of well-being relies heavily on the everyday objects that surround them. Things that surround us, our

DOI: 10.4018/978-1-5225-0510-5.ch008

external environment, are an expression of how the older generation lived and how we want to live today. The objects and products that an individual, family, community or society designs, manufactures, and choose and accepts, and how the relation between many of them evolves, reflects equally both individual and social values.

These individual and social values constitute the culture of a civilization and generate behavior patterns, customs and knowledge. Culture is a recurring issue, pointed out by several authors in the conceptualization and in the design definitions. After all, the design operates in the cultural universe, giving shape and establishing the culture.

Designers and researchers set a consonant voice on this subject, and indicate that becomes increasingly important

(...) think the activity and the role of the designer as the result of a global relation, which includes the midst, where the set object is inserted, the collective and the subjectivity, derived of the culture, which is present in the subject's relation with the object (...) (Couto & Oliveira: 1999, 9).

In other words, it can be said that design is culture because it creates and develops an artificial and symbolic universe to the society in which it operates. On the other hand, design is production of culture because it establishes products that are the result of cultural analysis and interpretation.

Gustavo Amarante Bomfim (1953-2005) is considered a pioneer in Brazil in the dissemination of scientific issues and the relation between philosophy and science in the design. On the way of his inquiries this researcher has developed an important study in an attempt to propose guidelines for the construction of a theory of design. Are some of these reflections and thoughts that are presented below in order to elucidate the relation between design and culture and to reflect on the fashion design or the design and the fashion.

This author justifies the intrinsic relation between design and culture from three main arguments. First, design is culture because:

(...) is an activity that configures using objects and information systems and, as such, incorporates part of the cultural values that surround it, that is, most of the objects of our midst are first of all the materialization of the ideals and inconsistencies of our society and its cultural manifestations as well as, on the other hand, announcement of new ways. Second, because the design, understood as conformed matter, participates in cultural creation, that is, the design is a praxis that confirms or challenges the culture of a given society, which features a dialectical process between mimesis and poiesis. In other words, the design community expresses the contradictions of this community and will be as perfect or imperfect as it. The Design has thus essentially a specular nature, either as ad, either as a complaint (Bomfim: 1999, 150 and 151).

And ends with the third question:

(...)culture is the expression of self, which manifests itself in all its works and activities is the poetic condition of the spirit in the act to confirm the matter. Design, is therefore not the activity protagonist of the configuration of a complex backdrop composed of political, economic, social, technological variables etc., but rather one of the possible interpretations of the various possibilities offered by these

variables. Design is more than the specification of the parts of a whole as the cartesian tradition. It is not a universal rule of configuration, but an interpretive action, creative, that allows diverse forms of expression (Bomfim: 1999, 152).

When Bonfim, as well as some more contemporary authors, refer to the design, they are not mentioning one or other segment design, but instead they act in the epistemological construction from a broader thinking and global. To them, there is no separation between design and the segments and areas that have been determined over time. They focus on design as a large area, a generator field of knowledge, values and customs that includes different forms of expression and configuration of objects and systems. And in this regard, fashion is a relevant field to the universe of the design as it configures using objects and also information systems.

OBJECTS AS DESIGN SIGNS

The design is a field in which the habits and lifestyles of human beings were always revealed, as well as the aesthetic production in a crisscrossed process over time and that is proclaimed through the objects, spaces, systems or strategies developed and produced by this area by the different historical approaches.

Several authors have discussed and reflected on the object, among them, we can mention Moles (1981), Aicher (2001), Coelho (2006) and Bonsiepe (2011).

An object is the result of creation, expression and production. Element elected to compose the symbolic set of human beings, which in this relation can be also called user or consumer. The object allows us to establish a broader approach and consider expanded relations in its significance. The objects incorporate approaches brought by other nomenclatures, such as products, parts or works.

Generally, when we refer to products we carry the sense that they are made directly to the consumer. In turn, the word part indicates something that is part of a set, a line or collection, normally destined for the market in order to attend changes in tastes and target audience options.

Works carries in itself the meaning of the untouched and at times, of immutable, what at first seems can not be questioned because it belongs to the universe of distinct and elected. But objects incorporate other meanings.

An object is a constructed, something that is put before us, exists outside of us and establishes a relation of mediation between man and society. An object builds the environment carrying values because they convey signs. Objects arouse thoughts, desires, and actions and are revealers of our way of life, our habits and our attitudes. So it mediates between situations, actions and people. Carries a dynamic that transforms, that generates new thoughts and relation, raise questions, evoke and provoke new and other looks, new and other relations of being with the thing itself.

The object is an essential element in the human environment. The environment involves the individual and the object is man's action product to conform the matter in the construction of the everyday environment and of the media system. The performance of the object occurs in the mediation between man and the world, reveals the society.

The search of the custom values on the opposition between private and public, artificial and natural generates the revaluation of the material elements that set the everyday life of man, occurring the promotion of the object on goods and products. This transformation includes the search of the satisfaction of desires in the relation between people, situations and acts replacing the collective spirit by the individual, where objects and individuals act in a cathartic relation between desire, satisfaction and compensation in the individual's reaffirmation in a consumer society. Because of the importance that the object has in our daily lives, marks the time in which we live.

According to Julier (2006)*Design Culture* is the study of the interrelations between design artifacts, in all its possible manifestations, that is, the work of designers and production of the object, - including marketing, advertising and distribution - and its consumption. Therefore it refers to the formal characteristics, how these are created and / or arise, and the various meanings and functions that the design represents.

The contemporary objects portray and reveal environments, ways and lifestyles and are expressions of conceiving, creating, designing and producing. They help to interpret the man, time and culture. They refer to the universe of appreciation and give another realm, another place and another value to design objects in their relation to other fields of knowledge.

DESIGN PRODUCTS: CULTURE AND INDUSTRIALIZATION

Even with huge distortions and simplifications, the design today is a recognized activity as essential for anyone looking to qualify their products and make them more competitive. In the modern world, the responsibility of the relation between industry and culture lies in the shoulders of the design. The product is the mediator between the manufacturing and the consumer, and its design that contains the message. To see the design in this way, more than just the mediator between the instantaneous social status and style - what the propaganda would like us to believe - should put it in a much broader context than the media usually presents.

Since the eighteenth century the dominant support for the design has been the production and mass consumption system based on joint forces of industry and trade. One of the themes to which design is inevitably connected is the growth of the capitalist economy. Are evident in the evolution and development of the design of objects of desire, and business contracts to attract new markets. Fashion and design, for example, have been intrinsically linked since a long time until today. The design can also become a tool to promote a new lifestyle. Design as communication meaning is capable of converting so many meanings as many as we are able to transmit.

The democratization of design has become a reality that affects virtually every consumer choice. The acceleration of stylistic redundancy has made the designer needs to constantly provide new styles to replace the existing in the market. With the advantages of automation, industry has organized its production methods in order to serve this great stylistic variation. The old fordist system, which was equivalent to standardized mass production, is dead, as well as the Modernism equation of mechanization and universality.

DESIGNERS: DREAMS INTERPRETERS, TRANSLATORS AND PRODUCERS OF SIGNS

The designer's responsibilities are several. These include not only designing the components of the object itself, putting them together properly, but also includes being the interpreters of our dreams, aspirations, anxieties and also of identities, creating formal symbols in everyday objects appropriate to do so. Because we form our identities, individual and / or collective, through the things that surround us, and designers have to take this into account.

Since the 80s of the twentieth century the design term has been appropriated by the mass media and by the advertising industry as the "path" to add "desirability" to the products. Style consciousness has become sine qua non not only for the young but for an entire group of consumers, and even the most sophisticated differentiation between products has been labeled as a stylistic choice possibilities, determined by gender, age or class. Thinking the activities of production and consumption, which needs and desires people (consciously or not) will find through visual images, materials and devices entering the market and help define who we are. In the culture of interfaces with the design, the relation is significant at both levels, either "elitist" or "popular". If the consumer culture makes the design necessary, the technological progress makes it possible. So the design carries a technological message within a socio-cultural context.

With the globalization of the free market economy, also the design has become a truly global phenomenon. Across the industrialized world, manufacturers of all types increasingly recognize and implement the design as an essential way to reach a new international audience and to gain competitive advantage. More than ever, the design of products form a global material culture and influence the quality of our environment and our daily lives. At the same time, the market is generating an excessive number of similar products with the same technology, the same price, the same performance and the same characteristics. This avalanche of options ends up confusing the consumer who has difficulty in understanding these differences, and attribute them their proper value. Thereby, and in an increasingly way, many industries and brands use the design as a way to differentiate themselves from the others.

The psychological aspects of the design are also widely mentioned with unprecedented importance. There is a general consensus that the product must be in addition to considerations of form and function if they intend to become "objects of desire" in an increasingly competitive market. To achieve this, the products need to establish pleasant emotional attachments to its users through the pleasure of handling and / or the beauty of its form. The emotionality is considered by many designers not only a powerful and essential way to facilitate better and more meaningful connections between the product and its users, but also an effective way to differentiate their solutions from those of their competitors. We live in a world of signs and symbols, and this reality is also one of the design supporting pillars. In a globalized society, is acquired not only the object, but the speech of the object.

As the designers clearly play a key role in determining the nature of manufactured products, there is little doubt that they exceptionally influence the expectations and buying habits of consumers. There is therefore a growing moral imperative for them to plot a new and better guidance for the design, particularly one that focuses on the development of real solutions of necessity, humane and sustainable. By choosing the advanced materials and technologies, while engaging in providing simplified design solutions with easier emotional connection with consumers, designers are able to create the kinds of ethical and relevant products that will be needed in the future.

MATERIAL CULTURE

The quality of the overall material culture is being determined by the actions and choices we make now, and so it must be right that every individual - creator, manufacturer and consumer - understand the need for a culture based on responsibility and must share the collective goal of building a better tomorrow.

The nature of industrial products is determined by a more complex design process that is subject to many influences and different factors that exaggerate the restrictions imposed by social, economic, political, cultural, organizational and commercial contexts, within which new products are developed, and the character, thought and creative capacities of individual designers or teams of designers, aligned experts and manufacturers involved in its production.

The discussion about the culture and identity of individuals and social groups is crucial for understanding the role of design in the development of the artifacts that set the material culture, and thus become the references that contribute to the knowledge/recognition of people and of society who develop them.

The plurality and cultural dynamics of societies require an equally plural and dynamic concept of identity in the design, which is considered the cultural heritage of individuals and social groups, their transformations and interrelationships.

In this context, an approach that considers the fundamental cultural values is necessary, but that is also open to contributions of the world, which can be absorbed by local cultures to build its own way. Cultural identity is based on diversity, and therefore has a dynamic and multidimensional nature, can not be understood as something immutable.

The same way that the discussion on cultural diversity as identity expression is fundamental to understanding the role of design in product development for the society, facing the standardization of requirements and diversification and the local and global issue that refer to the dialectic between cultural standardization and the diversity of identities in the context of material culture.

In this way, it is up to the designer - in product development for the global society - contribute to the development of a global community, eliminating the rivalries between nations, and at the same time preserving different cultural identities.

"Think globally, act locally", the globalization of products requires them to communicate with the markets in order to overcome the cultural barriers. This is, globalization, although unavoidable, does not occur in a homogeneous way, and therefore, unlike submit passively to imposed standards, seek creative ways to reduce the most of the social costs.

DESIGN AND EMOTION

Understanding the development of theories of usability and attitude in the design, taking into account the relationship of man with artifacts, is of great importance in understanding what is the emotional design. According to Mont'Alvão and Damazio (2008), in the trajectory of the design from the beginning of industrialization, there was confusion about the formal conception of the product, which sought in the artistic movements a solution with strong aesthetic value. In the first decades of the twentieth century, the quality stands out as synonym for functionalism, contributing to the elevation of the result of the design project. Still according to the authors, with the growth of ergonomic studies, after World War II, in addition to functionalism, adding to the design projects the need to adapt the product to the user.

Nowadays, the supply of technically well resolved products, characteristic of a globalized society, makes the functionality a differential virtually in disuse.

Consumers yearn for an effective consumer experience, with the satisfaction of their needs. In this relationship between consumption and consumer, the design strengthens the obligation to fulfill their expectations considering aspects functional of usability and subjectives.

In this way changes in posture of design projects started to be considered to meet the new needs that have settled with postmodernism, and the sensitive human values make the experiences with products less impersonal. In the present scenario, emotions, experiences and memories are directly linked to the consumption process.

The advances from the rise of flexible manufacturing technologies and marketing niches lead to projects and to a production able to attend the needs and desires of the individual users (FREIRE, 2009).

Therefore rescuing the subjective values and of the human sensitivity, making the material relations with the individual less impersonal, is a feature of the contemporary society to escape the routine, coming from the exhausting work and strict quality of life based on leisure time. "Thus, the experiences will acquire the characteristics and social functions that the goods have today" (Neves, 2010, p. 2650), and it is in this scenario that fits the new role of design objects and communication systems.

New emotional experiences that come from the relationship with the products and environments, induced by the project, may mean new possibilities to extract the man from this alienating framework, characteristic of today's society. This new man's relationship to their daily lives has been studied with certain constancy in the design, focused on the subjective aspects of users and focusing on emotion and its consequences (ALCANTARA; MACHADO JUNIOR, 2010).

According to Almeida (2010, p. 1905), the "[...] products often act as mediators of social relations and message communicators about their users," that is, the products will identify which group the user belongs, or which ideology he is adept.

It can be observed the continued growth of studies on Emotional Design, which, in addition to understanding the emotions of product users, are aimed at the development of techniques and tools that facilitate this process, and that can be of great importance to user-centered approach. Is necessary to understand the behavioral responses of users in the choice of objects to direct its creation process.

The objectives are always the same, so that, although the authors have different views, they believe that products can affect people's emotions and for that, we need to know and develop ways to help designers to achieve the emotions of users.

In this respect, the implementation of the Emotional Design can provide benefits with regard to the relation with the user and understanding of certain brain processing patterns, elements that allow the understanding of the product's users responses, helping in design projects conception that go beyond the mere satisfaction of functional experience.

The existence of the emotional component in the products and its use, however, is still unknown and unconscious for most people. But, throughout human evolution, they have developed muscle systems that expose their emotions, at the same time that it is possible to interpret the emotions of others, through perceptual systems, ensuring the survival of man for millennia.

Emotion and Cognition

Emotions play a fundamental role in everyday life. Damasio (1996, p. 168-169) points out that

[...] emotion is the combination of an evaluative thinking process, simple or complex, with operative responses to this process, mostly directed to the body itself, resulting in an emotional state of the body, but also directed at the brain itself (neurotransmitters nuclei in the brain stem), resulting in additional mental changes.

The emotions come from stimulus that act associated with the repertoire of each individual, but are also dependent on the environment in which he lives. Norman (2008, p. 30) points out that one of the ways that emotions act is by the "[...] neuroactive chemical substances that penetrate certain brain centers and change the perception, the decision-making and the behavior."

Since ancient Greece, it was common for men seek the separation between reason and passion, thought and feeling, cognition and emotion. However, the study of cognition, that is, of thought, has evolved significantly in recent years, providing a conceptual framework and a methodology capable of addressing all aspects of the mind, including emotion, once considered on the fringes of rational thought (LEDOUX, 2001).

The cognitive Science, defined by Silva and Marar (2010, p. 5860) as "[...] the generic designation for a set of interdisciplinary efforts to understand the mind and its relation to the human brain", deals with the understanding of the way how to acquire knowledge of the world. When designing the showcase, is necessary to know how the public interprets the information in order to adequately provide this information to them.

Reason and emotion are equally important, even if in some cases one is more prominent than the other. As Damasio (1996) points out, the emotional behavior distinguishes people, and the reason, by itself, does not know when to start or end to see what is good or bad, in decision making, being the frame of reference of the emotions that makes it. Surviving is in the foreground, so, according to Norman (2008), we react emotionally to a situation before evaluating it cognitively because the understanding and interpretation of what surrounds us is in charge of cognition, fitting quick decisions to emotions.

The creation of affection between the product and the individual can also be used as a tool to stimulate the consumers. Can even pre-exist a direct relationship between the user and a specific product to be explored.

It is important to emphasize the relation established of feelings with emotions. From the perspective of Damasio (1996, p. 173), "[...] all the emotions originate feelings, if you are awake and alert, but not all feelings come from emotions." Universal emotions like happiness, sadness, anger, fear and disgust, in action with the body, originate feelings when people feel happy, sad, angry or afraid.

Therefore, feelings, together with the emotions that they generate, guide and interfere in communication and in the objective decisions.

Positive and Negative Emotions

In contemporary culture, positive or negative emotions are used to awaken the viewer's attention. Damasio (1996, p. 177) explains that we have reactions according to negative and positive bodily states:

In conjunction with the negative bodily states, the image creation is slow, its diversity is small and reasoning ineffective; in conjunction with the positive bodily states, the image creation is fast, its diversity is wide and the reasoning may be fast, though not necessarily efficient.

The experience of consumption, today, is configured generally as a combination of different types of positive and negative emotions, and understanding of the emotional content and the nature of the relationship between the product and the consumer is of great importance to the designer in the development of the project. In a society that offers many options of products, the one that manages to transmit an effective message can reach best selling.

The emotions contribute as consumer behavior indicator, so that the adoption of the concepts about human emotion to design projects can be an important differential in creating products that generate positive emotions of people, which can guarantee its success in the market. However negative emotions are inevitably harmful.

Theoretical Approaches

There are several theoretical approaches on Design and Emotion. The approach that assumes that the three processing levels of the human mind, proposed by Norman (2008), wherein it gives rise to emotional reactions, resulting from three different levels of brain structures:

- **Visceral Level:** The automatic layer, pre-programmed, which governs the answers through direct, pre-conscious awareness. Are the physical aspects, such as appearance, touch and sound that dominate and give rise to first impressions.
- **Behavioral Level:** The part that contains the brain processes that control everyday behavior, involving automatic reactions, but also captures affective reactions. It is directly associated with the use and experience with a product, including: function, performance and usability.
- **Reflexive Level:** It is the contemplative part of the brain, involving affective reactions due to conscious thought. Only level at which consciousness and feelings, emotions and cognition reside, allowing the experience of both thought and emotion. Can be influenced by culture, experience, education level and individual differences.

In relation to the product, determines its full view, that is, makes you remember of a product, reflecting on its overall appeal and the experience of using it. Its role is to understand, interpret and find reasons and assign causes. As Norman (2008, p.165), the "[...] most of our richest and deep emotions is that in which we assign a cause to an instance."

In line with these levels, Norman proposed three design strategies: appearance design (visceral design), for ease of use (behavioral design) and reflexive sense (reflexive design) (DEMIR; DESMET; HEKKERT, 2009).

- **Visceral Design:** Corresponds to the initial moment of contact with a product, how they look, touch and feel, where the shape, texture and weight of the materials are important. Visceral design is related to the immediate emotional impact, therefore, needs to transmit a pleasant feel and look good. Through its aesthetic, is intended to arouse on the consumer an emotional and instinctive impact, resulting in a corresponding action.
- **Behavioral Design:** The main focus of behavioral design is the human being, that is, understanding and satisfying the needs of people who actually use the product, because of its experience with the product, as well as on usability and performance offered by it. Therefore investigate consumer needs and their satisfaction is the main focus of the behavioral design.
- **Reflexive Design:** It has to do with long-term relations with feelings of satisfaction produced by have, display and use a product. It is at this level that takes the interaction between the product and its identity, reinforcing the sense of self-identity of a person. Projected for the reflexive level, the design may be subject to the time factor, because this level is sensitive to cultural differences, trends in fashion and the continuous fluctuation.

The visceral and behavioral levels are immediate, while observing and use the product, but the reflective level can be influenced by time, according to Norman (2008, p. 58), "[...] through reflection you remember past and consider the future.".

In the level of process approach, responses are different, according to the processing levels of the brain.

The decision making process usually is related to thinking, idealizing to put aside emotion to decide logically, what, in practice, is not always possible. This decision implies that the individual is informed about the subject, is aware of the choice alternatives and what will lead each option to be chosen in a given situation. It is a process often impracticable, when it should be done in short time, that is, seconds. However, are perceived first the emotional aspects, as highlighted by Iida & Mühlenberg (2006) and, of them, it can be "[...] decided if we like or dislike a product, just with a visual exploration, olfactory or auditory for 30 seconds". If the decision of a person about a product, in its first emotional contact, is negative, it will be unlikely your change later, during use, by the functional benefits.

If the user decisions are also based on emotions, the design needs to pay special attention to this aspect, since the emotional aspects can, in many cases, become the main factor in purchasing decisions.

Emotion as the Designer Tool

In the relation between the individual and the product does not exist the consciousness on the part of this, of an emotional component in how products are conceived and arranged to use. The attention of the authors already cited demonstrates the growing interest of designers in the study of emotions, given its relevance in decision-making, overcoming in many cases the rational aspects, traditionally considered as the first deciding factor in the design, however, to Iida & Mühlenberg (2006) "[...] as the market will saturating and competition intensifies, manufacturers tend to value the emotional factors.".

Valorizing the subjective aspects of the user, the Design and Emotion, is based on the perspective that understand what motivates the consumer to desire to get involved in living experiences, what are their values and aspirations, becomes the focus in the selection process and use of the products. Therefore, it should be taken into account as important project requirements to be heeded in the design. Once understood the user's emotions, the designer can anticipate the positive emotional effects and thus avoid the aspects considered undesirable (DESMET; HEKKERT, 2009).

However, the challenge for the design is to adapt the user in a line of thought that influence your attitude or behavior, so that, in contact with the design project, are convinced that it is emotionally desirable and important to their life.

Mont"Alvão e Damazio (2008) ratify this aspect when considering that even essential, emotions are uncertain for designers because it is complex apprehend what is this emotional adaptation, since the emotional response from people to products is not something set. In most cases, emotions are caused by derived meanings of the product, not the attributes of the product itself. The author argues that (p. 57) a '[...] designer can not depend merely of their own constructs because there is absolutely no guarantee that they are similar to those of the receiver'.

It is not up to the designers, as a result, define an experience, but rather provide the conditions and optimal levels so that the desired experience comes true. Emotional responses are not specific, but understanding the context in which they happen and how the design project can act as a stimulus, in this context, creates possibilities for design aiming at a joint effect in this knowledge as a stimulus to create a new experience or fostering a existing one. It is therefore necessary to plan the interaction of three elements: context, people and products.

It is difficult to anticipate the emotional responses that people have because they can respond in different ways to the same project. Understand the emotional responses to products allows you to have knowledge of the individual anxieties and make it a tool for the project.

DESIGN SPEECHES

The design never had, and still does not have a fixed definition of limits, or an ideal to be followed. The conceptual changes are constant, reflecting a practical scenario influenced by changes in ideologies. Se um discurso projeto pode ser desenvolvido ele precisa ser aquele em que pode ser reconhecido o elevado nível de relativismo, pragmatismo e contextualização que deu o último conceito, e, como tal, sem dúvida, continuar a afectar o seu futuro. Design will continue to be influenced by consumption, by fashion system and identities of all types, by production, whether industrial or artisanal basis, as well as supported by ideologies and discourses beyond their control. That is, a continuous movement, though his future is, like everything else, greatly influenced by the past.

The idea of things and impregnated images of aesthetic and functional characteristics as meaning attraction and meeting the needs of consumers and users have a long history, and this is closely linked to the development of what has been called "modern" society.

In addition to the transformation of the design process itself, new manufacturing methods were introduced in the last decades of the twentieth century, making a wide range of consumer goods available in the market.

In this mass market context of the late twentieth century the term "designer" implied a level of individualism and taste that was resurfacing to people who wanted to be different. While only marketing strategy, one of the effects was the popularization of the concept of design and designers. Designers became celebrities in the same order of North Americans industrial designers of the 1930s. Not only in Europe but also asian countries were learning the benefits of having an advanced culture in design.

Despite these developments bring to the designer greater prominence, ensuring a place in the cultural hierarchy alongside architects and artists, it also had the effect of aligning the design concept very close to the advertising and marketing. The effect was not to give importance to the most key role as an element within the production process and as part of the material everyday environment. When culture in design, along with the economy, had a low in the early 1990s, the designers had to think of ways to overcome the superficiality that had characterized the 1980s.

Above all the world design at the late twentieth century has gone global, with an awareness that it was no longer possible to think only in terms of local or national identities. From a modest beginning in the hands of a few individuals seeking to create and improve the material world that reflected the modern age, design flourished into a force that affects virtually every aspect of modern life.

DESIGN IN THE PROCESS OF COMMUNICATION

The design for a long time was interpreted as qualifying factor for social distinctions. However, currently, this view has been changing with the change of attitude on the part of the market, professionals and consumer, making it more accessible. There is also the belief that investing in design adds value to the company, products and services. However, just keep the design from the perspective of value is already a common discourse and historically exploited. Its use must involve broader and current issues, face of so many new responsibilities imposed upon it. Including the need to distinguish which projects are related to the design area.

Once the design project involves discussions, reflexions and questionings about its use and especially about what the market is intended, it is crucial that designers and companies have a common goal, aiming at better integration between the product and the user. In that speech, the designer must start the project with the needs of man as major reference center, however, comprising all the circumstances involved in determining these needs.

Extending this approach, that returns the activity to the interests of the user and of the community, according to Moraes (1999), it is possible point out the aspects and new characteristics that challenge the role of the designer, divided into material, cultural and behavioral of the users. The materials would be the rapid development of industrial technology; the emergence of new materials; the advent of computer technology; and the emergence of interactive products. The cultural: the post-industrial era; the new possibilities of consumption of the electronic world; the increasing interaction between industrial, graphic and visual design through products and interface services; the design not only as project activity, but of management and strategic business decisions; and the design as a philosophical and political gesture.

And behavioral: the rapidly changing human social behavior; new user needs; the consideration of psychophysiological, cognitive, psychological and subjective aspects; design as reference of the cultural identity of a region and / or territory of a generation and of a time.

For the design remains the lesson that all that is projected also reflects a societal project. This trend is confirmed by Moraes (1999, p.114):

[...] in the exercise of designer social function - as conceptor element - along with industry - as a producer element - it is our task, increasingly, try to reduce the error margins [...] in an attempt to build a artificial world more interactive and smarter for humanity.

Therefore, although it is observed the importance of new usage factors as a consequence of change in human behavior, it is necessary observe the social cause. However, this view must not differ from market needs, turning to the decoding of pluralism itself originated in the existing different cultures in the country. Futhermore, according Moraes (2006, p.193): "[...] this new reality also requires of the designers to consider the three aspects: design, culture and territory". That is, should seriously contemplate the environmental issue, the excess of products on the market and the unbridled problem of consumption and production, in short, the sustainability.

Returning to the question of the user as business focus, one should also consider a new area under discussion, the emotional design, which is included in the several designer responsibilities. According Landim (2009, p. 15), these include not only "[...] designing the components of the object itself, putting them together properly, but also includes being the interpreters of our dreams, our aspirations and our anxiety, creating formal symbols in everyday objects suitable for such. "

This factor can be seen in the way the user form their identity, individual or collective, through the objects that surround it.

In this sense, in addition to the users, it is up to the designer also equally serve the interests of the holders of the means of production, generating the demand for a reflexion on their activity. Niemeyer (2000) comments that the growing importance of the designer in the construction of the material culture, should generate a reflexive and critical attitude in relation to their formation, development and action in design.

Many factors are transforming the scenario of the formation of design and also their behavior and performance. This posture change provides an opportunity to reflect, in addition, about teaching with the formation of professionals for a global market, with new routes that open, but with attitudes that include the environment, subjectivity and collective vision.

In the construction of the showcase, the responsible professional must have the domain of visual communication process, one of the design areas. The importance of visual perception in this process is vital to the structure of the visual message, its planning and configuration in order to obtain a proper understanding of the receiver.

DESIGN AND VISUAL PERCEPTION

There are many definitions for the term *perception*, from which is possible to make use of to understand its importance in the relationship between man and the world around him. As Coelho (2008, p. 44), more "[...] generically, perception relates to cognition or to immediate or intuitive judgment .".

The relation between the different settings is evident, emphasizing the fact that the word corresponds to the process, act or faculty of perceiving, become aware of objects in general, or to be aware of something. According to the same author, it is seen "[...] as well as the fact of perceive through the senses and emotion, or from REPRESENTATION aware of sensations."

Visual perception happens through external stimulus that create an individual visual repertoire or subject to cultural patterns throughout life. It is a process in stages of information that is seen (AUMONT, 1995). Therefore, unlike the instinctive notion that has that color or brightness is part of the object seen, they arise from the reactions of the human visual system, being a result of its perception.

Thus, it is pertinent to assume that one of the most widespread streams of visual perception is grounded on the results of experiments conducted by Gestalt psychology or gestaltism.

According Dondis (1991), with her studies and experiments on perception, the Gestalt psychology contributed with significant information for the understanding of the visual patterns that help to decipher the vision process based on the human organism. The author also stresses the importance of the work done by Gestalt psychologists, with the interest in the principles of perceptual organization, the configuration process of a whole from the parts. Considers that all visual occurrences is a form with content, and this content is influenced by the amount of parts that structures it, such as color, tone, texture, dimension, proportion, as well as their compositional relations with the meaning. Thus, these interacting parts can be analyzed separated from each other, completely free, and then combined in whole, but alter any part of this structure means changing the whole.

When receiving an information by means of a composition, the viewer sees the whole, and is subject to their stimulus, although the parts that constitutes it not always be separately identified by it, that is, the basic elements of the composition. The domain of knowledge and manipulation of these elements belongs to the designer, who must be able to work its structure. Being the design an area that works the information by manipulating these elements, attributes significant importance to studies from Gestalt.

The principles of Prägnanz of the shape, worked by Gestalt, also serve as parameters for visual analysis. Considering that the Gestalt provides procedures for an objective evaluation of visual perception.

Visual Message

Arnheim (2002, p. 39) states: "Seeing is understanding." This is a spontaneous process that occurs in fractions of a second, and involves receiving, retaining and understanding of a large number of information units.

For an effective communication it is necessary for the viewer quickly understand the coherent arrangement of the elements that set the visual message in order to understand its contents. Have knowledge and mastery of the principles and laws governing the psychology of visual perception, of the composition techniques, combined with semiotics, ergonomics, emotion, and other areas that reinforce this structure is the necessary multidisciplinary character that can energize the expressive intentions and controls responses to be achieved, so that the visual elements emphasize the content and form, allowing direct the meanings.

Dondis (1991, p. 18) states that there is "[...] guidelines for the creation of compositions. There are basic elements that can be learned and understood by all investigators of visual media [...]".This knowledge combined with manipulative techniques, enables the development of easily understandable visual messages. But one should be aware that the logic of languages, with an organized and simplified structure, is not organized in such a simple way for visual messages whose biggest characteristic is the complexity. In fact, they may change on the psychological states and cultural contexts, or even by environmental perspectives.

Thus, it is worth considering that, besides the organization of the elements for transmitting the visual message, is necessary to evaluate the environment in which it will be inserted, that is, a space with particular characteristics and that involves persons or groups with specific histories and behaviors. Você quis dizer: Ter conhecimento do público e do ambiente que o cerca são fatores aliados ao processo da comunicação visual, uma vez que deles depende também a compreensão de qualquer informação.

Have knowledge of the public and the environment that surrounds them are factors allied to the visual communication process, since it also depends on their understanding of any information.

This is a communication process in which the compositional decisions determine what the viewer receives as message. The chosen elements - color, shape, texture, tone, proportion - are related to each other to transmit through the composition of objects, the goal of those who projects them: for this study, the designer. "Careful planning, intellectual inquiry and expertise are needed in the design and in the visual pre-planning" (DONDIS, 1991, p. 136).

The importance of technical knowledge is evident, however, must be considered that the professional, on designing, needs to be focused for those who are projecting for, and that his knowledge and technique should be to the user's service, which often can not distinguish the elements of the composition, but have preferences for stimulus that receives. The external stimulus that the human organism gets, by the planned composition, generate psychophysiological actions as the author emphasizes:

[...]the tendency to organize all visual cues in the simplest shapes possible; automatic association of visual cues that have identifiable similarities; the uncontrollable need for balance; the compulsory association of visual units born of proximity [...] (DONDIS, 1991, p. 137).

Always working together, the elements are never isolated and the perception of some affects the perception of others. So with the ability of the visual system to recognize them, they "[...] are like the basic units of our perception of objects and of the space" (AUMONT, 1995, p. 29).

Therefore, the attention to people and how they interact with the visual message, from a planned configuration process, focusing on the experiences and emotions that are evoked by them, can influence and assist the consumer in the evaluation and in the process decision of purchase.

Configuration

In the design, which has the design activity as the essence, the term *configure* differs from the common meaning attributed to it. According to Coelho (2008, p. 169), *configure* "[...] means defining how should be the product, which should be its characteristics and attributes", that is, is a design activity. Arnheim (2002, p. 89), in a simpler way, stresses that "[...] the configuration serves, first of all, to tell us about the nature of things through their external appearance."

In short, the configuration is the means that you have to represent in the form of content what you have in mind, being done consciously or unconsciously. A "[...] conduct basic line of Gestalt psychology, says that for the sense of sight, any visual pattern will tend to the simplest configuration possible in given conditions" (ARNHEIM, 2002, p. 403).

Although the principle of simplicity clarify why the elements are unified or split up, some have volume and depth and others are flat, some are incomplete and others full, the parts are the whole and at the same time have individual values, the simplicity does not justify all that is seen and should be balanced with other principles also active.

In this process, the contrast is a key strength for being a rich instrument of expression. The contrast, as opposed to this biased state of stagnation of human desire, stimulates and boosts its existence.

However, the search for the balance is constant in the visual sightings and, both the sender and to the receiver, the lack of balance and regularity is a disorienting factor (DONDIS, 1991). The simplest way to achieve balance in the composition is by symmetry, but it is also common to work with mutual compensation between the forces that set a system.

According to Arnheim (2002), the weight and direction are the properties that influence the balance, whereas the factors that influence the weight are: location, spatial depth, size, isolation, and configuration. Thus, as a result, a visual object placed higher will carry more weight, that is, the balance in the vertical direction is not achieved with equal objects in different heights.

The farther from the center, the greater the weight of an element, the same is true for the largest depth of an area of the visual field, that is, larger the weight. Still according to the author, in size, the largest object is the heaviest, when other factors equalize; isolation is another factor which underlines the weight. Figures with simple geometric shapes seem heavier, while the vertical forms have more weight than inclined, that is, the configuration also influences the weight.

As for the spatial configuration, the three-dimensional provides greater freedom that the two-dimensional, and this provides greater flexibility in the composition. In the projects of showcases, the three-dimensional does not have the same force, despite its presence, because the depth, that is, the third dimension, is disproportionate to the height and width measurements.

The techniques of visual communication, proposed by Dondis (1991), are forming the visual solution that the communication process needs. The use of theoretical knowledge in a practical way allows that the receiver has an easy assimilation of the perceptual content. The author recognizes that there are many techniques, however, points out the most used and their polarities.

The well-structured configuration, with proper organization of the elements making use of proper techniques, through its planning and organization, is responsible for the communication efficiency with the full understanding of the message by the receiver.

CONCLUSION

The design as a tool to carry the power of nations has come a long way from the views of the nineteenth century, who filled pavilions with the first mechanical products.

National identity has become a sort of showcase on the market. This comes embedded within the mass media and experienced in everyday bases. As individuals, consumers have learned to negotiate their identities and their sense of "belonging", aware that it was composed of various elements defined by multiple categories including gender, age, ethnicity, lifestyle and location. Helped by the consume of goods, services, spaces and projected images available at any time in the market, individuals can negotiate their own identities, and play a role as an active part in the construction of the everyday modern, postmodern and contemporary life.

REFERENCES

Aicher, O. (2001). *Analógico y Digital*. Barcelona: Gustavo Gili.

Alcantara, C. P. & Machado, J. H. C., Jr. (2010) Arte e Design: para a construção de uma relação mais afetiva entre homem e objeto. *Proceedings of: Congresso Brasileiro de Pesquisa e Desenvolvimento em Design*.

Almeida, R. R. de. (2010). Emoção na Pesquisa de Público-alvo: Estudo de caso em disciplina de projeto de produto. *Proceedings of: Congresso Brasileiro ee Pesquisa e Desenvolvimento em Design*.

Arnheim, R. (2002). *Arte e percepção visual: uma psicologia da visão criadora*. São Paulo: Pioneira.

Aumont, J. (1995). *A imagem*. Campinas: Papirus.

Bomfim, G. A. (1999). Coordenadas cronológicas e cosmológicas como espaço das transformações formais. Formas do Design. Rio de Janeiro: 2AB.

Bonsiepe, G. (2011). *Design, Cultura e Sociedade*. São Paulo: Blucher.

Coelho, L. A. (2006). *Textos Selecionados de Design* (G. C. Lima, Ed.). PPDESDI UERJ.

Couto, R. M., & Oliveira, A. J. (1999). *Formas do Design*. Rio de Janeiro: 2AB.

Damásio, A. (1996). *O Erro de Descartes: Emoção, Razão e o Cérebro Humano*. São Paulo: Companhia das Letras.

Desmet, P. M. A. & Hekkert, P. (2009). Special Issue Editorial: Design & Emotion. *International Journal of Design*, *3*(2), 1–6.

Dondis, D. A. (1991). *Sintaxe da linguagem visual*. São Paulo: Martins Fontes.

Freire, K. (2009). Reflexões sobre o conceito de design de experiências. *Strategic Design Research Journal*, *1*(2), 37-44. Retrieved November 20, 2010 from http://www.unisinos.br/sdrj/pdf/63.pdf

Iida, I. & Mühlenberg, P. (2006). O bom e o bonito em design. *Proceedings of*: *Congresso Brasileiro ee Pesquisa e Desenvolvimento em Design*.

Julier, G. (2006). From Visual Culture to Design Culture. *Design Issues*, *22*(1), 64–76. doi:10.1162/074793606775247817

Landim, P. C. (2009) *Design/Empresa/Sociedade*. (Unpublished associate professor dissertation). São Paulo State University.

Ledoux, J. (2001). *O cérebro emocional: os misteriosos alicerces da vida emocional*. Rio de Janeiro: Objetiva.

Moles, A. (1981). *Teoria dos Objetos*. Rio de Janeiro: Tempo Brasileiro.

Mont'Alvão, C., & Damazio, V. (Eds.). (2008). Design, ergonomia e emoção. Rio de Janeiro, Mauad X: FAPERJ.

Moraes, D. (1999). *Limites do design*. São Paulo: Studio Nobel.

Neves, J. D. (2010). Sobre o Design e Consumo de Experiências à Luz do Consumo de Bens. *Proceedings of*: *Congresso Brasileiro de Pesquisa e Desenvolvimento em Design*.

Niemeyer, L. (2000). *Design no Brasil: origens e instalação* (3rd ed.). Rio de Janeiro: 2AB.

Norman, D. A. (2008). *Design emocional: por que adoramos (ou detestamos) os objetos do dia-a-dia*. Rio de Janeiro: Rocco.

Silva, H. F. J. & Marar, J. F. (2010) Emoção, Percepção e Design: Aspectos Teóricos e Práticos da Ciência Cognitiva e Neurociência no Projeto de Produto. *Proceedings of*: *Congresso Brasileiro de Pesquisa e Desenvolvimento em Design*.

KEY TERMS AND DEFINITIONS

Design: Defined as an actuation field, constituted by social, economic, cultural transformations and at the same time responsible by these same transformations, and also as a cultural expression.

Design Culture: The interrelations between design artifacts, in all its possible manifestations, that is, the work of designers and production of the object.

Designer: as maker of wishes, responsible for creating objets that arouse emotions in users.

Emotional Design: Creates objects that allow the evocation of emotions in users.

Mass Culture: It can be understood as the everyday objects that reflect the cultural production of a social group.

Perception: Possible to make use of to understand its importance in the relationship between man and the world around him.

Visual perception: External stimulus that create an individual visual repertoire or subject to cultural patterns throughout life.

Section 3

Mapping and Designing Cognition:

Artificial and Natural Perspectives of Learning and Memory in Design and iCinema

Chapter 9
Qualia and Extended Field of Contemporary Design

Leila Reinert
Anhembi Morumbi University, Brazil

ABSTRACT

This study presents the concept of qualia, studied by neuroscience as one of the important aspects for the development of a conscious mind, and its association with the contemporary design focused on the subject. With the expansion of the field of design in understanding and extension, the meaning of the word design makes it necessary to review new relations that are formed among the artifact, user, and projected environment. The advancement of neurosciences combined with digital technologies provides a profound transformation in the understanding of human subjectivity when searching to unveil, empirically, the mysteries of the human mind. This transformation is directly linked to the manner in which designers design because they are responsible for redrafting the collective life on Earth before the forthcoming ecological crisis. Design and qualia are a necessary articulation to cautiously improve our surroundings.

INTRODUCTION

Human consciousness is formed from subjective feelings. According to Damásio, "the feelings of pain or pleasure are the foundations of the mind" (2004, p. 11). Differentiating between the affections that occupy the body and the mind as well as the emotions and feelings is necessary to effectively understand the mechanisms that regulate the simplest to the most complex sensations in the brain of a living person. Damásio believes that "a state of consciousness that covers a sense of self [...] is indispensable for survival"(2000, p. 384). According to the author, consciousness consists in states and processes that have a first-person subjective character. It is a private process called mind and is linked to the subjective states of our experience—the *qualia*

The Latin word *qualia*, plural of *quale*, was used for the first time in the modern era by Clarence Irving Lewis (1929), a renowned American logician, and designates the phenomenal properties of subjective experience, such as the sensory data captured by the senses. Smells, sounds, flavors, and

DOI: 10.4018/978-1-5225-0510-5.ch009

colors are sensory experiences irreducible to any definition. This means that, for example, possessing the knowledge of the physical properties of light and the different wavelengths is worthless compared with experiencing the subjective sensations provided by colors.

The study of qualia opens a wide field of research to the design activities of designers, given that design expanded its areas of operation, in addition to the daily artifacts, designers design landscapes, food, bodies, culture, as well as nature. Designers design ways of life and consequently are active agents in the processes of subjectivation of the contemporary subject.

In the articulation between qualia and design, we can understand how the projects develop, prioritizing the interaction between product, user and environment, and no longer think about the product "in itself." In other words, Latour (2008) states that we surpass the "matters of fact," relating to the eminently objective concerns of the modern era, and turn them into "matters of concern", focused on the human being, in care for the environment, in the subjective involvement with the environment being designed. In this manner, designers do not design objects but "things," in the original, etymological meaning of the word thing—both the English 'thing' and the High German 'ding', meaning a meeting convened to deal with a matter in question (Heidegger, 2006). Designers design artifacts, spaces, communications, interfaces, services; in short, this implies things designed as complex sets of matters of general interest.

Considering the things and the qualities of our experienceable perception, this study proposes a reflection on how the association between qualia and design can leverage our existence, our being, and living in the world.

QUALIA AND CONSCIOUSNESS

Throughout the 1980s and 1990s, numerous researchers in the field of the philosophy of mind have discussed regarding the concept of qualia, issues of subjectivity, and neurobiological processes that cause consciousness. In particular, focus was placed on issues triggered mainly by the new instruments that science and contemporary technology introduced to scientific research concerning the relation between the mind and the brain, or in traditional terms, between the physical and the mental. Moreover, according to Searle (1997), the origin of consciousness has not been adequately explored because Western thinking finds it difficult to exclude obsolete categories or assumptions inherited from philosophical and religious tradition.

How do we get away from the polarities: duality or monism, materialism or idealism? How can we conceive the qualitative and subjective mental states like our joys, affections, memories and moods as being identical to the chemical processes and electrical discharges of a physical matter? What differentiates, then, an emotional thought from a stomach pain? On the other hand, as consciousness is conceived as a completely separate phenomenon of its materiality, as if it were a separate entity of the brain, what is the possible correlation for the existence of an 'I'? In short, issues that seek scientific principles to understand the phenomenological and subjective qualities of our being and awareness in the world were considered to be epistemological problems without a solution for centuries.

Several controversies have emerged on the existence of qualia and the conscious mind in the conflict between philosophers such as Dennett (1988), Chalmers (1996), Searle (1997), Delacour (2001), and scientists such as Ramachandran (1998), Damásio (2010), Mogi (2013), among others, without exhausting the issue nor, in fact, completely clarifying the function of brain. Though the advance of research on this topic was considerable, much more scope exists to decipher human subjectivity.

One way to understand why these notions are so controversial, even though, apparently, the phenomenal characteristics felt in our experience are obvious—there is no denying the delicious sensation of the smell of freshly ground coffee in the morning—is the fact that they are experienced in the perspective of direct observation of the individual conscious mind, of the first person singular. According to Dennett (1988), a philosopher of the functional materialist mind, any subjective experience that cannot be objectively analyzed by a third person is an insurmountable obstacle and invalid for a scientific study. The whole argument developed by Dennett (1988) contains denying the existence of qualia and consciousness in favor of a brain function in which mental states are states of information processing. A pain, for example, is an *input* stimulus that causes a physical state in the nervous system, that is, in the perceptual information-processing tool, and results in a behavior of physical output. Dennett (1988) considers qualia and consciousness to be false problems that can be disqualified as objects of study as these are not legitimate scientific issues.

On the other hand, according to Searle, "we are blinded to the natural, biological character of consciousness and other mental phenomena by our philosophical tradition, which makes 'mental' and 'physical' into two mutually exclusive categories" (1997, p. xiii). It is necessary to admit that consciousness is, at the same time, a subjective and qualitative phenomenon, i.e., a mental and physical phenomenon, which belongs to the natural world. Searle (1997) defines consciousness as an emerging property of the brain—causally explained by the behavior of the components of a system, but not found in individual properties of these components nor explained by the sum of them—a biological phenomenon, because biological brains have the incredible capacity to produce subjective experiences that are perceived by a mind aware of the first person. Searle (1997) does not distinguish between qualia and consciousness; both are qualitative subjective experiences, and although we do not understand exactly how our biological system produces conscious minds, the challenge is to transform what was understood as a mystery into a problem. It is because consciousness is our condition of existence and because we are conscious beings that things have meaning and seem sensible.

These aforementioned discussions were made by philosophers based on the research by scientists from different fields of science, such as physics, medicine, biology, computing. Consequently, a relatively hermetic discussion for common sense was proposed in the field of language and in the history of Western culture. When the scientific research starts progressing in the study of the brain, the neuroscientists, or the new scientists of the mind, take on the argument, the focus becomes more empirical and relies on case studies with the aid of cerebral imaging techniques that enable us to view and map the brain activities during the implementation of different actions. Currently, through positron emission tomography (PET) or functional magnetic resonance imaging (fMRI), it is possible to see the brain areas affected when we become emotional, feel pain, dream, imagine something or look at certain images. Our subjectivity is being investigated in laboratories by more sophisticated and less invasive tests and demands of a more comprehensive knowledge of human nature and their modes of subjectivation in modern life. Excessively rhetorical arguments, which were often difficult to counter, are now verifiable and explained through advances obtained in the research on the functioning of neurons in molecular biology. It is very likely that the most valuable discoveries concerning the human mind today have their origin in the biology of the mind, or neurobiology, and not in the traditional disciplines related to the mind, such as philosophy, psychology, or psychoanalysis (Kandel, 2006).

However, before a new controversy is established, it is worth looking again at the properties that define qualia, those that are a consensus for most of the authors, to best exemplify the issue. There are four fundamental traits that, although questioned, underpin all discussions:

1. **Ineffable:** Sensations or sensory qualities that print affective shades from lived experiences and are difficult to analyze and are non-quantifiable and indescribable.
2. **Private:** Experienced in the first person, without objective confirmation of an outside observer.
3. **Intrinsic:** An event in itself (non-relational), without comparison and is irrevocable.
4. **Immediate:** Comprehensible from the awareness of the here and now. Feelings and sensations that occur in the immediate present.

If we can view the workings of the body at the molecular level (PET or fMRI) and locate the areas activated in the brain to investigate cognitive functions by the induction of emotions (Busatto, 2000), would it be legitimate to ask if we can see the sensorial phenomena experienced by a person through cerebral tomography images? Course it not, because the brain imaging results are a graphical representation, a translation of electrical impulses and neurochemical reactions that occurred in the body as a result of a subjective sensation. It is impossible to establish a correspondence between a qualitative sensation and its graphical representation. However, if the sensation experienced corresponds to a color, a red room, for example, the brain areas of the visual system located in the occipital lobe, will be activated and mapped. If the red room is also loaded with affective memories, in addition to the visual areas, there is a coordinated activation in other brain regions affected by the subjective sensations, such as the pre-frontal, insular cortex and amygdala, which are known for giving the emotional tone of our affective experiences. Therefore, the new techniques of cerebral imaging neither explain nor reproduce the qualia, but make observable the reactions of the body toward subjective experienceable phenomena.

There is a rather interesting aspect related to the graphical representation of cerebral activity and the experience of qualitative subjective phenomenon that gave rise to it because it highlights the difficulty of translation from one language to another, which is in the origin of the controversy about the existence of qualia and the conscious mind. According to Ramachandran (1998), a color-blind scientist who cannot detect color, knows everything about the brain and the physical properties of the color. He states, “the problem is that I can tell about my qualia (my experience of seeing red) only by using the spoken language. But, the ineffable ‘experience’ itself is lost in the translation. The current ‘redness’ of red will remain forever unavailable to you” (1998, p.148). We can, of course, defend writers and poets for carving out the language and making it plausible to fathom the sensorial experience of another. Notably, if we understand the issue of qualia as being a problem of translation, as Ramachandran (1998) suggests, then the division between physical and mental phenomena is just a result of existing barriers between different languages, surpassable only when the means to translate into another are found. In this sense, the viewpoint of Ramachandran is close to that of Searle, which proposes “a kind of ‘epistemic pluralism’ that does not imply a metaphysical thesis; a plurality of forms of language to which it is possible to speak significantly, according to different conceptual registers, from the ‘same’ world”, as Bento Prado Jr. states, in the introduction to the Brazilian edition of the book ‘The Mystery of Consciousness’ (Searle, 1998).

Notably, of the four fundamental traits submitted, two lose their strength of evidential reasoning of the disqualification of the qualia as objects of scientific study. First, the new cerebral imaging techniques make observable the experiences lived in the first person singular. Second, it is difficult to describe the redness of red; however, the translation of sensory data and subjective experiences has been an exercise of constant creation in the language of artists from the most diverse areas. This does not imply that the qualia are not private or ineffable, but that they are not as inaccessible to objective knowledge as was assumed by the arguments of the functionalist philosophers.

Currently, for neuroscience, no more doubts exist about the importance of qualia in the creation of the conscious mind. It clarifies ways to empirically assess the manifestations of the phenomenological qualities of subjective experience. They have also been examined as a fundamental aspect for learning. Recent studies have investigated correlations between being conscious of subjective experiences (phenomenal properties of consciousness) and cognitive skills from the concept of metacognition (Mogi, 2013). It is a widespread concept in neuropsychology, cognitive psychology, and education, which describes the mental activities of planning, monitoring, and control of the cognitive processes considering the subjective experiences inherent to the process of knowing one's own cognition, although the term still lacks greater accuracy in the approaches that have been proposed (Corso, Sperb, Jou & Salles, 2013).

The relation between qualia and learning questions the other two characteristic traits of qualitative experienceable phenomena: to be an immediate and intrinsic detail. The here and now, which determines a subjective experience, is not necessarily a single event nor an event "in itself," non-relational. If we take Proust's *madeleines* as an example, the sensations and feelings experienced by the character are repeated every time the cake is tasted. Flavors and smells are seized by memory and relived when some fact functions as a trigger for the stored affective memory. When there is repetition, there is learning there is a new recognition of the feeling being experienced, no matter how unique, since they are separated by time (Delacour, 2001).

For our discussion, the role of qualia in the modes of subjectivation of the contemporary subject and its effective participation in the actions that determine their existence become evident.

CONTEXTS: QUALIA AND DESIGN

Qualia, What Are Their Advantages?

My brain discards several sounds that it still does not know, even if I receive all the "messages" that are conducted by my ears, through the Individual Sound Amplification Device (ISAD) within the largest possible gain; in these recent months, with the new hearing aid, I have noticed these "wastes." As I capture the sounds, increasingly, my brain filters them, i.e., it distinguishes them depending on the amount of different noises emitted by the environment. As soon as I hear them, I pay attention, repeatedly, until my brain registers them, and the next time my intellect recognizes them more quickly! (Witt, 2013).

The story of Patrícia Rodrigues Witt (2013), bearer of a profound and bilateral congenital deafness, on learning of auditory perception with the help of hearing aids, introduces the issue that mobilized the writing of this study. What benefits do qualia offer to us? What is the importance of subjective qualities in designers' design activities? However much we may contest the existence of qualia, or their relevance for action in the world—how the phenomenal features felt in our experience reflects our choices and actions—when we are faced in our day-to-day lives with the many relations established with the apparatus that surround us, we realize how many times we are affected by qualia sensory data stimulators. Designers are responsible for designing the connection between the "artificial" world and us, and it is this relation between qualia and design that the foundation for our reflection is to be found.

Learning, refining, and leading to consciousness of sensorial perceptions (sensitive), in the way Patrícia (2013) tells us, is part of the learning of feeling for all of us. Certainly, the comparison is not intended to be simplistic, given its proportions we have developed our perceptions by being alert and attentive to the phenomena that surround us. It is also known that we filter and discard all incoming messages, non-stop, in our brains, this amazing biological system capable of producing subjective conscious minds. Notably, as we have already shown previously, qualia are not experiences "in themselves," as if they were pure sensations but the result of the relations established between perception and things, i.e., relatively stable realities, such as an apple, a chair, or an event, and are articulated to the cognitive aspects of the mind (Delacour, 2001).

In this sense, Patricia (2013), in her narrative about the apprehension of sounds that were unfamiliar to her, or not yet identifiable, like the brakes on a bus, car alarms, birds in closed environments, meows, etc., draws attention to a very special sound: "and finally the interesting mystery of the sound of my grandmother's kettle when it whistles, which, in fact, I still cannot hear clearly." The high-pitched sound of the whistle of the kettle interfered in Patricia's tone of voice, which, at first, seemed to her to be a problem with the hearing aids. Only later did she realize that the sound information captured by the ear was sent to the brain but was discarded as it did not correspond or have any usefulness to what she was hearing, showing the necessary link between the auditory channels and the process of cerebral mental maps triggered by sound. Deciphering the sound of the kettle whistle and focusing on the experimentation of the perceived sound sensation, a hearing-impaired person will extend their qualia at the same time as they expand their knowledge of the world that surrounds them.

However, the kettle and the sound of its whistle has lot more to tell us. In addition to it being a kitchen appliance used to boil water, the whistling kettle is an artifact of communication that informs you when the water has reached the boiling point. This information prevents you from forgetting the kettle on the burner, which could cause a serious accident. In the case of coffee lovers, the sign of boiling water anticipates the aroma and taste of the drink that will be enjoyed soon, with pleasure, and transforms the kettle and its whistle into an artifact of memory. Technically, the whistling noise, caused by the steam passing through the spout of the kettle is the example of a system called "hole tone" or "bird call," a physical phenomenon observed for the first time in 1854 by the German physicist Karl Sondhauss. The hole tone is a system basically made up of a cylindrical tube with two parallel plates positioned close to one another, which contain similar holes through which a shot of steam that generates, with the impact, acoustic waves as a result of the interaction between the passage of the fluid and the structure (Henrywood & Agarwal, 2013). This system has been under research that sought, as one of its objectives, the control of frequency and amplitude of the whistling noise not only in reference to a kettle whistle but also in similar conditions in industrial duct systems.

There is one more important characteristic to the whistling kettle; it is the result of a design project. The link between qualia and design began to be developed from the role of design in contemporary times, i.e., from kettle 9091 by the German designer Richard Sapper, shown in Figure 1. Why choose his kettle in particular, among so many others? The reason is quite simple: unlike the effort invested to make the sound produced by the whistle of the kettle nozzle correspond to the high-pitched musical notes of "mi" and "si," the sound was inspired by the sound of the Italian steamboats and barges that sail down the Reno river (Norman, 2004). To the primary function of boiling the water efficiently, which provided the reason for the existence of kettles, many others were added in the course of time, and it became an artifact of multiple functions, including producing qualia, in so far as the designer proposes not only a unique design but a musical sound for the whistle coming out of the spout. The kettle that whistles musical notes is far beyond the original object that is designed to boil water.

We are far from a design project that solely focuses on seeking the best way for an object to perform a specific function, as in the modern era. The contemporary designs project ways of life, more than

Figure 1. Illustration of Kettle 9091 Alessi, 1983 by designer Richard Sapper
© [2015] [Leila Reinert] Used with permission.

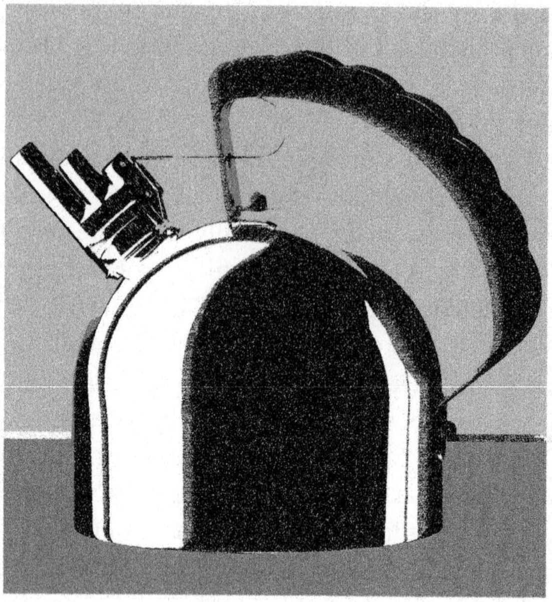

appropriate autonomous shapes that reflect the "taste" of their users. Currently, design, art, and architecture project affectivity (Costa, 2010). But what does this mean? Taking our kettle as an example, it becomes clear to understand that the relation between the user and object raises qualitative subjective experiences, which were previously unthinkable. When listening to the sound of the whistling of the kettle, the user will be affected by it in the most varied way and according to their experiences. If they are Italian and frequently saw the steamboats navigate through the Reno, memories will probably be triggered; if it is a child, it is possible that a magical world will be instilled in their imagination; if it is a person with a hearing impairment, maybe a lower-pitched sound than that of conventional kettles is easier to capture and less enigmatic; and so on. Each experience built into the fact that the water boils and the kettle whistles is another affection being created.

What is the reason for such a magnification of senses that the example presents to us? The culture and society today reflect, in their productions, the transformations that have occurred since the second half of the

20th century. We have transformed from an industrial society of design focused on the production of consumer goods, oriented by the industrial product, and "controlled" by a market economy to a society based on systems of communication and information networks, whose most valuable capital is human capital. It is the ways of life, the ways of seeing, feeling, and thinking that are consumed more than material goods (Pelbart, 2003). Thus, design is focused on the subject and their way of life and now also considers the relation that the user establishes with the designed environment. Aligned to the technological advances of distinct areas, the designer has gained new areas for work and, above all, a broad recognition of their professional activities.

The contemporary thought, of course, is much more comprehensive and complex than how it was presented here, as well as the history of design, but for our purpose, we simply need to understand the displacement of the sense of design from the historic changes that have occurred over the last 60 years. One of the relevant factors for the proposed example is the exploitation of nature, caused by the idea of progress of the modern era and the indiscriminate use of natural resources, which requires a profound change in the course of production, consumption, and disposal of products to enable new habits in our day-to-day lives, as well as raising one of the most currently discussed topics in design: design and ecology. Designers are social actors, directly responsible for everyday interactions between human beings and their artifacts, and play a major role in the transition toward the sustainability needed for the maintenance of the quality of life for humans on the planet (Manzini, 2008). The question is centered around the modern subject's way of life, not so much as an isolated individual, but mainly in the collective arrangements that make sustainable ways of life possible; in other words, the design, or the designers, are triggers and creators of qualia as they design experiences with subjective qualities that transform behaviors.

Design in the Extended Field

Not so long ago, the word design was not very familiar to people in general, and they possessed less knowledge about the areas of knowledge assigned to designers. Historical reasons have contributed to the difficulty of establishing definitive parameters to clearly delimit and define the concept of design, or even the specific allocations of this professional activity. The first is the word "design" itself, which was not incorporated by the Latin languages immediately, and neither did its translations clearly correspond to the assignments of the profession. Besides, in English, the word design can be used as a noun or verb, and it was also turned into an adjective in several languages (perhaps in all) for the additional aesthetic value added to the design products.

The English word "design" is derived from the Latin word *designare*, which can mean indicate, represent, mean, and note. According to Flusser (2007), as a verb—to design—it means to project, plot, configure, and proceed in a strategic manner. As a noun, it means plan, project, drawing, goal, conspiracy, form, or purpose. In other words, the term 'design' carries, already in the etymological sense of the word, actions related to both material issues of achievement and the intellectual process of planning; therefore, design acts at the junction of concrete and abstract design (Cara, 2010). The problem is that these aspects have become contradictory throughout the 19th century, when there is a conceptual distinction between art and technique and from which industrial society starts understanding the production of aesthetic objects and the production of common objects as being opposite practices. Le Bot (in Ars, 2008) states that the rupture that develops with the advent of the industrial society between these productions derives from changes in the conceptual field and refers to the processes of the division of work imposed by a mode of capitalist production that begins to consolidate economic, political, and ideological practices in all sectors of the society, especially favoring the law of productivity.

We must remember that design, or *industrial design*, was directly connected to the ideals of the modern movement when, in 1959, the International Council of Societies of Industrial Design (ICSID) established the functions of professionals who produced industrial objects and, still at that time, craft objects were made on a production line. With the objective of defining a specific field of knowledge in design and, mainly, to establish the differences between the design and other areas of knowledge, in particular, the art and craft industry, professional functions and the conceptualization of the notion of design were and have been revised and redefined several times by the ICSID (Cara, 2010). However, this was not the only reason for so many systematic revisions to this professional activity. Given that design is directly connected to the existing socioeconomic systems, i.e., the development of productive forces, techniques, and technologies available in production processes of companies in different contexts, in different times and spaces, it is impossible to give definitions without considering the flexible, mutant, non-linear, and relative character of this professional activity. There have been many changes in the world since the first parameters were formulated, outlining the areas of performance and knowledge in the field of design. Currently, design fully accepts the diversity and amplitude of its functions and invests in research that places the key role played by design in the necessary transformation of social behaviors and envisages more sustainable futures (Manzini, 2008; Tonkinwise, 2014; Irwin, Kossoff, Scupelli, & Tonkinwise, 2015). The expansion of the field of knowledge of design, not only in its understanding but also in its extension, is breaking down typically modernist ideological barriers and indicating how far the designers' design activities are important for the necessary reconstruction of our collective life on Earth (Latour, 2008).

But in what way can we better understand this statement? For Latour, the modernist utopia, rooted in the concept of humanism, tends to separate material and objective constraints on one side and symbolic and subjective constraints on the other. For a long time, humanism dismissed objects, as it did not understand that human beings do not exist by themselves, because their existence depends on the supports of material life and objectives that protect and maintain them (Sloterdijk, 2009). The best understanding of the concept of the term "design"—design actions related both to matters of material execution and intellectual planning—which operate in the symbiosis of objective and subjective aspects of a project, redefines the notion of building, designing and making and links the designer to the strange figure of a cautious Prometheus. Or, in the words of Latour:

In design, there is nothing foundational. It seems to me that to say you plan to design something, does not carry the same risk of hubris as saying one is going to build something. Introducing Prometheus to some other hero of the past as a "designer" would doubtlessly have angered him. Thus, the expansion of the word "design" is an indication (a weak one to be sure) of what could be called a post Promethean theory of action. This theory of action has arisen just at the moment (this is its really interesting feature) when every single thing, every detail of our daily existence, from the way we produce food, to the way we travel, build cars or houses, clone cows, etc is to be, well, redesigned. It is just at the moment where the dimensions of the tasks at hand have been fantastically amplified by the various ecological crises, that a non – or a post-Promethean's sense of what it means to act is taking over public consciousness (2008, p. 3).

This other look into methods of production does not disqualify the objects, the material world, in so far as they are not understood as being inert materials but can be things, i.e., matters of interest (Latour, 2004). Here, understanding how Latour differentiates an object from a thing is necessary to determine the interactions of humans with their artifacts and the manner in which the matters of fact become increasingly understood as matters of interest.

An object is a material, passive being that is addressed by the subject, or that is in opposition to the subject and that does not do anything by itself. Objects generate matters of fact, objective realities, which one cannot doubt nor draw anything from beyond what has happened. A glass of water that is dropped as an object has no other meaning other than the insignificance of the fact that a glass of water has been dropped. When Latour (2008) says that objects can be things, he recalls the etymological sense of the word thing, through Heidegger—which establishes a dichotomy between object and thing, to recover our relations with the objective material world. The word "thing," and "ding" from High German, means a meeting convened to address a matter of interest, an issue in dispute (Heidegger, 2012). Still following Heidegger (2012), "thing" and "ding" better reflect the Roman word "res," which means "that which pertains to man," "the matter in question." The Latin word that corresponds to "res" is precisely the word "cause," in the sense of the case or subject of interest in general. Hence, the neo-Latin terms are derived from *la cosa, a coisa, la chose*. In this manner, things generate matters of interest, and a glass of water that falls, as a thing, can become an event. The glass of water that fell can be the trigger for a memory, a conversation, a spot of ink on paper. It is a thing that becomes a trigger for a myriad of much more complex situations than the fact that occurred.

Objects can be things. Designers design things—everyday artifacts, environments, bodies, food, spaces, etc. Consequently, these things are actively participating in our processes of subjectivation. In other words, if designers are responsible for designing a large part of the interactions among humans, the environment, and its artifacts, then designing is dreaming up human beings' modes of action through their experience with the designed world. If designing modes of action means producing subjectivity, designing is also configuring the subjective aspects of our being in the world—qualia.

THINGS THAT CREATE QUALIA

"Anyone who knows the soil and subsoil of life knows very well that a section of a wall, a stool, a carpet, an umbrella, are rich in ideas or feelings, as are we, and that the reflections of the partnership between men and things make up one of the most interesting phenomena on earth," wrote Machado de Assis in the novel 'Quincas Borba' in 1891.

This excerpt from the novel Quincas Borba was cited as an epigraph for Chapter "The Life and Speech of Shapes" in the book "Design for a Complex World" by Rafael Cardoso (2013, pp. 101-168). However, Cardoso does not, in fact, indicate the reflections of partnerships between humans and things, although there is a comprehensive discussion that includes our relationship with the appearance of objects (their material configuration), including commercial and affective values, with the formal and informational aspects (ability to communicate something) that result from the specificities of a project elaborated through design. The design will create objects embedded of meanings that transform, resist, and survive beyond their initial purpose. In short, the matters addressed by Cardoso are broad in terms of thinking about the relation among the artifacts, their users, and their meanings—both the designed meanings and those introduced by cultural, regional, or intellectual characteristics, unique to each user. However, the constraints are still bound to objects and the matters of fact. Rather, they tend to separate the objective aspects on one side and the subjective on the other, although there are a few moments in the text that allow glimpses into more symbiotic relations between human beings and their artifacts.

At this moment, it is interesting to present an excerpt from Machado de Assis's writing, not included in the epigraph for the chapter mentioned above: "The expression: '*Conversar com os seus botões*' [Portuguese for 'talking with one's buttons', meaning 'thinking to oneself'] appears to be a simple metaphor, but is in fact a real and accurate sense. The buttons operate synchronously with us; they form a sort of Senate, comfortable and inexpensive, which always votes for our motions". To the attentive reader, the connections between the experiences of qualitative subjective phenomena—qualia and design—elaborated from designing matters of interest (things) seem to make "real and accurate sense" when we talk with our buttons, do they not? Despite the playful tone, we have long daily conversations with artifacts and devices, more or less, technological, that are surrounding us. For instance, like the kettle that whistles every morning, which has already been mentioned.

We influence and are influenced by the things that involve us in many ways, in which the partnership of senses arises, like Machado de Assis says. The creative processes for projects in design are simultaneously, empirical and reflective, and the predominance of one or of the other is related to the purpose of the project. The more comprehensive the project, the greater the need for reflection and systematization of the proposed solutions; the more limited the field of action, the less conflicts in it, the more empirical the solutions will be. In any case, the precise definition of purpose, and even the rigor of conscious intentions, does not exclude the subjective aspects of experienceable qualitative phenomena that appear in design solutions, whether they are conscious.

A simple and beautiful example is the communication design placed at Schiphol Airport in Amsterdam. In the beginning of the 1990s, under the management of Aad Kieboom, the team responsible for the expansion of the terminal at Schiphol implemented the project of engraving an image of a fly in each of the urinals in the men's room at the airport. A fairly simple action, but which resulted in an improvement of 80% less splashing and consequently considerable savings in maintaining the cleanliness of the bathrooms. It seems that, even without paying attention, men keep and update their primitive instincts of the hunter, who watches his prey to launch his attack (Figure 2).

The success of the urinal fly design from Schiphol airport was huge, as was their propagation to several other bathrooms around the world, whether institutional. The idea of the etched flies in urinals came from Jos Van Bedoff, one of the cleaners responsible for bathroom maintenance at the airport. When Bedoff served in the Dutch army in the 1960s, he noticed that small red dots had been placed in the urinals in the bathrooms at the headquarters to direct the men's actions, whose urine flow is normally poorly directed. Instead of a simple and discreet red point, Bedoff suggested they used the image of a fly because, in his view, men aim at the insect to try to kill it, or immobilize it, with ease. What is interesting is that what were designed were a context and a consequence: the male hunter aims at his prey and directs the flow of urine at the urinal, avoiding undesirable splashing. They did not design a more appropriate and elegant urinal to avoid splashing or an ultra technological material capable of absorbing and evaporating the urine without leaving any residue. What was dreamed up, or recovered, was an instinctive behavior of primitive man through his experience with the designed "thing." The design endorsed a condition of use in the symbiotic link between the subjective and objective constraints that the situation demanded. It exploited a qualitative subjective phenomenon intrinsic to the operation of brain mechanisms and the neurobiological processes of the human mind.

More interestingly, the central idea for the project came not from a designer but a co-designer, in this case, a cleaner. This points to the increasingly more frequent emergence of what Manzini (2008) calls "design networks," i.e., groups of people who work collaboratively to generate solutions in complex sets of matters of general interest. We are living in a period of transition, in which there is an extreme need to constantly redesign our life projects to come up with everyday actions that can generate social innovations to generate more sustainable conditions for living on planet Earth.

The interactions between humans and things, or as Machado de Assis would say, our "talks with our buttons", configure the sensory qualities of our subjective experiences, our qualia. The example of the fly in the urinals is one of many other projects conceived through design that clarifies how our pro-

Figure 2. Illustration of engraved fly in the urinals in Schiphol Airport

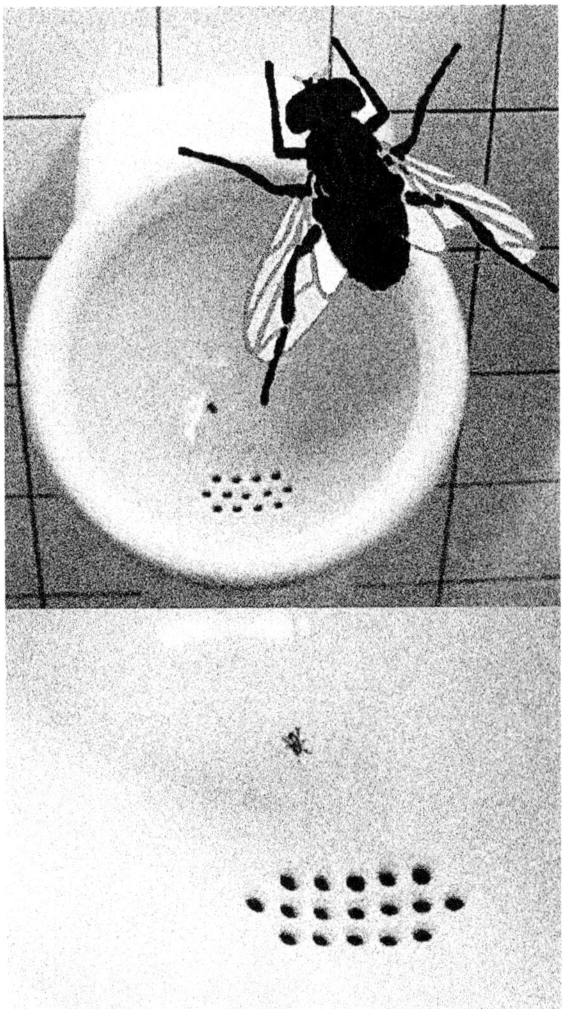

cesses of subjectivation are entangled in things, but it is our attention to our surroundings that actually configures our sensory perception. The manner in which we deal particularly with each thing or affect and are affected by them determines our existence.

DIARY OF A CLOTHESPIN

Of the many possible things we deal with on a day-to-day basis, we would like to present quite a peculiar artifact commonly found in the laundry of any home: the clothespin. But our clothespin is not made of wood, like the old-fashioned ones, nor does it have a metal spring so typical of clothespins. Taciana Silva and Marcela Albuquerque, two Brazilian designers designed our multipurpose clothespin in 2002 (IF Design Award 2004). They where design students at PUC in Rio de Janeiro at the time. Its major difference is that it looks nothing like a clothespin, which makes it an unidentifiable object at first sight, although it is very pleasant to the eye and extremely seductive to touch—it awakens the desire to immediately pick it up. It is molded in one piece by thermoplastic injection; it is light, colorful, malleable, and flat. Measuring 4 cm × 9 cm, neither too large nor too small, it is easy to handle. It is really a peculiar artifact (Figure 3).

We could go on for hours describing its texture, temperature, or the elliptical dynamics of its shape; however, it is impracticable to experience all its combined sensory qualities and learn about its multiple uses without knowing, seeing, and holding this clothespin in the same way as a color-blind person will never know the redness of red.

Just as we "talk with our buttons," we also talk to our multipurpose clothespin. If we take it out of the context of laundry, if we alter its field of action, we invent and experience unprecedented usages: a bookmarker, a hairclip, a memo holder, or even a simple artifact that we hold in our hand when we feel nervous and anxious when faced with a given situation. We create new relationships, change our original direction by proposing functional operations that are distinct from what they were designed for. However, none of this happens if we are not attentive. It is our state of alertness in the face of our

Figure 3. Multipurpose clothespin. Designers: Taciana Silva and Marcela Albuquerque, 2002

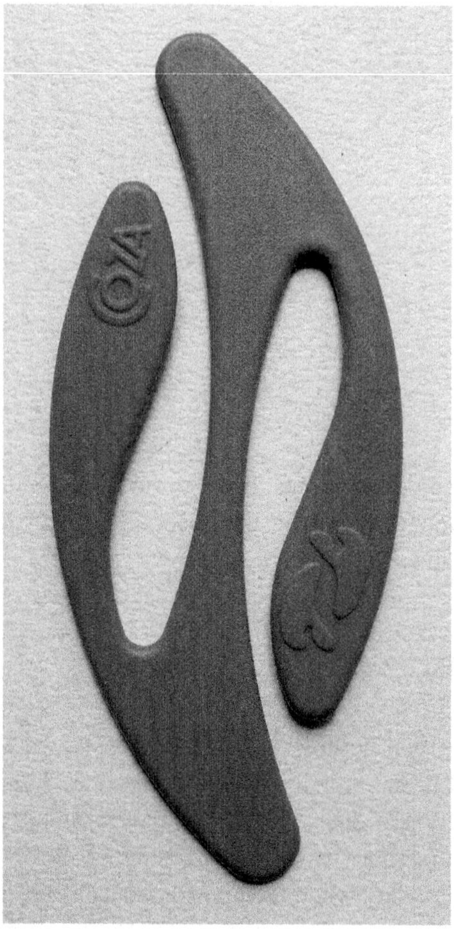

environment that makes our partnership with things possible and creates multiple forms of action, i.e., a perceptive alertness, a sort of "almost" permanent wakefulness, and, why not, a state of caution. In the moment our little perceptions are unconscious, diffuse, obscure, hazy, or hallucinatory before they become conscious and pragmatic perceptions (Figure 4). It is worth remembering the case of Patrícia: "and finally the interesting mystery of the sound of my grandmother's kettle when it whistles, which in fact I still cannot hear clearly."

The mental processes of learning and memory organize our sensory perception, and as we have previously seen, recent studies have investigated correlations between the awareness of qualia (phenomenal properties of consciousness) and cognitive skills from the concept of metacognition— the process of knowing one's own cognition. It seems important to us, from now on, to bring research on learning and

Figure 4. Multipurpose clothespin. Variations
© [2015] [Leila Reinert] Used with permission.

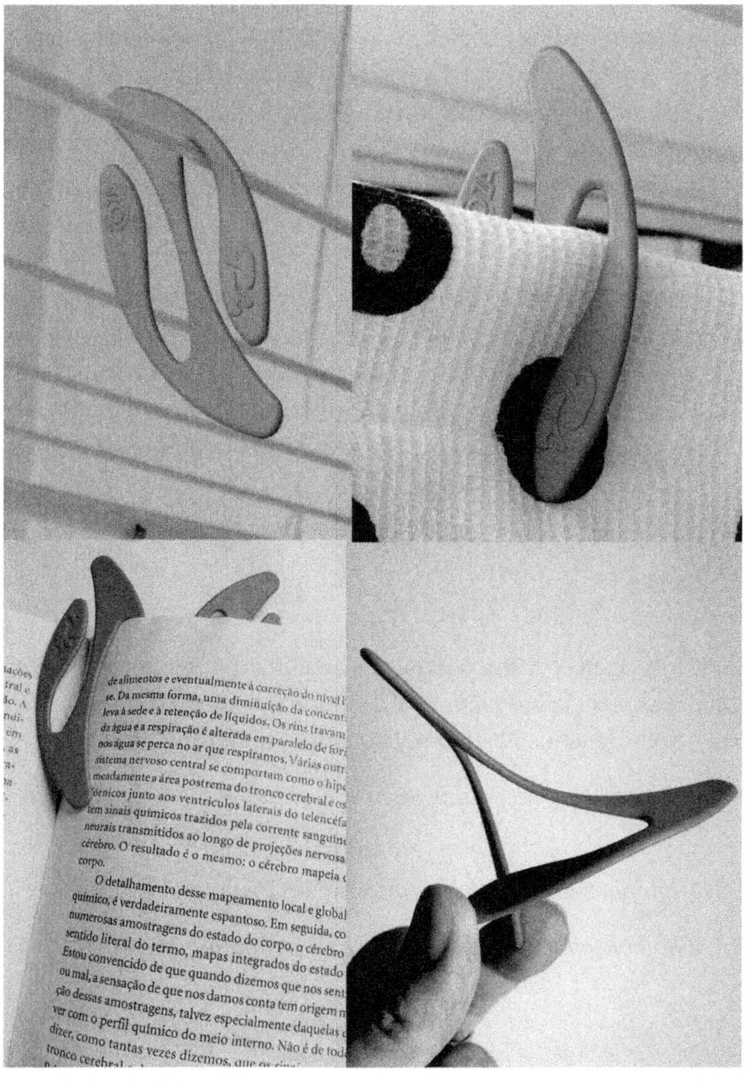

memory of the links between qualia and design closer together to further enhance the reflections of how much, and how, things (complex sets of matters of general interest) participate in the processes of subjectivation in contemporary times.

In the research developed in molecular biology, or neurobiology, Eric Kandel (2007) elucidated neural reactions occurring from different patterns of sensory stimuli that deliver the assimilation of knowledge and its storage. Kandel's (2007) research is based on the principle that there are various forms of learning produced by different patterns and combinations of stimuli and these variations create differences in the storage of memories. The creative processes in design are also made possible by different patterns and combinations of stimuli and creative action results from the variations of these stimuli in our sensory perception. Our multipurpose clothespin, this peculiar artifact, is really stimulating, is it not?

CONCLUSION

Scope for research always exists when we work with content that can be extended to different fields. Research in the field of design and neuroscience are in full swing and share matters that redefine our way of understanding the world in which we live and ourselves. If it is necessary to redesign our collective life on Earth, one must understand the potential powers of qualia to create sustainable subjectivities resulting from the experience with the designed world.

More than any conclusion or summary of the arguments presented, the words of Machado de Assis, from long ago, rightly approach the "the real and direct sense" of this study. They deserve to be remembered,

Anyone who knows the soil and subsoil of life knows very well that a section of a wall, a stool, a carpet, an umbrella, are rich in ideas or feelings, as are we, and that the reflections of the partnership between men and things make up one of the most interesting phenomena on Earth. The expression: "talking with one's buttons" appears to be a simple metaphor, but is in fact a real and direct sense. The buttons operate synchronously with us; they form a sort of Senate, comfortable and inexpensive, which always votes for our motions.

REFERENCES

Busatto Filho, G. (2000). Imagens do funcionamento cerebral durante tarefas cognitivas e emocionais: aplicações da técnica de ressonância Magnética funcional em psiquiatria. *Revista de Psiquiatria Clínica, 27*(3). Retrieved from www. Hcnet.usp.br/Ipq/revista/vol27/n3/art164.html

Cara, M. (2010). *Do desenho industrial ao design no Brasil: uma bibliografia crítica para a disciplina*. São Paulo: Blucher.

Cardoso, R. (2013). *Design para um mundo complexo*. São Paulo: Cosac Naify.

Chalmers, D. (1996). *The conscious mind: in search of a fundamental theory*. Oxford, UK: Oxford University Press.

Corso, H. V., Sperb, T. M., Jou, G. I., & Salles, J. F. (2013). Metacognição e funções executivas: relações entre conceitos e implicações para a aprendizagem. *Psicologia: Teoria e Pesquisa*, *29*(1), 21-29. Retrieved from http://www.scielo.br/pdf/ptp/v29n1/04.pdf

Costa, C. Z. (2010). *Além das formas: introdução ao pensamento contemporâneo no design, nas artes e na arquitetura*. São Paulo: Anablume.

Damásio, A. (2004). *Em busca de Espinosa: prazer e dor na ciência dos sentimentos*. São Paulo: Companhia das Letras.

Damásio, A. (2000). *O Mistério da consciência: do corpo e das emoções ao conhecimento*. São Paulo: Companhia das Letras.

Delacour, J. (2001). *Conscience & cerveau: La nouvelle frontière des neurosciences*. Bruxelas: Éditions De Boeck Université.

Dennet, D. C. (1988). *Consciousness in Modern Science*. Oxford, UK: Oxford University Press.

Flusser, V. (2007). *O mundo codificado: por uma filosofia do design e da comunicação*. São Paulo: Cosac Naify.

Heidegger, M. (2006). *Ensaios e conferências*. Petrópolis: Editora Vozes.

Henrywood, R. H., & Agarwal, A. (2013). The aeroacoustics of a steam kettle. *Physics of Fluids*, *25*(10). Retrieved from http://scitation.aip.org/content/aip/journal/pof2/25/10/10.1063/1.4821782

Irwin, T., Kossoff, G., Tonkinwise, C., & Scuppelli, P. (2015). *Transition Design Overview*. Carnegie Mellon School of Design. Retrieved August 25, 2015, from https://www.academia.edu/13122242

Kandel, E. R. (2007). *In search of the memory: the emergene of a new science of mind*. New York: W. W. Norton & Company.

Latour, B. (2008). *A cautious Prometheus? A few steps toward a philosophy of design (with special attention to Peter Sloterdijk)*. Keynote lecture for the Networks of Design meeting of the Design History Society. Retrieved from http://www.bruno-latour.fr/sites/default/files/112-DESIGN-CORNWALL-GB.pdf

Latour, B. (2004). Why has critique run out of steam? from matters of fact to matters of concern. *Critical Inquiry*, *30*(2). Retrieved from http://criticalinquiry.uchicago.edu/past_issues/issue/winter_2004_v30_n2/

Le Bot, M. (2008). Arte/Design. *ARS Revista do Programa de Pós-graduaçnao em Artes Visuais*, *6*(11), 7–21.

Lewis, C. I. (1929). *Mind and the World Order: Outline of a Theory of Knowledge*. New York: Charles Scribner's Sons.

Manzini, E. (2008). *Design para inovação social e sustentabilidade: comunidades criativas, organizações colaborativas e novas redes projetuais*. Rio de Janeiro: E-papers.

Mogi, K. (2013) Cognitive factors correlating with the metacognition of the phenomenal properties of experience. *Scientific Reports*, *3*(3354). Retrieved from http://www.nature.com/srep/2013/131127/srep03354/full/srep03354.html

Ramachandran, V. S., & Blakeslee, S. (1998). *Phantoms in the brain: probing the mysteries of the human mind*. New York: Willian Morrow and Company.

Searle, J. R. (1997). *The mystery of consciousness*. New York: The New York Review of Books.

Sloterdijk, P. (2009). *Esferas I: burbujas*. Madrid: Ediciones Siruela.

Pelbart, P. P. (2003). *Vida capital: ensaios de biopolítica*. São Paulo: Iluminuras.

Tonkinwise, C. (2014). *Design for Transition - From and to What?* Retrieved April 25, 2015, from https://www.academia. edu/11796491

Tonkinwise, C. (2015). *Committing to the Political Values of Post-Thing-Centered Designing*. Retrieved August 2, 2015, from https://www.academia. edu/14560093

Witt, P. R. (2013). *O mistério da chaleira*. Retrieved April 18, 2015, from http://surdezsilencioemvoodeborboleta.com/blog/2013/06/03/o-misterio-da-chaleira/

KEY TERMS AND DEFINITIONS

Consciousness: Consists in states and mental processes that cover a sense of self. It is a private process called mind and is linked to the subjective states of our experience.

Design: Actions related to both material issues of achievement and the intellectual process of planning; therefore, design acts at the junction of concrete and abstract aspects of our life–experiences with subjective qualities that transform behaviors.

Metacognition: Process of knowing one's own cognition.

Matter of Concern: Some questions much more complex than the fact that occurred. Matter of interest is a trigger for reflections and thinking.

Matter of Fact: A partial approaches that are given in objectives experiences (based on or in accordance with reason or logic).

Process of Subjetivation: Practices and procedures of the self be. Subjective conducts to transform themselves.

Qualia: The phenomenal qualitative properties of subjective experience, such as the sensory data captured by the senses.

Thing: Complex sets of matters of general interest.

Chapter 10
Conceptual Experiments in Automated Designing

Petr Ivanovich Sosnin
Ulyanovsk State Technical University, Russia

ABSTRACT

The success of designing Software Intensive Systems (SIS) and other objects may be improved by incorporating conceptual experimentation to be part of the design process. This paper presents a scientific approach to conceptual experimenting with objects that are units of designers' behavior and is aimed at solving project tasks in conceptual design. The proposed approach is based on specifying the behavior units as precedents and pseudo-code programming of experiments' plans. Reasoning used by designers in the experiments is registered in a question-answer form. Experimenting is supported by a specialized toolkit.

INTRODUCTION

Five years ago, the group of well-known researchers and developers in software engineering had initiated a process of innovations, which have been named SEMAT (Software Engineering Methods And Theory). The main reason for this initiative was *an extremely low degree of success in developments of Software-Intensive Systems.* Last twenty years, this problem is steadily registered in statistics (success approximately in 35% of cases) that are presented in reports (Chaos, 2013) of the company "Standish Group."

It is necessary to notice that, in normative documents of SEMAT, a way of working used by a team of designers is marked as a very important essence (Jacobson, 2012). There "way-of-working" as a notion is defined as "the tailored set of practices and tools used by the team to guide and support their work." Also, the specification "tailored" indicates that designers of any SIS should choose and apply appropriate means in accordance with situations arising in real-time. By other words, designers have to build and use the technology of their work in the design process. Therefore, investigations aimed at the search of effective ways-of-working are perspective and can lead to an increase of success in developments of SISs.

DOI: 10.4018/978-1-5225-0510-5.ch010

Told above suggests that designers should have an opportunity for real-time interactions with "libraries" of practices and tools and their units. Such interactions have to support creating and using the technology T(t) of working that is based on the accessible experience applied by designers.

We are deeply convinced that the efficiency of ways-of-working can be greatly enhanced by increasing the efficiency of interactions with the experience in the process of solving the project tasks. The basis of this conviction is our research of the place and role of experience at the conceptual stage of designing the SISs. It should be noted that this stage is estimated as an inevitable source of costly mistakes due to the problem of personal and collective understanding.

Our research interests have concerned real-time interactions of designers with their natural experience and its models in collaborative designing. The main result of the research is the question-answer approach (QA-approach) to conceptual designing that is oriented on the constructive use of question-answer reasoning (QA-reasoning) for interactions with the experience at this stage of actions. Investigations of interactions of this kind have led us to the development of an instrumentally-simulation environment WIQA (Working In Questions and Answers) and a corresponding system of workflows "Interactions with Experience" (Sosnin, 2013 a).

It should be noted that reports of Standish Group were "guides" for planning of our research, applying of QA-approach and evolving of the toolkit WIQA and workflows "Interactions with Experience." Step by step the QA-approach concerns criteria and factors that facilitate increasing the degree of the success in designing of SISs.

This paper presents a step of a research that focuses on interactions with the accessible experience in conceptual experimenting of designers in the work with project tasks being solved. A conceptual experiment is understood below as a thought experiment, the content and process of which are operatively reflected on a semantic memory of a question-answer type, and results of reflections are applied in the process of experimenting with useful purposes. Conceptual experiments are similar to automated thought experiments.

BACKGROUND OF INTERACTIONS WITH EXPERIENCE

Naturally-Occupational Experience of Designing

Any modern SIS is developed with the use of the accessible experience E^A a part of which is accumulated in specialized standards. Such sources of experience help to decrease the level of complexity in interactions of designers with the system that is otherwise not constructed. For example, at any step of the conceptual stage of designing, a number of standards helps to simplify the work with the SIS that exists in a form of a system of conceptual models M created by a team of designers in a collaborative development environment. An important subset of these models and corresponding technological practices can be inherited from the following standards:

1. Capability Maturity Models Integrated for Developing (CMMI, version 1.3, about fife hundred practices) that offers not only an important subset of best practices for process P of designing but also prescripts the ways for its continuous improving;

2. People Capability Maturity Models (P-CMM, version 2.0, more six hundred practices) includes descriptions of recommended practices that provides the rational use of human resources and effective work of the designers' team G;
3. Project Management Body of Knowledge (PMBOK, version 5, several hundred practices) that collects the best practices for all areas of management actions in CDE;
4. IEEE Recommended Practices for Architectural Descriptions of Software Intensive Systems (IEEE, version1471) that specifies some viewpoints on the architecture of the SIS and practices for its architectural creation and use;
5. ISO/IEC 9126 Software engineering- Product quality that helps to model requirements to quality with the use of empirical means.

Thus, developing the SIS, designers creates the appropriate way-of-working WW by the inclusion of useful best practices and corresponding tools in the applied technology T implemented in CDE. In this part of their work, designers orient on indicated standards and other sources of best practices for building the technology T.

It should be noted, the use of best practices $\{P_i\}$ leads to solving the corresponding normative technological tasks $\{Z^N_i\}$ the definite subsets of which are implemented in coordination as workflows $\{W^T_j\}$. Any workflow W^T_j can be also interpreted as the task Z^W_j of workflow type.

In implementing the technology, tasks of Z^W_j-type and corresponding tasks of Z^N_i-type are applied for solving the tasks $\{Z^N_i\}$ that belong to the subject area of the created SIS. Such work is impossible without interactions with the experience E^A combined units of the natural experience E^N with models E^M of its components. Successful solving of indicated project tasks essentially depends on:

- Effective integrating of personal components of E^N in a collective wholeness;
- Occupational maturity of E^N that is confirmed, first of all, by mastering of best practices included in the technology T applications of which embody in a component E^{NT} of the experience E^N;
- Measure in which the designers are experienced in the SIS domain or, by other words, sufficiency of them experience E^{NS} in the subject area of the developed SIS for successful designing;
- Effective modeling and simulating of components E^{MT} and E^{MS} combined in integrated model E^M used by designers in parallel with the experience E^N in their real-time work;
- Measure of naturalness of E^M-units or, by other words, the degree of their conformity and compatibility with "units" of E^N;
- Effective means of designers' interactions with E^A.

It should be noted, that enumerated dependencies can be used as a source of requirements for a system that intends for real-time actions of designers with the experience and models of its units. This source has been used for the development of the QA-approach and its embodiment in workflows "Interactions with Experience" and the toolkit WIQA. The use of these means showed that this approach can be applied as for designing of SISs so for automated designing the other objects, for example, for designing the configured templates of airplane parts (Sosnin, 2015).

Typical Structure of Experience Units

In the QA-approach, the conformity and compatibility among units of E^N-type and E^M-type are achieved on the base of their unified understanding as models of precedents. The unified understanding is based on the hypothesis that units of the human experience are generated on the base of intelligently processed human reactions as conditioned reflexes. Intelligent processing is aimed, first of all, at the stable reuse of reactions by the human in similar conditions. Any result of such processing is a "model" of a corresponding precedent in the brain structures.

No one knows how these models are embodied in human brains' structure. But, there is a possibility for controlled influencing on such "models" by their models that are built outside brains. Below models of "models" will be called as models of precedents.

In the design processes, models of precedents should be created with using of intelligent processing for the solutions of project tasks. Such way of their creating will facilitate the conformity and compatibility among models of precedents and corresponding "models" localized in designer brains. Therefore, interactions with any model of the definite precedent will be transferred on the corresponding "model".

Thus, controlled interactions of designers with units of the natural experience can be implemented through interaction with models of corresponding precedents. Such understanding of "interactions with experience" is used in the QA-approach and will be used below. Moreover, in the approach, modeling of precedents is controlled by a typical model P^M that integrates some subordinate models reflecting precedents with some important points of view.

Basic features of the QA-approach are bound with the logical model Pr^L of the precedent Pr that has the structure shown in Figure 1.

The model Pr^L fulfills the role of a kernel for Pr^M and applies for specifying other its components. First of all, the subordinated model Pr^L helps to formulate and structure the textual model Pr^T as a statement of a task for the corresponding precedent. Such help is additional for the work with the statement Pr^T which, in the general case, begins with its initial state $Pr^T(t_0)$ describing the new task. The help is additional for a question-answer analysis (QA-analysis) of the state $Pr^T(t_0)$ and other states $Pr^T(t)$ which step by step evolve the model P^T till its adequate description.

The model Pr^T generalizes results of the QA-analysis conducted on the base of QA-reasoning that can be effective used as a question-answer model (QA-model) of the task (Sosnin, 2014), or, that the same, the question-answer model Pr^{QA} of the corresponding precedent.

Figure 1. Logical model of the precedent

Name of precedent Pr_i:

c h o i c e	while [logical formulae (LF) for motives $M = \{M_k\}$] as [LF for aims $C = \{C_l\}$] if [LF for precondition $\mathbf{U'} = \{\mathbf{U'}_n\}$], then [plan of reaction (program) $\mathbf{r}_q$], end so [LF for post conditions $\mathbf{U''} = \{\mathbf{U''}_m\}$] -------------------------------- there are alternatives $\{Pr_j(r_p)\}$.

An effectiveness of understanding in the work with the model P^M can be increased if this model will include subordinated model Pr^G that presents block-and-line descriptions of the precedent task. In the QA-approach, the model Pr^G is specified so that it supports diagrammatic reasoning of designers in their interactions with the model Pr^M. Also, built diagrammatic reasoning can be aimed at a simulation of a necessary perception coordinated with QA-reasoning.

The model Pr^T includes the reaction r_q that should implement if the corresponding precedent Pr is chosen. In the QA-approach, any reaction r_q can be defined as a conceptually algorithmic model Pr^I in the form of the executable pseudo code program that presents a conceptual solution of the corresponding task. If the model Pr^I fulfills the role of the precedent prototype for its programming in another executed version Pr^E then, this version also includes the integrated model Pr^M the typical scheme of which is presented in Figure 2.

The model Pr^M is shown in the activity context when the designer solves the task Z as the reuse of the precedent Pr. Any of its subordinated models or a definite their group can be interpreted as a projection of the precedent model. One of the important projections is the QA-model of the task.

It should be noted, the model Pr^M and its subordinated models, Pr^L, Pr^{QA}, Pr^G, Pr^I have a common feature. They are created in a semantic memory that is oriented on the implementation of question-answer processes (Sosnin, 2014).

Question-Answer Processes in the Semantic Memory

The choice of the QA-approach based on interactions with the experience is explained that consciousness has a dialog nature the existence and process of which are being opened via the use of the natural language L^N. The natural intension of the language L^N is an intelligent processing of conditioned reflexes that underlie human life. Used in processing of conditioned reflexes, language L^N helps to create and remember items of human behavior in forms that are "prepared" for the reuse as precedents.

Figure 2. Typical scheme of the precedent model

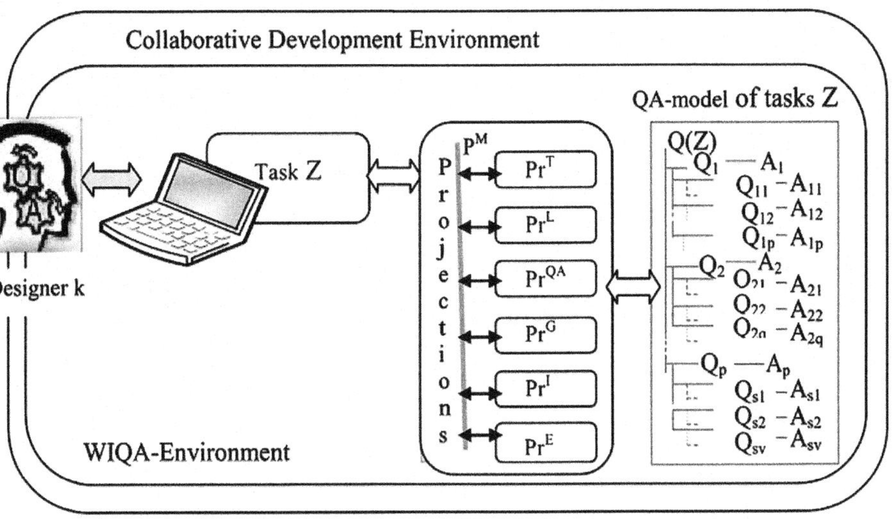

The described case supposes that intelligent processing applies for preparing the solutions of project tasks for their reuse as precedents the typical model of which is shown in Figure 2. Such models are created by designers who intelligently process the solved tasks with using the naturally occupational language. For conformity and compatibility of processing with dialog processes in the consciousness, designers use a reflection of their actions on the semantic memory that intended for the personal and collective work with question-answer processes (QA-processes).

Solving the task, any designer has the possibility for real-time registering and reusing of applied QA-reasoning as its protocol (QA-protocol) in the semantic memory. As told above, any action or interaction of the designer with registered QA-reasoning initiates the dialog process in his/her consciousness for useful planned or situational aims. Also, QA-reasoning is registered by the one of designers is accessible for other members of the team and initiates similar processes in the consciousness of each of them. Acting with QA-protocols in the WIQA-environment, designers have possibilities for various variants of their perceiving and processing. Moreover, useful variants can be programmed in the pseudo code language built into the WIQA. That was the reason to call the semantic memory of the toolkit WIQA as “QA-memory”.

By told above, the QA-memory can be interpreted as the memory intended for the controlled transfer of QA-processes from brain structures of designers to the collective use in designing. This intention demonstrates the scheme presented in Figure 3 where the task solved by one of the designers and mapped on the QA-memory is accessible for another designer.

Thus, QA-processes mapped to the QA-memory simulate processes in the consciousness of designers. Furthermore, simulated QA-processes participate in the feedback of their influence on processes in consciousness.

Figure 3. QA-memory as an intermediate environment

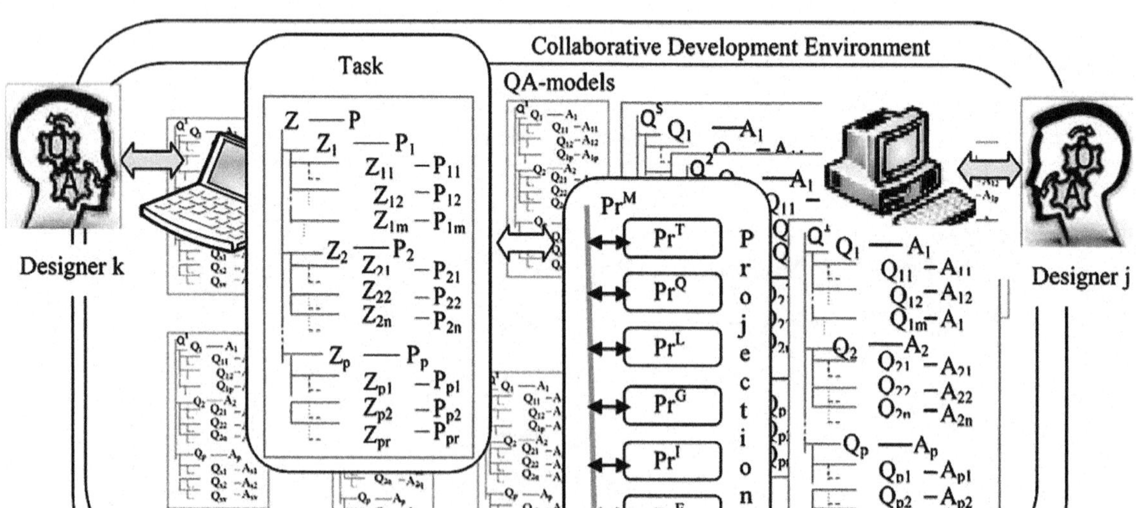

Feature of Conceptual Experimenting

In conditions of a project company, at the conceptual stage of designing, the semantic memory of WIQA is used by designers in following processes:

- registering of a set of created projects each of which is presented by the tree of its tasks in the real time;
- parallel implementing the set of workflows $\{W_m\}$ each of which includes subordinated workflows and/or project tasks $\{Z_n\}$;
- pseudo-parallel solving of the project tasks on each workplace in the corporate network;
- simulating the typical units of designers' behavior with using the typical precedent model.

These details of designers' activity are opened on the scheme presented in Figure 4 where relations with the collaborative development environment are shown only for one of the designers.

The scheme underlines that intelligent processing of the solved tasks leads to the creation of an Experience Base E^{Pr} that unites built models of precedents in their library for the collective. It should be noted that any built precedent evolves the accessible experience by inclusion additional components $\{\Delta E_k\}$ to it.In the general case, intelligent processing can be the source of following kinds of additional components:

- A new model of the precedent;
- A new projection of the existing precedent model;
- A modified projection of the existing precedent model;

Figure 4. The place of the semantic memory in processes of designing

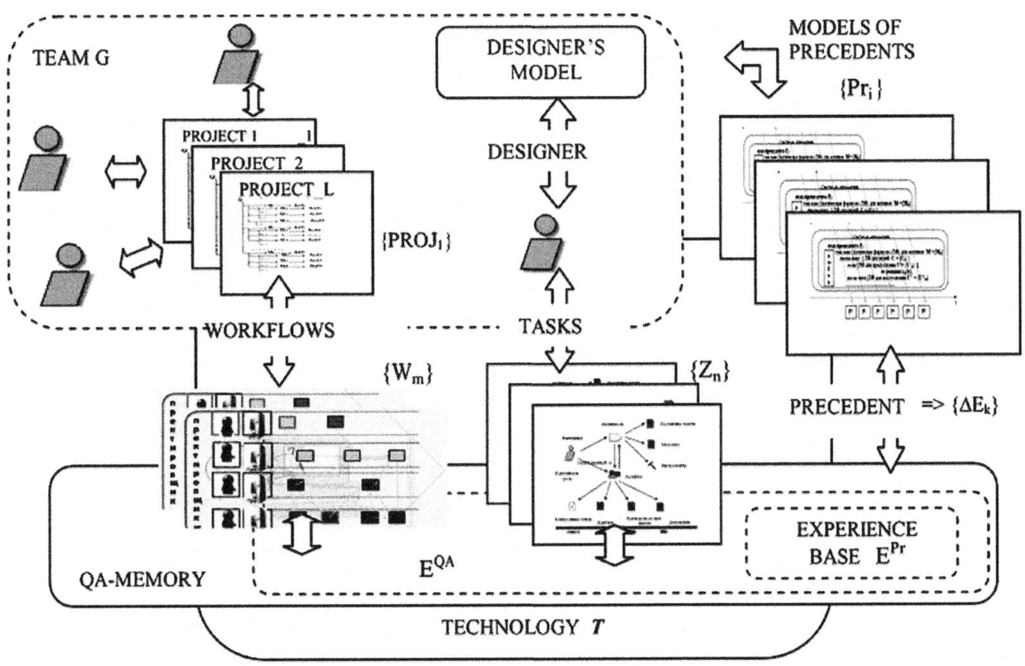

- A modified version of the existing model of precedent;
- A new concept that evolves an ontology of the Experience Base;
- A modified concept of the ontology

In the QA-approach, conceptual experimenting is the basic way for discovering these additional components and their substantiated inclusion into the Experience Base. The offered version of conceptual experimenting has following features:

1. Conceptual experiments are conducted in the conceptual space (C-space) of the QA-memory that supports mapping on it the operational space of designing a system of SISs.
2. The C-space combines models of components of the operational space, including models of designers and units of the used experience. Such C-space is interpreted as the space of designers' activity that is aimed at the conceptual development of SISs. Developing of SISs, designers create the C-space in the QA-memory, and they interact with this space for achieving the useful purposes.
3. In the C-space, its processes are managed by interactions with the experience the typical units of which are models of precedents that are defined above.
4. In the In the C-space, its processes are managed by interactions with the experience the typical units of which are models of precedents that are defined above., the models of precedents obviously or implicitly reflect the regularities of this space or, by other words, "laws" of this space. These regularities reflect regularities of the operational space (regularities of the subject area of SISs, norms of the used technology, requirements and restrictions managed by developing of SISs).
5. A set of useful purposes that can be achieved by conceptual experimenting includes:
    a. Discovering the regularities of the C-space that exists in QA-memory;
    b. Evolving the C-space by an inclusion of the additional regularities into a system of the space regularities;
    c. Reusing the definite regularity or their composition in designing;
    d. Adjusting the definite regularity or their composition in designing;
    e. Understanding the definite regularity or their composition in designing;
    f. Learning the definite regularity or their composition in designing.

It is necessary to remember that processes in the C-space activate processes in the consciousness of designers who participate in these processes. All these processes are being implemented on the base of interactions with the accessible experience.

Conceptual experiments are automated thought experiments that are conducted in the C-space and implicitly supported by processes in the consciousness of designers who participate in experimenting.

Intellectual Processor

Indicated interactions with the accessible experience are a kind of the specialized designer activity which can be implemented as the role "Intellectual processor" (or, shortly, I-processor) that plays by any designer in parallel with other appointed roles.

Any modern technology used for designing the SIS includes modeling the work of designers with the help of roles. For example, the current version of Rational Unified Process (RUP) supports activities of designers playing approximately 40 roles. RUP is a "heavy" technology therefore for small teams the quantity and specifications of used roles are decreased and simplified (Borges,2012).

In existing technologies, behavioral actions provided interactions with experience are being fulfilled by designers without its explicit specifying and special instrumental support. The role of I-processor has been defined for its use by designers in WIQA-environment (Sosnin, 2014) but the role can be materialized by the other way.

The essence of I-processor is being defined by features enumerated below:

1. First of all I-processor is a role being played by any designer solving the appointed tasks in conceptual designing of SIS.
2. In any case, the role is a special version of a designer behavior that satisfies the definite set of rules. Role specifications depend on appointed tasks and tools supporting their preliminary solving and reuses of solutions.
3. I-processors are intended for experimenting with tasks the solving of which are problematical without explicit real-time access to the personal experience and/or collective experience and/or models of useful experience.
4. The role of I-processor is oriented on the behavior of a scientist who prepares and conducts an experiment, registering its results in the understandable and checkable form. Experimenting, the designer investigates units of the behavior that have led to solutions of corresponding project tasks in conceptual designing.
5. In its activity, any I-processor interprets investigated tasks as precedents and interacts with experience units as with models of precedents.
6. The real-time work of any I-processor is being accompanied by QA-reasoning and its models being registered in QA-memory of the toolkit WIQA.
7. To re-used tasks, experimenting, the designer creates the model of the corresponding precedent, which is loaded into the Experience Base and included in the Personified Model of the Designer (PMD).
8. The kernel of the PMD is a model of the corresponding personal experience that consists of models for mastered precedents.
9. Personified models are interactive dynamic objects that are stored in the one of sections of the Experience Base.
10. Any PMD is accessible for any designer who executes the role of the I-processor in the team of designers.

Thus, in the general case, the activity of I-processor can be understood as the experimental activity of the designer who creates the model of the task for conceptual experimenting with the prototype of its solution.

Related Works

The nearest group of related works concerns the subject area "Conceptual Development and Experimentation (CD&E)."The specificity of this subject area is presented in the report (MCM, 2010) where the role and place of conceptual experimentation in military applications are indicated in detail. The publication(Jaros, 2011) demonstrates some specialized solutions that are focused on a behavioral side of experimenting. In this subject area, the greater part of published works concern thee strategic level of the experimental activity.In published works, analogies with thought experiments and reflections on the semantic memory are not used.

The idea of the designer model as an I-processor is derived from publications (Karray, 2008) in which the Model Human Processor (MH-processor) is described as an engineering model of human performance in solving different tasks in a real-time regime. It is necessary to emphasize that I-processor is similar to MH-processor and includes the cognitive processor. However, the existence and work of this component of I-processor are derived from another technique that is based on reasoning of the question-answer type.

The EPIC version of MH-processor uses programs that are written in the specialized command language Keystroke-Level Model (KLM). A set of basic KLM actions includes the following operators: K – key press and release (keyboard), P – point the mouse to an object on screen, B – button press or release (mouse), H – hand from keyboard to mouse, or vice versa, and other commands. Operators of the KLM-language and their values are used to estimate temporal characteristics of human interactions for alternative schemes of interfaces. KLM-programs do not correspond exactly to the used reasoning, and, therefore, they do not reflect interactions with the accessible experience. It is necessary to note that processors of both indicated types are oriented toward experimenting with units of human behavior.

The version of the offered experimentation is coordinated with basic principles of the SEMAT Kernel, which is described in (Jacobson, 2012). The version is oriented towards ways-of-working that focus on the real-time activity of designers. The process is what the team does. Such processes can be formed dynamically from appropriate practices by current situations. The informational cards and queue mechanisms are being used for managing of ways-of-working (Jacobson, 2007)

It is necessary to note that this version applies units of task types for structuring the activity of the designers because solutions of tasks facilitate the enrichment of the accessible experience by scientific experimentation. By experimenting, the scientists solve specific tasks creatively.

In the offered approach, the scientific viewpoint correlates with two faces of the software engineering described in (Cares, 2006), where functional paradigms and scientific paradigms are discussed. In the context of this paper, the approach means are oriented towards scientific paradigms used by software engineers.

Therefore, the important group of related studies includes publications that present empirical viewpoints on software engineering. In this group, we note the following works (Jeffery, 2002) and (Sjoberg, 2007), which present the domain of empirical software engineering; papers (Basili, 2001) and (Southekal, 2011)], which define the Goal-Question-Metrics method (GQM-method) and Experience factory, which includes the Experience Base. All of the indicated studies were taken into account in the offered version of scientifically experimental ways-of-working.

One more group of related publications concerns the use of question-answering in computerized mediums, for example, papers (Xu, 2005) and (Webber,2010). In this group, the closest research presents experience-based methodology BORE (Henninger, 2003), in which question-answering is applied as well, but for the other aims, this methodology does not support programming of the creative designer activity.

A set of typical kinds of ontologies, according to their level of dependence on a particular task or point of view, includes the top-level ontologies, domain ontologies, tasks ontologies and applied ontologies. All these type of ontologies is defined in (Guarino, 2009) and (Guarino, 2008) as means that are used in different systems.

For the occupational activity, the adequate type of ontologies is applied type, which must be expanded using the other types of ontologies. By the publication (Bullinger, 2005), the theory and practice of applied ontologies will require many more experiences yet to be made. It is necessary to notice that the project ontology as a subtype of applied ontologies is fundamentally essential for the personal ontology. Project ontologies in the greater measure are aimed at the solving the complicated tasks and their families with using the experience base (Eden, 2007). The specificity of project ontologies is indicated in some publications. In the technical report (Ikeda, 1998) the main attention is concentrated on people, process and product and collaborative understanding in interactions. Investigating the possibility of the ontology-based project management is discussed in the paper (Iqbal, 2013).

It should be noted, all papers indicated in this section were used as sources of requirements in developing the set of instrumental means provided conceptual experimenting in automated designing. Any of these papers concerns only a part of the QA-approach to conceptual experimenting.

IMPLEMENTATION OF QA-APPROACH

Reflection on Question-Answer Memory

As told above, the QA-approach supposes that components of the operational space S are reflected on the QA-memory. The generalized scheme of the reflection is shown in Figure 5 where any component and other elements are marked by their name and symbolic designations.

Figure 5. Operational space

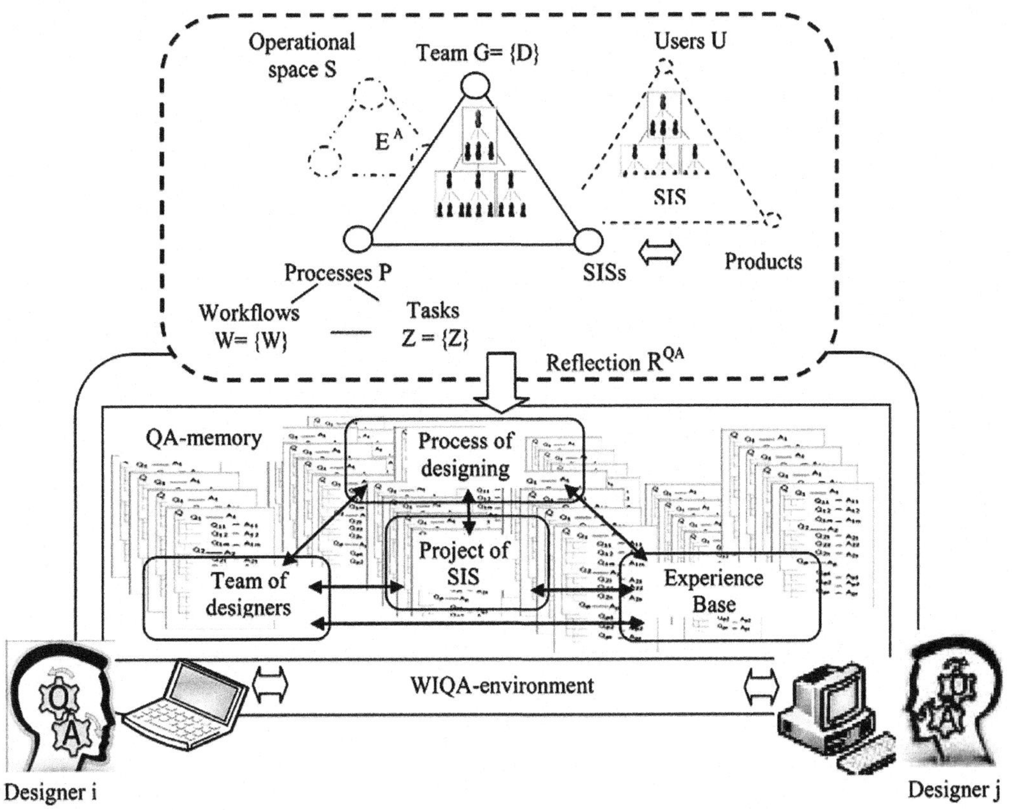

On the scheme, the reflection $\mathbf{R}^{QA}$ shows that all what involving in designing of the family of SIS is found their expression as models in the QA-memory. It can be written by the following expression:

$$WW(S, P, G, E^A, SIS, t) \xrightarrow{R^{QA}} S^{QA}(t+\Delta t) \cup P^{QA}(t+\Delta t) \cup G^{QA}(t+\Delta t) \cup E^{QA}(t+\Delta t) \cup SIS^{QA}(t+\Delta t), \quad (1)$$

where ***WW*** is a Way-of-Working used by designers, $S(t+\Delta t)$ presents relations among components of the operational space ***S*** and other symbols correspond to designations of essences in Figure 1. Any expression $X^{QA}(t+\Delta t)$ presents the state of the model for the corresponding essence after the next step of the reflection. It should be noted that any component of S is a composition the units of which also are reflected on the QA-memory. Below, specifications of subordinated units will be described by rules of a generative grammar GR^{QA}.

The choice of grammatical specifications is caused by the specificity of the reflection $\mathbf{R}^{QA}$ that implements as a step by step generating of the question-answer objects (QA-objects) and their uploading in the QA-memory. Such work is fulfilled by designers who use a part of means included into the toolkit WIQA. After uploading, QA-objects can be interactively accessible for designers or can be processed as program objects.

Specificity of QA-Memory

The QA-memory is specified and materialized for storing the conceptual description of the operational space ***S*** in memory cells in the understandable form. For this reason, cells are oriented on registering the textual units in the form of communicative constructions any of which can be divided on "theme" (predictable starting point) and "rheme" (new information) that should receive additional meaning in answering the corresponding question. Such orientation has led to the solution of using two types of corresponding cells. The cell of the first type is intended for registering the simple question (Q) while the cell of the second type registers the corresponding answer (A).

Cells of both types are specified equally because any question includes a part of awaited answer, and any answer includes components of the corresponding question. Therefore, a pair of corresponding cells with a question and answer presents the description unit in details.

QA-memory with its cells is intended for registering the conceptual content of reflected units with taking into account the semantics of their textual descriptions. The necessary semantics is fixed in basic attributes of the cell and in additional attributes that can be added by the designer if it will be useful for a simple object stored in the cell. The potential structure of a simple object is presented in Figure 6.

A set of basic attributes of the cell helps to register for the stored object its unique identifier (address in QA-memory), type, description, name of the creator, time of storing or last modification, name of the parent object, quantity of "children" and a number of other characteristics. These attributes with their values and the definite subsystem of operations (commands) support interactions of designers with visualized object stored in the corresponding cell. Additional attributes are attached to the definite simple object for enriching its computerized potential, for example, to enrich the semantics of the object representation in QA-memory.

The designer has the possibility to attach to any cell some useful files, pictures and references to informational sources for enriching the expressiveness of used cells.

Figure 6. Specification of the interactive object

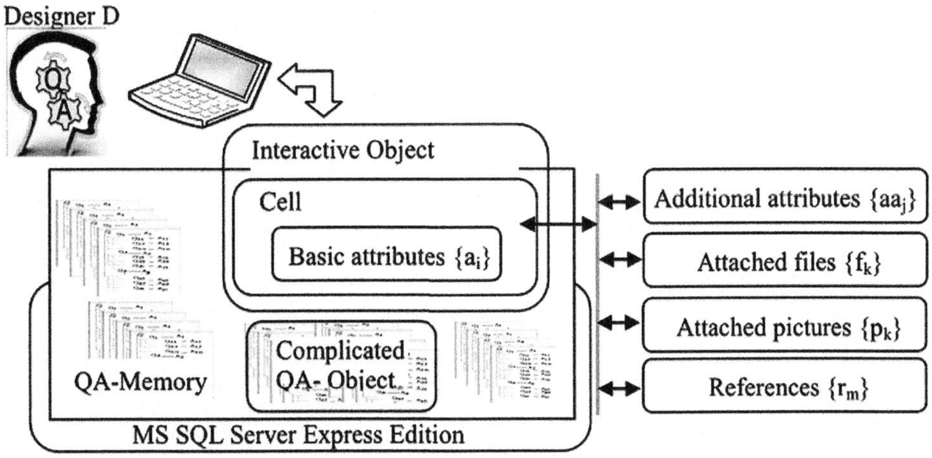

In the formal grammar GR^{QA}, generating of QA-objects implements by following rules (written with using the extended BNF-notations):

QA-Memory = {QA-object};

QA-object = Question, "←", Answer;

Question = Q | (Q,"↓",{Q});

Answer = A | (A, "↓", {A}); (2)

Q = ({a}, {[aa]}, {[f]}, {[p]} {[r]});

A = ({a}, {[aa]}, {[f]}) {[p]} {[r]};

a = (address, type, description, time, the others),

where "Q" and "A" are typically visualized objects stored in cells of QA-memory, symbol "↓" designates an operation of "subordinating."

Conceptually Algorithmic Programming

The QA-memory has an algorithmic potential because cells of this memory can be adjusted on emulating of any type of data and coding of any operator an interpretation of which will implement the necessary processing of corresponding data. This potential has been investigated and tested in the work with some practical tasks. The mastered experience has allowed creating the original pseudo-code language programming L^{WIQA}, built-in the toolkit WIQA. In its current state, this language has the structure presented in Figure 7.

Figure 7. The structure of the pseudo-code language

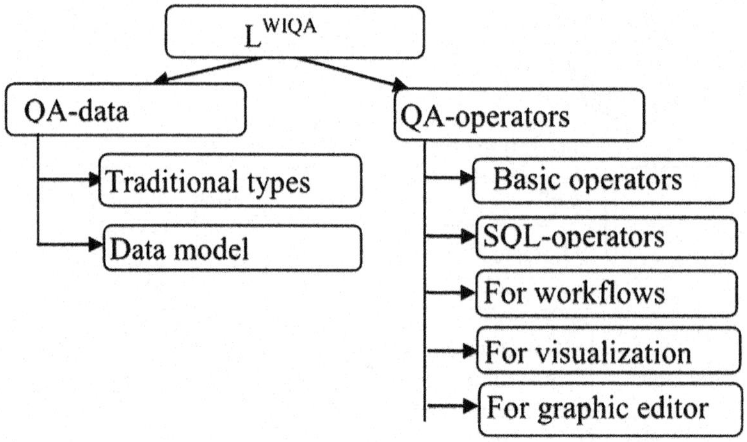

The main feature of the language L^{WIQA} is caused by the controlled inclusion of the necessary semantic as in memory cells so in objects stored in these cells. Such stored objects inherit the semantic value of the corresponding cell. Also, emulated data and operators allow interpreting them with using of questions and answers (Sosnin, 2014). This interpretation opens additional possibilities for processing as data so operators. Real-time switching on their semantic attributes helps in interactions with the natural experience and its models. Told about features was a reason to qualify described version of pseudo-code programming as the version that is distinguished from known versions of pseudo-code programming. Therefore, this version was qualified as conceptually algorithmic programming of the question-answer type or, shortly, as QA-programming. To distinguish them from other types of programs, programs encoded in the language L^{WIQA} have been called QA-programs.

Called features have been taking into account in the development of an environment of pseudo-code programming in the language L^{WIQA}. This environment includes:

1. A step-by-step interpreter that executes any operator the code of which is written in the cell attribute "description" as at the punch card. At the same time, in the execution of this operator, the interpreter can process not only the code expression but other attributes and information indicated in Figure 6. By other words, the operator can be adjusted in the condition of its execution. Moreover, the adjustment can be implemented in the real-time of the operator execution. That is very useful for conceptual experimenting.
2. Means of adjusting the step-by- step interpreter on executing the chosen group of operators or all operators of the definite QA-program.
3. Means for the transformation of the diagrammatic scheme in the corresponding QA-program the execution of which draws this scheme in the field of the specialized graphical editor.

All described features of the work with operators and QA-programs were used in conceptual experimenting/

Reflection of Processes

Details of the operational space that are presented in Figure 2 not only indicate reflections of projects on the QA-memory, but it also demonstrates structures that should find their presentations in such memory. For example, models of the project and its components in QA-memory correspond to the following rules of GR^{QA}:

PROJ = ZP;

ZP = (Z, "↓", {Workflows };

Workflows = ZW | (Workflows, "↓",{ZW});

ZW = {Task};

Task = Z | (Task, "↓", {Z});

Z = QA-model | (Z, "↓", {QA-model}); (3)

QA-model = {QA} | (QA-model, "↓",{QA});

QA = (Question, Answer),

Question= Q | (Question, "↓",{Q});

Answer = A | (Answer, "↓",{A}),

where "Z", "Q" and "A" are typically visualized objects stored in cells of QA-memory, symbol "↓" designates an operation of "subordinating." These objects have the richest attribute descriptions (Sosnin, 2014). For example, a set of attributes includes the textual description, index label, type of object in the QA-memory, the name of a responsible person and the time of last modifying. Any designer can add necessary attributes to the chosen object by the use of the special plug-ins "Additional attributes" (object-relational mapping to C#-classes).

Reflections of Workforces

As told above, the success of designing the family of SISs depends from the occupational maturity of used human resources in the essential measure. Therefore, modeling the team of designers as a whole and members should play an important role in the project activity. In the toolkit WIQA, modeling of the team is supported by the specialized plug-ins "Organizational structure" (Sosnin, 2014).

In the proposed approach, a job description (JD) is used as a kernel of the personified model of the designer. "Job descriptions are written statements that describe the duties, responsibilities, most important contributions and outcomes needed from a position, required qualifications of candidates, and reporting relationship and coworkers of a particular job" (http://humanresources.about.com/od/ job descriptions/a/develop_job_des.htm). In the described case, JD-components have the constructive references on models of precedents stored in the Experience Base. The framework of the PMD with such a kernel is presented in Figure 8.

The scheme demonstrates the PMD in the context of its forming and creating. In grammar G^{QA}, the PMD is described by the following rules:

PMD = (DName, JD, {List of Features});

JD = {Section};

Section = Text -{[Role]} -{[K]}- {[O]} - {[L]}; (4)

List of Features = (Type, {F}, {[AF]});

F = ZP | G | Z | O | L | V' | FPR';

AF = Additional Feature,

where a number of references indicate appointed ***Roles***, a number of ***K*** reflect mastered competencies, units of ***O***-type present the results of estimating the designer actions, units of ***L***-type register the results

Figure 8. Framework of the PMD

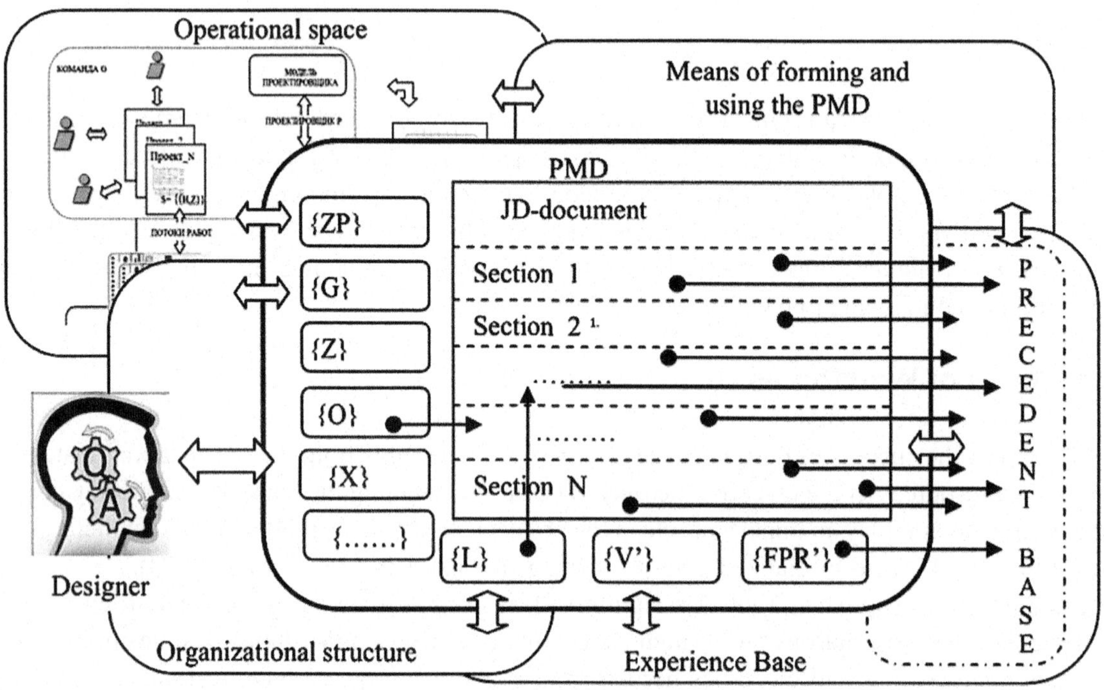

of experiential learning, lists of ***FPR'*** indicates the precedents' models built by the designer and units of ***V'***-type are the references on the other values that are created by one. Lists *{X}* and *{....}* indicate that PMD is opened for the extension.

Reflection of Experience

In the proposed approach, models of precedents are used for modeling of assets applied by the team in conceptual designing the family of SISs. Any asset is understood from the viewpoint of its application, and; therefore, it binds with the corresponding task of ***ZPr***-type.

Assets are divided on classes of different types the specifications of which are described by a set of attributes including "the name of type", "description", "kind of precedent version VFPr used for its storing in QA-memory" and the others. Some of the types specify the models of precedents for human resources, for example, they should include the type that specifies the personified models of designers.

Potential of L^{WIQA} is sufficient for QA-modeling and QA-programming the assets of following kinds: previous projects, valuable project solutions, prototypes, documents, interface samples, schemes of reports, standards, frameworks, guides, patterns, samples of different types, schemes of modeling, structure of the software, packages of the source code, tools, platforms, infrastructure and other valuable units.

Models of assets are registered in the catalog of Experience Base and allocated in the specialized area of QA-memory (Figure 9).

It is necessary to notice that only one part of assets is placed in Precedent Base. The greater part of assets are stored in corresponding libraries where they are presented in forms of precedent projections.

Figure 9. Assets in experience base

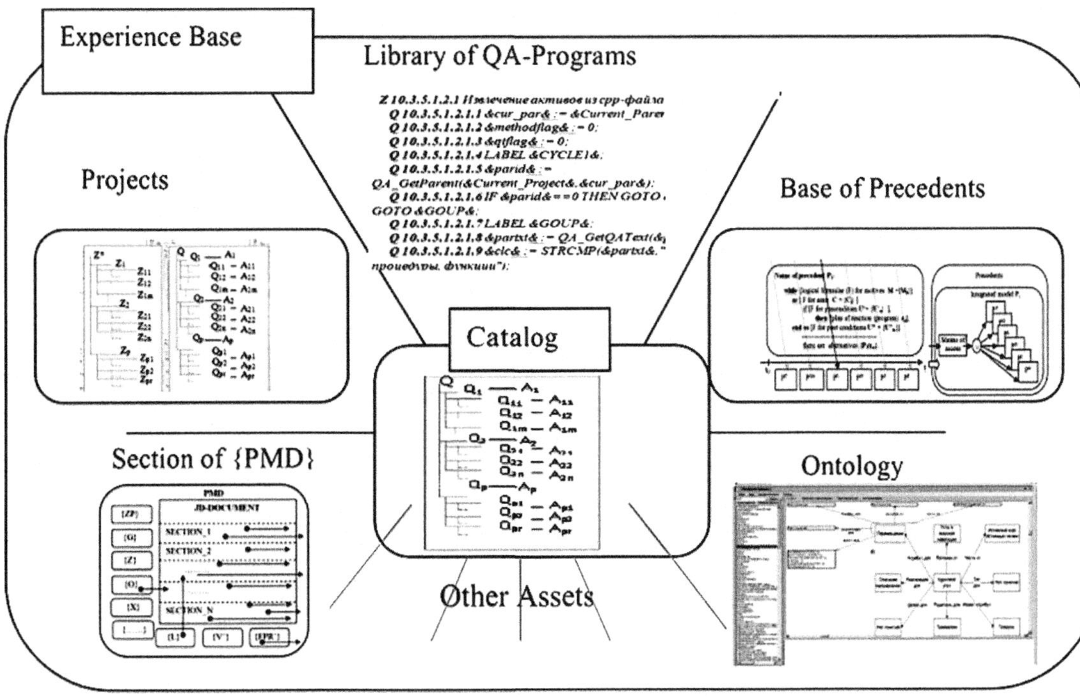

As told above, components stored in QA-memory can be bound with the attached files that are placed in corresponding libraries too. One of such libraries includes program written in C# that can be used in QA-programming.

The use of a large set of different assets (with orientation on precedents) leads to questioning about means of their systematization. In the describe case, this function is fulfilled by the project ontology, concepts of which are also stored in QA-memory.

Ontology

Means of the toolkit WIQA support the use of the integrated ontology, its project ontologies and their components till concepts for the following basic aims:

- Systematization of the Experience Base as a whole and its "projections" on any designed project that is implemented in the project organization;
- Access to the necessary assets the models of which are stored in the Experience Base;
- Extraction of requirements, their specifications and evolution in processes of designing;
- Creation of the controlled vocabulary and its applying in the frame of the occupationally natural language used in the project organization;
- Conceptual experimentation with the project solutions created by designers in the real time.

These aims are achieved in conditions where the ontology is created in the QA-memory with using the normative scheme of a concept presented in Figure 10.

Ontological relations specify a net of ontology units that are bound with the chosen concept. For the systematization of concepts, the designer has the opportunity for appointing relations such as "is-a", "part-of", "attributive", "causative", "associative" (temporal, spatial, synonymous, opposite, following) and relations of pragmatic types. The designer has the opportunity for extending the list of relations.

For the use of the controlled dictionary in implemented projects, the designer should indicate the type of materialization for any used concept. References to materializations fix relations of concepts

Figure 10. Concept reflection

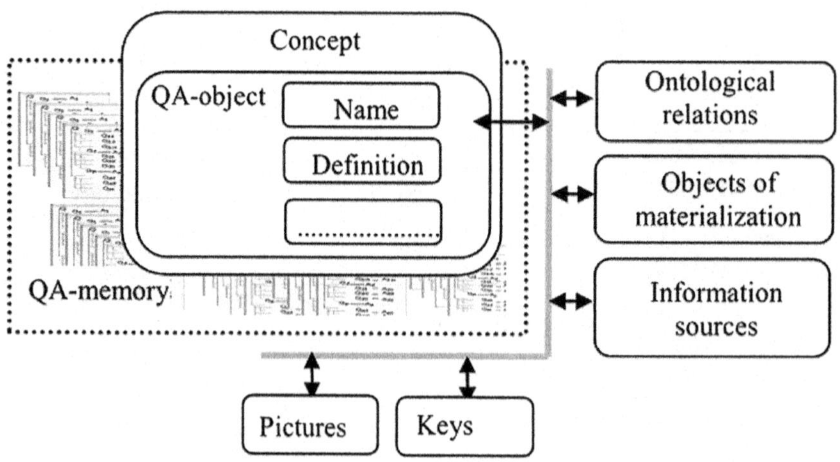

with the space of occupational interest of the designer. It allows preventing the use of words from uncontrolled vocabulary in solutions of project tasks. Furthermore, it helps to check the compatibility of used concepts in reasoning and texts.

Evolution of the QA-Approach

An adjustment of the QA-approach for conceptual experimenting is the latest step in its development. All passed steps of the development figuratively shown in Figure 11.

Steps are chosen with the orientation on the success factors registered in statistics of Standish Group (Chaos, 2013). Results of steps evolve the potential of the toolkit WIQA.

CONCEPTUAL EXPERIMENTING

Designing of Templates for Airplane Parts

The greater part of described above is used means of supporting the experimental work should help the designers create solutions for investigating tasks and opening the possibility for their confirmatory reuse in the designers' team. Any conceptual solution of the investigated task orients on its creating and reusable executing of the model for corresponding precedent by the designer or a group of designers who fulfill the role of the I-processor. Therefore, first of all, any conceptual experiment has a behavioral nature. Any precedent model helps in simulating the definite unit of the designer behavior. In the QA-approach, such model is understood as an "experimental setup" with instructions for this use that confirms the "existence" of the definite regularity or a group of cause-and-effect regularities in the operational space of designing.

Figure 11. Directions of the research

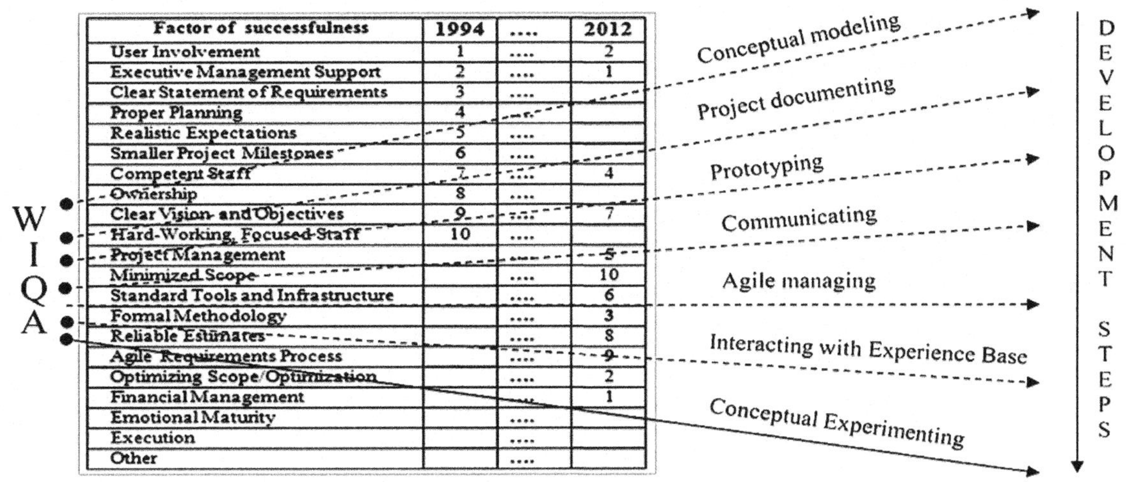

It should be recalled, once again, that the approach is aimed at the conceptual stage of designing because a system of SIS regularities should be formed, checked and confirmed during the design as early as possible.

As it is known, any experiment should be prepared, conducted and documented in the understandable form, so that is suitable for the reuse. In the described case, such form corresponds to the normative model presented in Figure 2. The form manages the actions the designer or designers who intelligently process the solution of the corresponding project task. This work is fulfilled in the semantic memory of the toolkit WIQA. Features of conceptual experimenting are described above.

The practical side of conceptual experimenting will be presented below on an example of designing the configured templates of aviations parts. This example will demonstrate the use of conceptual experiments as in designing the applied SIS (CAD-system for designing the template) so in automated designing the templates and their applications.

In the aircraft industry, for manufacturing the parts of a fuselage, wings and aileron, including details of their covering, technological templates are widely used. It is caused by such specificities of aviation parts as their large sizes, complicated configurations, low mechanical rigidity, high requirements for machining and linking in assembling. Also, for linking the parts included in each plane section of an aircraft construction, a system of rigid equipment is necessary for fixing the contours of the internal parts that make up this section. For example, large-largest parts have to be linked to the sheet metal control-contour templates.

Any template for the part is a kind of its machining attachment that supports definite technological operations, for example, manufacturing of the corresponding part, its checking on specifications and linking with another part in the frame of the definite plane section. For this reason, designing the template should take into account the required technological functions.

In the general case, the template is a plane figure that geometrically consists of the definite configuration of the corresponding part and additional elements that provide the implementation of technological functions. Any template is produced from a steel sheet with using of laser cutting.

For manufacturing the actual airplane, enormous number of various templates must be designed, produced and applied. This work is usually implemented by using the technology that is applied to manufacturing of the aviation parts indicated above. But the creation of configured templates is more complicated then the creation of the corresponding parts. It is necessary to note that manufacturing of the configured templates is based on computer numerical control programming (CNC-programming) oriented on the use of appropriate CAD-system and laser cutting machine.

The life cycle of template manufacturing includes the stages that require the creative-intensive activity of workers that participate in designing of the templates. It is caused by the need to take account of future ways to move the laser beam on all lines that represent a template for its steel billet. The projected CNC-program must specify and control the laser beam along the way.

There is some specialists (with different professional experience) who obviously or indirectly participate in designing indicated CNC-programs. Only some of them are programmers. In this case, conceptually algorithmic models of CNC-processes can help in reaching a common understanding on the projected CNC-processes and reduce the complexity of CNC programming. In the use of the toolkit WIQA, such collaborative work is achieved in the system presented on Figure 12.

The built ontology consists of the following sections: typical templates as concepts, a visualized classification of templates, templates in manufacturing of aviation parts, templates in a production control of parts, templates as models of precedents.

Figure 12. Life cycle of designing the template

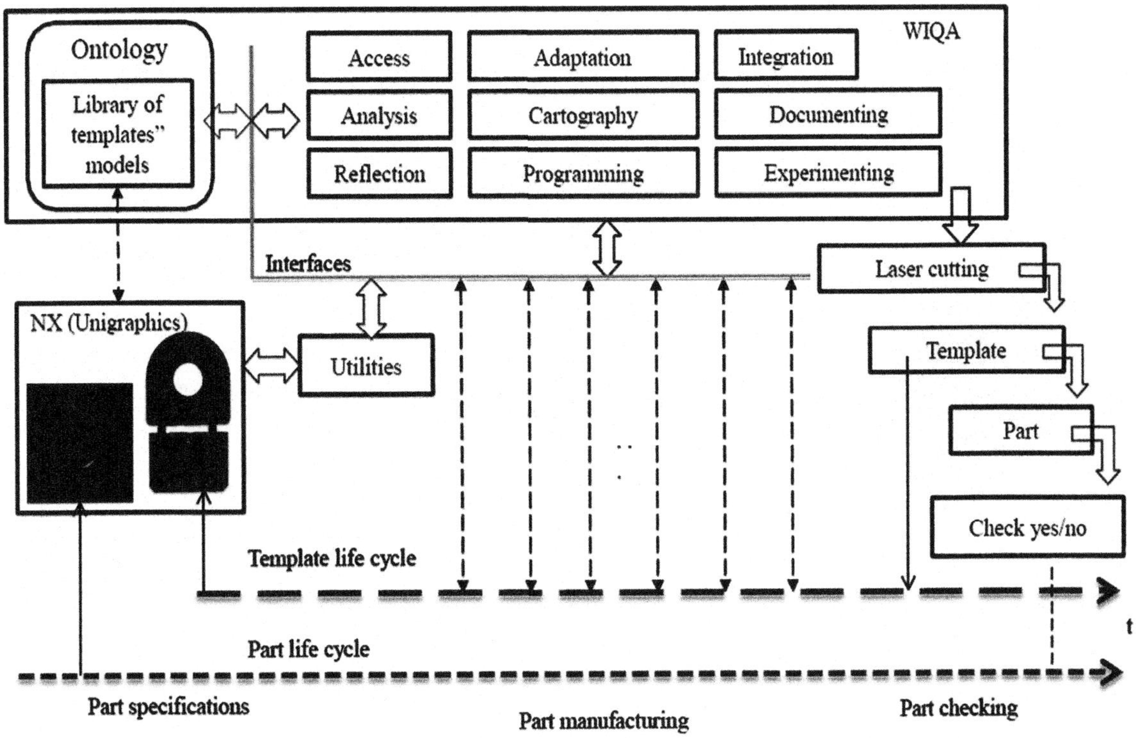

At the beginning of designing the necessary template, the designer uses the ontological (conceptual) support implements the following actions:

1. Choosing the useful template from the library (task "Access").
2. Analyzing the similarity and differences with the required template (task "Analysis").
3. Fitting the similar template to the requirements (task "Adaptation").

If the appropriate template is absent in the library, the designer should create its model in the NC-environment. There are some utilities that supports the transfer of models from NC to WIQA and back.

Models of Precedents for Template

The use of the precedent model helps to accumulate reuse results in designing the definite template. Such way is applied for building the model templates that are loaded into the library, shown in Figure 13.

In this figure, the example of the template model is presented partially only for demonstrating the forms of components P^T, P^G, P^I and P^E. The normative content of the logical component P^L must correspond to the structure opened above at Figure 1. The component P^{QA} defines the description of the solution way, and it includes the results of normative documenting (by attached files).

The developed extension (with other means of the language LWIQA) supports QA-programming of a set of useful tasks that arise in designing the templates. One of the important subsets of such tasks is

Figure 13. The structure and content of the template model

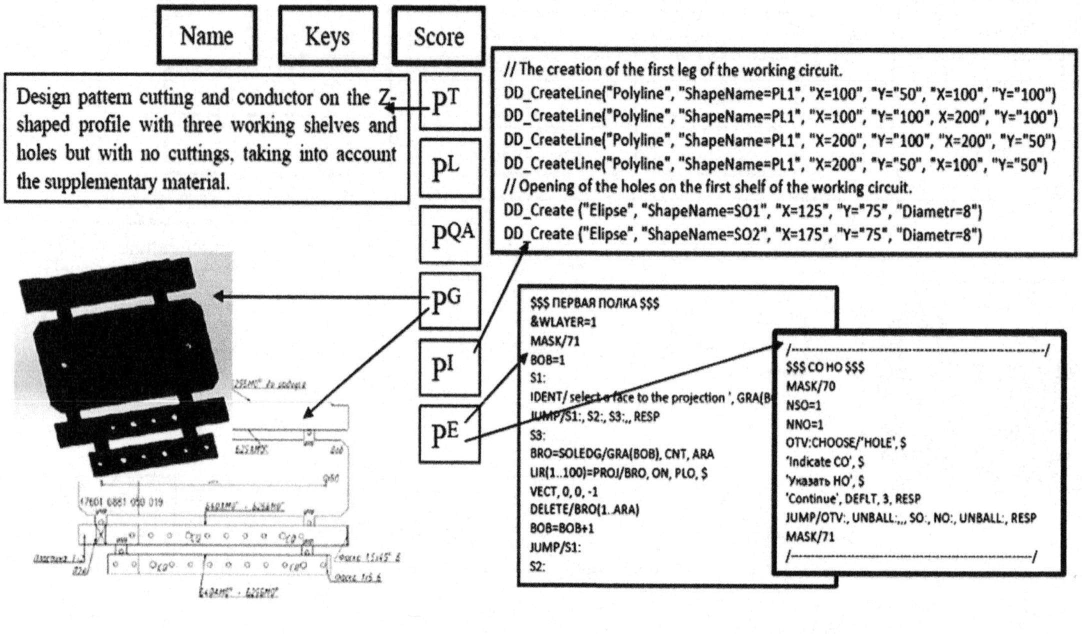

concerned with experiments aimed at the search of variants for technological components built-in the template. The other important subset of experiments can help in creating the ways of moving the laser beam on template contours or their fragments.

It is necessary to note that tasks of both subsets are oriented on experimenting with template contours that should be cut out by the laser beam. In a graphical editor of the toolkit WIQA, any of such contour or its fragment is described by the pseudo-code operators of the language LCNC. Operators are executed (step by step) with the use of the specialized interpreter that builds the trajectory of moving the laser beam. This process demonstrates the laser cutting in simulated time.

By other words, pseudo-code descriptions of controlled trajectory or its fragments are interpreted in actions aimed at their drawing in the plane of the graphical editor. For this reason, QA-programming of the simulation for the laser cutting relate to declarative conceptual programming.

It is necessary to note that the source code of any pseudo code program can be translated to the G-code executed on the CNC-machine. Consequently, in the offered approach, the CNC-processing of the template has two faces. The first face is concerned with QA-program for the simulation of laser cutting while the second face directs on G-program controlled the laser cutting in real time. This possibility is presented in Figure 14.

In the general case, the template has a complicated configuration. To reduce the complexity, the proffered approach uses a cartographic model of the geometry of the template. The cartographic model assumes that the template presentation is composed of a set of layers, each of which combines data collected from a particular feature. In designing, the order of the work with layers depends on the priority of design tasks associated with layers. It is necessary to note that the layer structure of the created template includes subordinated structural descriptions that correspond to template applications. Such applications lead to technological regularities that manage the automated manufacture of corresponding parts. In

Figures 14. Two faces of pseudo code programs

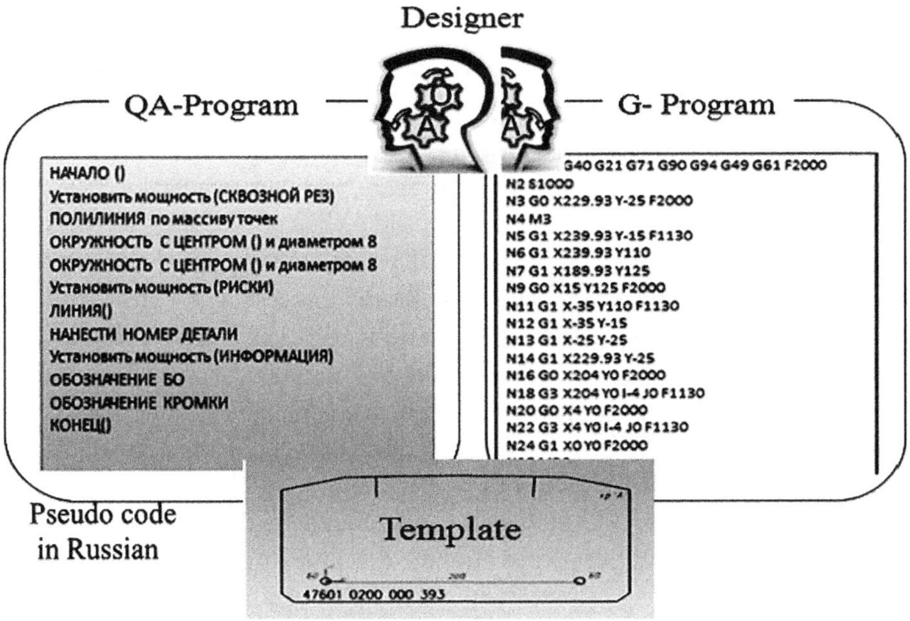

the described case, they open additional direction for creating the models of precedents and conceptual experimenting with subordinated structures of the cartographic models.

The use of described means opens the possibilities for experimental estimations of CNC-processes in questionable points of their life cycle. The set of such points is concerned, for example, with untypical characteristics of the controlled trajectory. Some such points are concerned with the use of templates for different technological applications.

The specificity of the described approach is defined by following features: the use of reflections of CNC-process on the semantic memory that opens the possibility of pseudo code programming actions of the designer and moving the laser beam on the controlled trajectory; the use of models of precedents for registering the results of designing for the future reuse; the use of the ontological maintenance of real-time actions in designing.

Offered versions of conceptual experiments are implemented in visual forms with using the specialized graphical editor that supports the conversions of the template geometry to the pseudo code program of its simulation in experimental aims. All achieved positive effects facilitate increasing the quality in designing of the CNC-programs.

FUTURE RESEARCH DIRECTIONS

Nowadays the main attention in evolving of the suggested means is focused on the agents-oriented solutions. Some such solutions are found, tested and embedded to WIQA in the form of program agents. Nearest plans include investigations aimed at the use of intellectual agents in modeling of precedents.

CONCLUSION

The approach described in this paper suggests a system of methods that support conceptual experimenting in automated designing objects and systems. In this kind of actions, designers use analogies with the work of scientists when conducting experiments. Moreover, they simulate such behavior with the help of pseudo-code programs that describe the plans for experimentation. Thus, designers investigate the programmed plans of the experiments that they prepare, and they conduct and describe the results in understandable and checkable forms, for later reuse.

In the conceptual experiments, the investigated behavioral units are modeled as precedents. Such a form for human activity is natural because the intellectual processing of precedents comprises the base of the human experience. In experimentation, the designers evolve accessible experience by using real-time interactions with its current state. This feature has found normative specifications in the role of "intellectual processor", which is played by designers and is being supported by the toolkit WIQA. In the collaborative way-of-working, this aspect can be used also to any other aspect of the technology that is applied in the conceptual design process.

This toolkit opens the possibility for the separate execution of any operator by the designer playing the role of the intellectual processor. Before and after the execution of any operator of any QA-program, the designer can check or investigate its preconditions and post-conditions. Moreover, the investigated operator can be changed and evolved syntactically as well as semantically, for example, with the help of additional attributes.

The possibility of experimenting is supported by the special library of QA-programs destined for cognitive task analysis, problem-solving and decision-making included in the named toolkit. Suggested means are used in one project organization in creating a family of SISs. They also have been used in designing the configured templates for airplane parts.

REFERENCES

Basili, V. R., Lindvall, M., & Costa, P. (2001). Implementing the experience factory concepts as a set of experience bases. In *Proceedings of the 2001 International Conference on Software Engineering & Knowledge Engineering.*

Borges, P., Machado, R. J., & Ribeiro, P. (2012).Mapping RUP Roles to Small Software Development Teams. In *Proceedings of International Conference on Software and System Process,* (pp. 190-199). doi:10.1007/978-3-642-27213-4_5

Bullinger, H.-J., Warschat, J., Schumacher, O., Slama, A., & Ohlhausen, P. (2005). Ontology-Based Project Management for Acceleration of Innovation Project. *Lecture Notes in Computer Science, 3379,* 280–288. doi:10.1007/978-3-540-31842-2_28

Cares, C., Franch, X., & Mayol, E. (2006). Perspectives about paradigms in software engineering. In *Proc. 2nd International workshop on Philisophiocal Foundations on Information Systems Engineering.*

Chandrasekaran, B., Josephson, J. R., & Benjamins, V. R. (1998). Ontology of tasks and methods. In *Proceedings of the Workshop on Applications of Ontologies and Problem-Solving Methods, held in conjunction with ECAI'98.*

Eden, A. H., & Turner, R. (2007). Problems in the Ontology of Computer Programs. In Applied Ontology (vol. 2, pp. 13–36). Amsterdam: IOS Press.

Guarino, N. (2009). *The Ontological Level: Revisiting 30 years of knowledge representation. conceptual modelling: foundations and applications, Essays in Honor of John Mylopoulos.* Springer-Verlag.

Guarino, N., Oberle, D., & Staab, S. (2009) What is an Ontology? In S. Staab & R. Studer (Eds.), Handbook on Ontologies, Second Edition. International handbooks on information systems. Springer Verlag. doi:10.1007/978-3-540-92673-3_0

Henninger, S. (2003). Tool Support for Experience-based Software Development Methodologies. *Advances in Computers*, *59*, 29–82. doi:10.1016/S0065-2458(03)59002-7

Ikeda, M., Seta, K., Kakusho, O., & Mizoguchi, R. (1998). Task ontology: Ontology for building conceptual problem solving models. In *Proceedings of ECAI98 Workshop on applications of ontologies and problem-solving model.*

Iqbal, R. (2013). An Analysis of Ontology Engineering methodologies: A literature review. *Research Journal of Applied Sciences, Engineering and Technology*, *6*(6), 2993–3000.

Jacobson, I., Ng, P.-W., McMahon, P., Spence, I., & Lidman, S. (2012). The Essence of Software Engineering: The SEMAT Kernel. Queue, 10(10).

Jacobson, I., Ng, P.-W., & Spence, I. (2007). Enough of process-let's do practices. *Journal of Object Technology*, *6*(6), 41–67. doi:10.5381/jot.2007.6.6.c5

Jaros, V. (2011). Concept Development and Experimentation in project STRUCTURE. Cybernetic Letters.Brno: Expertia, o.p.s., 2, 36-41.

Jeffery, D. R., & Scott, L. (2002). Has twenty-five Years of Empirical Software Engineering Made a Difference. In *Proceedings of the 2nd Asia-Pacific Software Engineering Conference*. doi:10.1109/APSEC.2002.1183076

Karray, F., Alemzadeh, M., Saleh, J. A. & Arab, M. N. (2008). Human-Computer Interaction: Overview on State of the Art. *Smart Sensing and Intelligent Systems*, *1*(1), 138-159.

MCM-0056-2010 NATO Concept Development and Experimentation (CD&E) Process. (2010). Brussels: NATO HQ.

Simperl, E., Michel, M., & Burger, T. (2010). Achieving maturity: The state of practice in ontology engineering. *International Journal of Computer Science and Applications &Technomathematics Research Foundation*, *7*(1), 45–65.

Sjoberg, D. I. K., Dyba, T., & Jorgensen, M. (2007). The Future of Empirical Methods in Software Engineering Research. In *Proceedings of Workshop Future of Software Engineering*. doi:10.1109/FOSE.2007.30

Sosnin, P. (2013). Scientifically Experimental Way-of-Working in Conceptual Designing of Software Intensive Systems. In *Proceedings of the IEEE 12th International Conference on Intelligent Software Methodologies, Tools and Techniques* (pp. 43-51). IEEE.

Sosnin, P. (2013). Role Intellectual Processor in Conceptual Designing Of Software Intensive Systems. In *Proceedings of the 11-th International conference on Computational Science and Applications* (LNCS), (vol. 7973, pp. 1-16). Springer.

Sosnin, P. (2015). A Precedent-Oriented Simulation in Designing the templates of Airplane Parts. In *Proceedings of 29th European Conference on Modelling and Simulation.*

Sosnin, P., & Maklaev, V. (2014). Question-Answer Reflections in a Creation of an Experience Base for Conceptual Designing the Family of Software Intensive Systems. Springer International Publishing.

Southekal, P. H., & Levin, G. (2011). Formulation and Empirical Validation of a GQM Based Measurement Framework. In *Proceedings of 11th International Symposium on Empirical Software Engineering and Measurement.* doi:10.1109/ESEM.2011.59

Webber, B., & Webb, N. (2010). Question Answering. In Handbook of Computational Linguistics and Natural Language Processing. Blackwells. doi:10.1002/9781444324044.ch22

Xu, S., & Rajlich, V. (2005). Dialog-Based Protocol: An Empirical Research Method for Cognitive Activity in Software Engineering. In *Proceedings of the 2005 ACM/IEEE International Symposium on Empirical Software Engineering.*

ADDITIONAL READING

Bass, L., Ivers, J., Klein, M., & Merson, P. (2005). *Reasoning Frameworks, Software Engineering Institute, Tech. Rep. Carnegie Mellon University* (pp. 2005–TR-007). Pittsburgh, PA: CMU/SEI.

Booch, G., & Brown, A. W. (2003). Collaborative development environments. In M. Zelkowitz (Ed.), *Advances in computers, 59.* San Diego, CA: Academic Press. doi:10.1016/S0065-2458(03)59001-5

Charette, R. N. (2005). Why software falls. *IEEE Spectrum, 4*(9), 36–43.

El Emam, K., & Koru, A. G. (2008). A Replicated Survey of IT Software Project Failures. *IEEE Software, 25*(5), 84–90. doi:10.1109/MS.2008.107

Guarino, N., Oberle, D., & Staab, S. (2009). What is an Ontology? In S. Staab and R. Studer (Ed.), Handbook on Ontologies, Second Edition. International handbooks on information systems. Springer Verlag, pp. 1-17. doi:10.1007/978-3-540-92673-3_0

ISO 9004-2009:Managing for the sustained success of an organization – A Quality Management Approach. 3rd edition. Geneva. Switzerland: ISO, (2009).

Karray, F., Alemzadeh, M., Saleh, J. A. & Arab, M. N. (2008). Human-Computer Interaction: Overview on State of the Art. *Smart sensing and intelligent systems*, 1(1), pp 138-159.

Karsai, G., ... (2009) Design Guidelines for Domain Specific Language, In Procedings of the 9th OOPSLA Workshop on Domain-Specific Modeling, pp. 7-13.

Pattit, J. M., & Wilemon, D. (2005). Creating high-performing software development teams. *R & D Management*, *35*(4), 375–393. doi:10.1111/j.1467-9310.2005.00396.x

Pew, R. W. (2007). Some history of human performance models. In W. Gray (Ed.), *Integrated models of cognitive systems* (pp. 29–47). New York: Cambridge University Press. doi:10.1093/acprof:oso/9780195189193.003.0003

Rational Unified Process, I. B. M. (RUP), Available at http://www-01.ibm.com/software/rational/rup

Sosnin, P. (2008). Conceptual Solution of the Tasks in Designing the Software Intensive Systems. *In Proc. of the 14th IEEE Mediterranean Electrotechnical Conference - MELECON 2008*, Ajacco, France, pp. 293 – 298.

Sosnin, P. (2010). Question-Answer Programming in Collaborative Development Environment. In Prioc. of CIS-RAM 2010, Singapore, pp. 273-278.

Sosnin, P. (2012) Pseudo-code Programming of Designer Activity in Development of Software Intensive Systems. In Proc. of the 25-th International conference on Industrial Engineering and other Applications of Applied Intelligent Systems (IEA/AIE 2012), Dalian, Chine, 2012, pp. 457-466. doi:10.1007/978-3-642-31087-4_48

KEY TERMS AND DEFINITIONS

Conceptual Experiment: The conceptual experiment is a thought experiment, the content and process of which are operatively reflected on a semantic memory of a question-answer type, and results of reflections are applied in the process of experimenting with useful purposes.

Intellectual Processor: Human who fulfils the pseudo-code programs by the way which is similar to the execution of computer programs with the help of the interpreter program (definition of the author).

Human-Computer Interaction: An interdisciplinary field focused on the interactions between human users and computer systems, including the user interface and the underlying processes which produce the interactions. http://encyclopedia2.thefreedictionary.com/human-computer+interaction.

Precedent: Precedents are actions or decisions that have already happened in the past and which can be referred to and justified as an example that can be followed when the similar situation arises. Available from http://dictionary.reference.com/browse/precedent.

Pseudo-Code: Pseudo-code is a compact and informal high-level description of a computer programming algorithm that uses the structural conventions of a programming language, but is intended for human reading rather than machine reading. Available from http://en.wikipedia.org/wiki/

Question-Answering: Question-Answering (QA) is a type of "an information retrieval in which a direct answer is expected in response to a submitted query, rather than a set of references that may contain the answers. Available from http://www.wordiq.com/definition/Question_ answering.

Software Intensive Systems: Systems in which "software represents a significant segment in the following points: system functionality, system cost, system development risk, development time" (Software, 2006).

Way-of-Working: The tailored set of practices and tools used by the team to guide and support their work" (Go, 2003).

Chapter 11
Can We Induce a Cognitive Representation of a Prosthetic Arm by Means of Crossmodal Stimuli?

Mateus Franco
Federal University of Sao Paulo, Brazil

Henrique A. Amorim
Federal University of Sao Paulo, Brazil

Tiago V. Ortiz
Federal University of Sao Paulo, Brazil

Jean Faber
Federal University of Sao Paulo, Brazil

ABSTRACT

The ownership feeling of our body occurs mainly due to feedback responses in real-time from environment stimuli to our own body. These constant feedbacks induce a neuronal arrangement, generating a representative map and, consequently, an ownership and unity feeling, named as a body schema. Although there is a relative well knowing of the sensorial mapping about each part of our body, there are still several gaps about how the integration of all this representation is indeed accomplished. Many researchers have shown high rates of prosthesis non-acceptance due to different reasons. Here, the authors discuss an experimental protocol to induce optimally the ownership feeling associated with upper limb prosthesis, by means of a crossmodal vibro-tactile stimulation over the individual's body. The main hypothesis is that through this procedure the participant will extend their proprioception and achieve an ownership feeling of the prosthesis.

INTRODUCTION

The development of prosthetics and orthotics are directly related to advances of micro-electronics, materials and automation engineering. All new devices have been using different technologies in order to expand functionality and improve the integration with the human body. However, some studies show that, despite of all these efforts, the actual prosthesis present high rates of non-acceptance. Statistical data indicate that this ratio may reach 70% of users. These users leave or use their prosthesis only spo-

DOI: 10.4018/978-1-5225-0510-5.ch011

radically along day life (Ostlie et al., 2012; Biddis & Chau, 2007). Among the main factors that justify these high rate of non-acceptance are: economic, because of the high costs; biological, due to uncomfort; function, due to the restrictions of movements and degree of freedom, when compared with the original limbs and; finally, psychological factors (Lisa et al., 2007; Ostlie et al., 2011). One of the most important psychological factors is the perception that the user has on the appearance of the prosthetic and also the lack of ownership feeling. That is, the user does not feel the prosthesis belonging to its own body, but as something which needs to be carried out, as burden. Despite the neural plasticity introduced by physiotherapy, training and by the natural adaptation, the subjects cannot cognitively represent the prosthesis in its totality; most probably because a new representation[1] depends on an adequate body interaction to its new object (Stephen, 1985; Ehrsson et al., 2008; Graziano & Botvinick, 2002).

The ownership feeling occurs mainly because of the real-time sensory responses from our own body – by external tactile and visual stimulus, especially. These constant feedbacks induce not only an effective neuronal arrangement, but also a representative mapping and, consequently, an ownership and unity feeling, say a body schema (Dhillon & Horch, 2005; Maravita & Iriki, 2004). Although these premises are necessary, the exact process under the neural-cognitive integration under the ownership feelings, regarding an external object, like some tool, is far to be totally understood. In spite of the relative well knowing about the sensorial mapping considering each part of our body and its correspondent senses, there are still several gaps in the way the integration of all these representations is accomplished. One of these gaps concerns how the brain maps spatial occupation and planning of our body movements (Vignemont, 2013; Kwok, 2013). Under these key aspects relies our prospective notion, or our predictive ability of future events. And, in a similar way of our memory working, our future prediction capacity, or prospective inference, is fundamentally relevant to construct our body schema (Luinge & Veltink, 2005; Hoffman et al., 2010).

Recently, different researches proposed tactile protocols to induce or optimize the sensation of ownership of a prosthetic arm. Most of these protocols are based on the classical Rubber Hand Illusion concept. In this illusion it is made a crossing tactile and visual stimulus, at the same time, on the artificial rubber hand and on the hidden real volunteer's hand, that promotes a sort of ownership feeling of the artificial rubber limb. With a correct tactile and visual stimulus the individual starts to consider the fake rubber hand as part of his/her body, and may even report pain (or some discomfort) when the rubber hand suffers some injuries (Luinge & Veltink, 2005; Hoffman et al., 2010; Kuiken et al., 2004; Marasco et al., 2011; Beeker et al., 1967).

Based on these experiments, here, the authors discuss a new multi-task protocol designed to improve the ownership feeling of an artificial limb. They propose a cross modal mapping which intends to transduce a set of predefined movements into a pattern of vibro-tactile stimuli, on different parts of the subject's body. In this context, this experimental procedure seeks to evaluate how strong and effective might be a cognitive representation of an artificial robotic arm induced by a simultaneous vibro-tactile stimulation with a predefined arm movement (Kim et al., 2010; Gohil et al., 2013; Tsakiris et al., 2007; Alimardani et al., 2013). The authors also describe the contextual conditions associated with behavioral conditioning by using the vibro-tactile stimulus, through electrophysiological recording to evaluate a possible optimal cognitive association. This protocol seeks to map a set of movements from an artificial arm and not its touch perception. The main hypothesis is that through this procedure an individual will be able to incorporate and extend cognitively the feeling of spatial location of his/her body, increasing his/her proprioception and thus the ownership of the prosthesis (Gallagher, 2006; Funk et al., 2005; Paillard, 2013).

BACKGROUND

How brain does represent our body? Although there are a vast knowledge about the mapping linking body limbs and senses to specific brain regions (Moseley et al., 2012), many points are still not clear. Actually, only having a map correlating brain regions with their "effectors" (say limbs and all motor expressions) and their "sensors" (vision, touching, audition, gestation and smell), it does not answer the causal relationship under brain circuits and each part of the body, neither how this relationship is integrated biologically and cognitively, in space and in time (Botvinick & Cohen, 1998; Armel & Ramachandran, 2003).

The answers that could explain our ownership feelings depend not only of a bi-univocal mapping linking parts of our body into specific brain regions, but also, these answers need to explain the intrinsical dynamics, with promotes a constant loop of self organization with environmental stimuli. By looking at this dynamic, considering the real-time feedback among environment/input-body-brain-body-environment/output, we may figure out an evaluation about the function, structure and cognition under properception and ownership feelings (Botvinick & Cohen, 1998).

In order to elucidate these issues the body schema theory presents an important approach considering a global and integrated perspective of how we see ourselves and how we interact with our environment; considering our movements and tools usage.

Body Schema

Body schema is the concept that describes the way our brain is linked to our body, in its whole dimensions and complexities. It represents our location as unity, as well as all parts of our body, dynamically, in time and space. In this way, our body schema represents a dynamical mapping, changing and being updated constantly, as we interact with the environment (Botvinick & Cohen, 1998; Armel & Ramachandran, 2003). The body schema describes the monitoring and measuring of our body in relation to different aspects of the environment. It provides us the necessary surveillance and attention with possible external threat, since it marks where our body starts and where the environment starts, in time and space. This constant measuring of body-environment interaction allows a constant checking of our physical body and its internal representation; and further, this real-time feedback allows the brain to produce an appropriate pain response, in order to guaranty the physical body integrity (Ehrsson et al., 2005).

The cognitive representation of the body-environment might be divided in two aspects. The first one is the body schema itself, which, as already said, provide a representation of the body and its parts, including its sensors and the way they are mapped onto the brain. Through this representation the brain is able to constantly monitor the position and body movements, avoiding and guiding into specific objects in order to achieve and execute planned actions. The second aspect is the so called peripersonal space which consist the representation of the immediate environment around the individual who perceives it (Borsook et al., 1998). There are many articles describing it from neurophysiology, neuropsychology, and psychophysics point of view, saying that there is an integrative representation of visual, somatosensory, and auditory of peripersonal space, in both, human and non-human primates. This representation has a predominance of the visual, somatosensory, and proprioceptive modalities. The peripersonal space concerns all aspects of the environment and objects around us that might be grasped, touched or manipulated. In general, all objects and environmental aspects that cannot be reached directly, without some movement toward them or toward to us, are classified as being in the extrapersonal space. In this way, it is expected that the brain must represent all objects situated in peripersonal space differently from those in extrapersonal space (Neuper et al., 2005). Objects located in extrapersonal space might be perceived by only few senses, most probably by vision, audition, and olfaction. On the other hand, objects located in peripersonal space, might impact upon all our sensory system, which may include touch and even gustation. Therefore, the brain information processing of objects in peripersonal space must be more complex, in the sense that involves more modalities of sensory information when compared to the processing of objects located in extrapersonal space, Figures 1-2. And in consonance with these results, different works have shown how the use of tools, and the viewing of the proper body or body parts in mirrors (or in display videos) might also modify the representation of peripersonal space in brain (Graziano & Botvinick, 2002; Maravita & Iriki, 2004; Yong, 2011).

It is important to state that there is different taxonomy to distinguish the concept of body schema and body image. The first one is a sensory-motor representation of the body used to guide movement and action, the second one is used to form our perceptions (body percept), conceptual (body concept) or emotional (body affect) judgments towards to our physical body (Funk et al., 2005).

Rubber Hand Illusion

The rubber hand illusion is one of the most impressive experiments that might help to elucidate on how we represent our body by using feedbacks from constant environment inputs. It also shows how

Figure 1. Human homunculus and cortex body representation. Each part of our body has a particular group of neurons, located mainly on the primary somatosensory cortex, that process all sensorial information coming from these parts. These local processing describes a mapping between our senses and limbs and our central nervous system

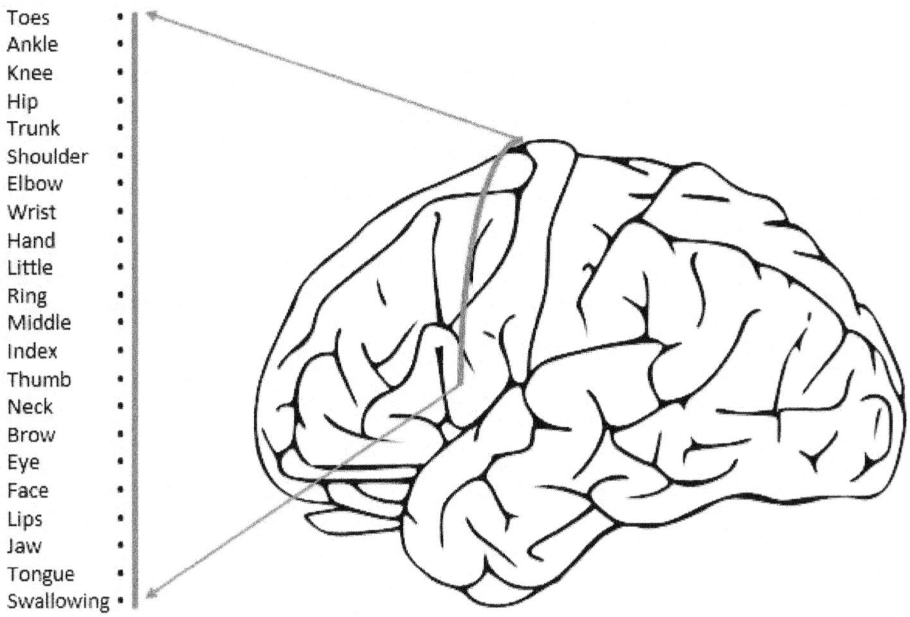

Figure 2. Body schema representation and visual field extended by tool use. This illustration shows the changes in the bimodal receptive field following tool-use. (1) Somatosensory Receptive Fields (SRF) of distal-type neurons. (2) Before tool use. (3) After tool use. (4) Passively grabbing the tool. (5) SRF of proximal-type bimodal neurons. (6) Visual Receptive Fields (VRF) before tool use. (7) VRF immediately after tool-use
(Scheme extracted from (Maravita & Iriki, 2004))

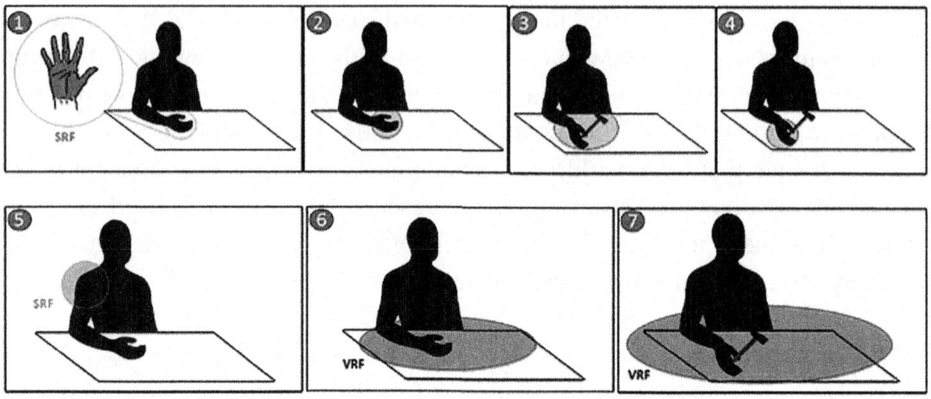

our body schema might be changed cognitively in a very fast simple way. This procedure suggests that most probably the ownership feeling of our body or external objects might be manipulated or inducted.

The procedure concerns in making an individual perceive an artificial hand as its own hand. An individual places one of its hands on a table and after a folding screen is put in front of the subject in such way he/she cannot see his/her own arm, but he/she can see the artificial arm in place of his own, simulating it (Figure 3). In sequence, the researcher applies on both hands, the artificial and the real, a tactile stimulus at the same time (scratching gently with a brush, for example). The consequence is that from the individual perspective the sensation of touch comes from the real hand and the visual stimulus of seeing the touching on the fake hand, cause a "feeling" on the prosthetic hand. Once the individual perceives the artificial arm as part of its own body, the prosthetic is striked up in such way we can measure the behavior reaction of the participant. In general, behaviorally it is commonly the participants withdraw their real arm from the table, and subjectively it is commonly a description of fear, awkwardness, anxiety or even pain. Despite not all participants respond in the same way and intensity, there is always a significant accordance with these results. Which might be resumed that, by applying a tactile stimulation, in congruence with a visual and proprioceptive feedback in real-time, there will appears an immediate reorganization of the body schema and of the sense of ownership body.

Based on this principle, the hypothesis of lack of ownership experience lies on a neural and cognitive failure representation of an artificial limb, Figure 4. Despite the neural plasticity introduced by training and by the adaptation process, the subjects cannot effectively represent cognitively the prosthesis because of a new representation depends on an adequate body interaction with its new object. The ownership feeling occurs mostly because of the real-time sensory responses about and from our own body – by external tactile and vision stimulus, mainly (Lewis, 2010). These constant feedbacks induce an effective neuronal arrangement, generating a representative map and, consequently, an ownership and unity feeling, or an effective body schema (Press et al., 2008).

Figure 3. Rubber hand illusion scheme. In this experiment there is a cognitive induction of ownership feeling of a false arm. A folding screen is put in front of a subject in order to hide his own arm. At the same time it is put a false arm nearby of his body in such way it simulates his own arm. After, a researcher applies some tactile stimuli simultaneously on his hand and on the rubber hand which the subject can see. By seeing the rubber hand and feeling the tactile stimulus there is a cognitive induction of ownership.

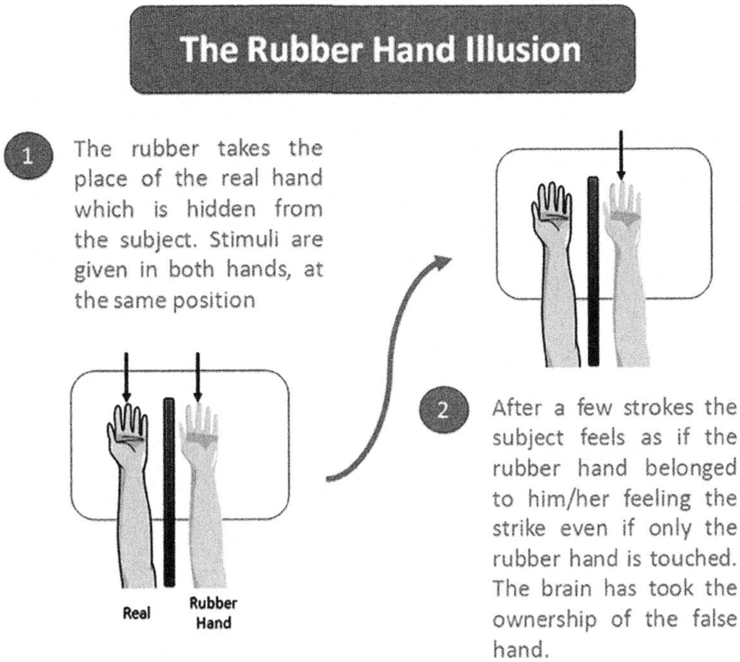

The brain plasticity, considering our cognitive body representation, also responds to physical changes occurred in the body along life. The best case is the adaptation in front of a limb loss (Ball et al., 2009; Kanayama et al., 2009). If, for some unfortunate reason, someone loses a limb (or even makes some hard body modification, like putting big piercing or hoops), the body schemas also change. There is a very common sensation reported by those who have lost some limb called 'phantom limb' sensations. Phantom limb is the sensation that a person who had some amputated or missing limb, or even an intern organ, feels. This sensation might be painful, or feeling itchy, twitch; and some patients even try to pick things up with their lost limbs. Many clinical techniques are based on the rubber hand illusion in order to try to heal this feelings, showing the efficacy and importance of the cognitive map of our body.

COGNITIVE ASPECTS ASSOCIATED WITH THE PROTOCOL

Different aspects of learning, memory and proprioception are directly linked to the protocol proposed here. Some of them deserve a short discussion in order to evaluate the deep and fundamental consequences of this protocol.

Figure 4.

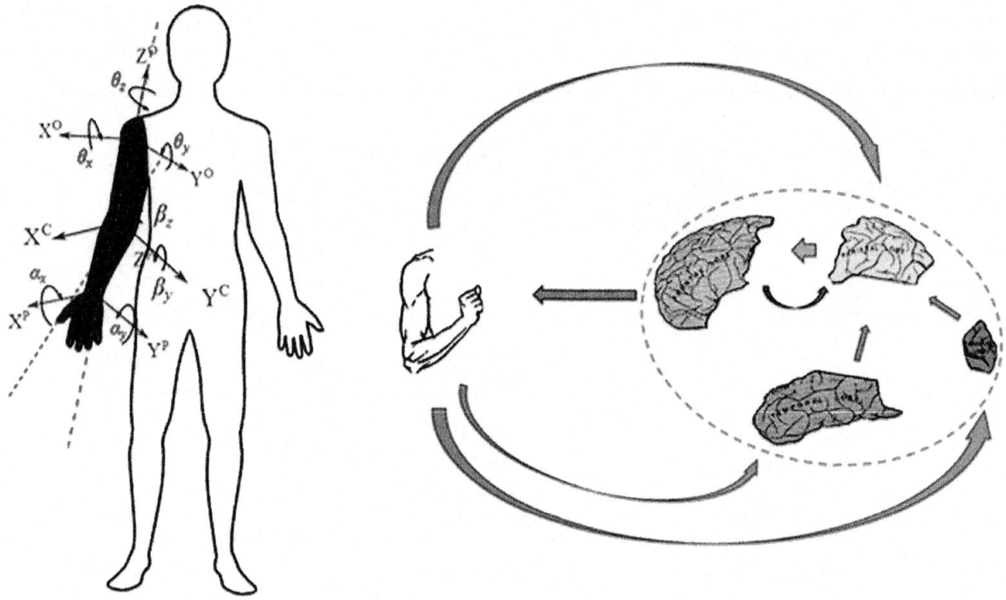

Conditioning and Associative Learning

An important cognitive factor linked to a crossmodal stimulus protocol is the associative memory capacity in a learning process. An associative memory is a learning principle stating that specific ideas and pratical experiences reinforce each other and can be linked to one another. In this way, here, in this protocol, through the repetition, the subject will be conditioned with the stimulus-response vibrotactile patterns, but it is also expected an associative learning among the real and robotic arm movements with the vibrotactile stimuli the subject receive. This learning takes place towards a reorganization and restructuring of neural network associated with the processing of each modular input and output (Paivio, 1971; Brasted et al., 2003; Mitz et al., 1991).

In the first step of the protocol the individual performs some pre-defined movement such that it triggers the stimuli patterns given by the array of vibro-tactile motors, spread out onto the individual's body. This procedure will impose an association for each movement with a specific configuration of intensity and frequency of vibration displaced on his/her body. This association provides an causal inference between the movement of his/her arm with the movements of the robotic arm. The individual then moves the robotic prosthetic arm by using sensors and effectors displaced onto his own arm in such way once he/she performs a movement, the same movement is realized by the robotic arm and, at the same time, the vibro-tactile stimuli is displaced. This triad provides the necessary elements for an associative learning, and when, even without seeing his own arm moving, he/she feels the tactile stimuli, already pre-configured, in order to evoke what he/she has learned.

Sensorial Cross-Modalities

One of the ideas behind this experimental protocol is to integrate different senses (vision and touch, in this case), using a crossed conditioning. Hence, by using vibro-tactile stimuli there is a cognitive induction and association considering the artificial arm movement. In other words, the individual "feels" the movement and position of a prosthetic arm through a vibro-tactile stimulus given by a matrix of miniature motors, spread along the subject's body. For each specific movement there is a code mapped onto a combination of intensity, frequency and position for each mini-motor. The assumption under this setup is that the individual will integrate and link cognitively the crossed events.

A similar protocol was already successfully used, in order to make a blind person to see through an electrical stimulus mapping on his/her tongue (Sampaio et al., 2001). Through fMRI measurements, they observed an intense activation of the subject's primary visual cortex (V1). It was concluded that this protocol provided not only an effective direct mapping between different stimuli and senses (sight and 'taste', in this case), but it also provided a cognitive impression and neural adaptation.

Prospective Inference

It has been proposed that, in a similar way of memories, there is another function helping us to mark and program our trajectory in daylife: our capacity of future inference, by simulating possible events chain. This type of *prospective inference* is fundamentally relevant to construct our body schema (Prata et al., 2011; McDaniel & Einstein, 2007).

At the same way we prospect and plan our physical position during our daylife scheduling directions, or even our future realizations in our social life, we also plan and predict each spatial occupation by each part of our own body, in different time scales. This prediction would be printed by expectations on what we plan, fitted by what we really did. As much more efficient and longer in time our performances are predicted, as stronger would be the reinforcement of our own behavior and proprioception. Surely, our predictability capacity is based on probabilistic predictions.

Therefore, associated with our past storage capacity of past events, given by memories, our capacity to predict future events, say our *prospective inference* ability, are fundamentally relevant to construct our body schema. In order to plan our actions we make a model of the environment and of ourselves. And through this model we prospect not only our possible spatial positions, but also our proper actions and movements, accomplished in different time scales. Analogously to our memory, our inferences could be divided into different temporal scales according to our behavior relevance, and to the degree of predictability: short, middle and long term prospection. The *middle term* would be associated with our daylife scheduling; the long term prospection would be associated with our social and existential plans of what we intend to be in life. Finally, the first one would be associated with plans of effective motor actions. This prospection, however, at this level, not necessarily would be aware but responsible for immediate actions, optimizing safety and efficiency. This idea might be corroborated by looking at performances of dance, music, sports or even in movie/theatre. In all of these sceneries the performance itself depends intrinsically of the repetition and learning in order to know what to do in each step along different scales of time. For all of these prospection terms the individual needs to simulate all possible future possibilities in order to make it effective, precise, soft and natural.

SOLUTIONS AND RECOMMENDATIONS

The protocol proposed here involves different stages where each of them needs a particular evaluation. And, due to its innovative aspect, it was necessary the manufacture of several devices in order to provide a proper individual interaction and measurements in real-time. In this way, the protocol objectives a different perspective of measurement, considering the interaction of the subject with the whole apparatus; it provides a simultaneous recording of the physiological signals, like ECG and EEG, and modulates the behavior of the individual. In order to evaluate all these aspects, we will focus on the representation of a robotic arm controlled by an electronic glove in different degrees of freedom and high temporal resolution.

The main idea under this protocol is to map a set of movements realized by the robotic arm into a specific vibration pattern produced by 20 mini motors displaced on specific sensorial points along the body of an individual. The hypothesis is that by means of sequential steps, this protocol induces not only a conditioning association, but also a cognitive mapping between the movement and the crossmodal vibro-tactile stimulation, which optimizes the ownership feeling of the artificial limb (Figure 5).

Metodology of the Experiments

In order to make easier the comprehension of the whole protocol we will divide in topics, showing each stage of the conditioning, learning and cognitive association. First thus, we will describe the experimental parts of the protocol of induction (how is made the induction of the cognitive prosthesis representation), and then, we will describe how each device was manufactured and its relationship with the whole procedure (how each device is constructed in order to optimize the ownership induction).

Tactile-Visual Feedback Stimulation

In this first step, there will be induced an association between a tactile stimulus, disposed in different parts of the subject body, by using a matrix of mini motors, and the movement of the natural subject arm. The subject will see his own arm moving in a set of predefined movements and at the same time he/she 'feels' the tactile stimulation (vibration) scattered at sensorial sensorial points along his/her body. The idea is to construct a mapping between each pre-defined movement and a configuration of vibration, modulated by a combination of intensity vs. frequency vs. position. With some sections, the individual will reinforce the association between visualization of each movement and tactile stimulus. This stage will represent an operant conditioning. But, through this conditioning there will be also a neuronal learning generated by the cognitive association (Figure 5 - A).

Robotic Arm Control by using its Own Arm Movements

In the second step, there will be induced an association between the movement of the subject real arm and the artificial robotic arm (say a prosthesis). The individual will be seated in order to restrict his movements only to one of his arms. Thus, by putting sensors of movement, such as accelerometers and/or gyroscopes, along the subject's arm it will be possible to monitor and measure each coordinate of it along time in a pre-defined restricted space. In this way, by constructing a proper electronic/algorithmic

Figure 5. A general scheme of the whole protocol. (A) Behavior conditioning and cognitive association between the specific set of arm movements and the vibro-tactile stimulus configuration. (B) Cognitive association among (real) arm movements, prosthetic arm movements and vibro-tactile stimuli. This step implies in an inference of causal correlation among the three main vectors of the protocol, by also using visual feedback. (C) The real arm is hidden and only the artificial arm is now visible. This step represents an extension of the rubber hand illusion where it is expected the ownership feeling and a remodulation of the subject body schema. (D) The prosthetic arm is controlled by EEG signals in a Brain Computer Interface system. In this stage the ownership feeling intensity is tested by modulating the temporal scaling, by delaying and advancing the prosthetic control response in order to differentiate expectation feelings and conflict coordination of the arm.

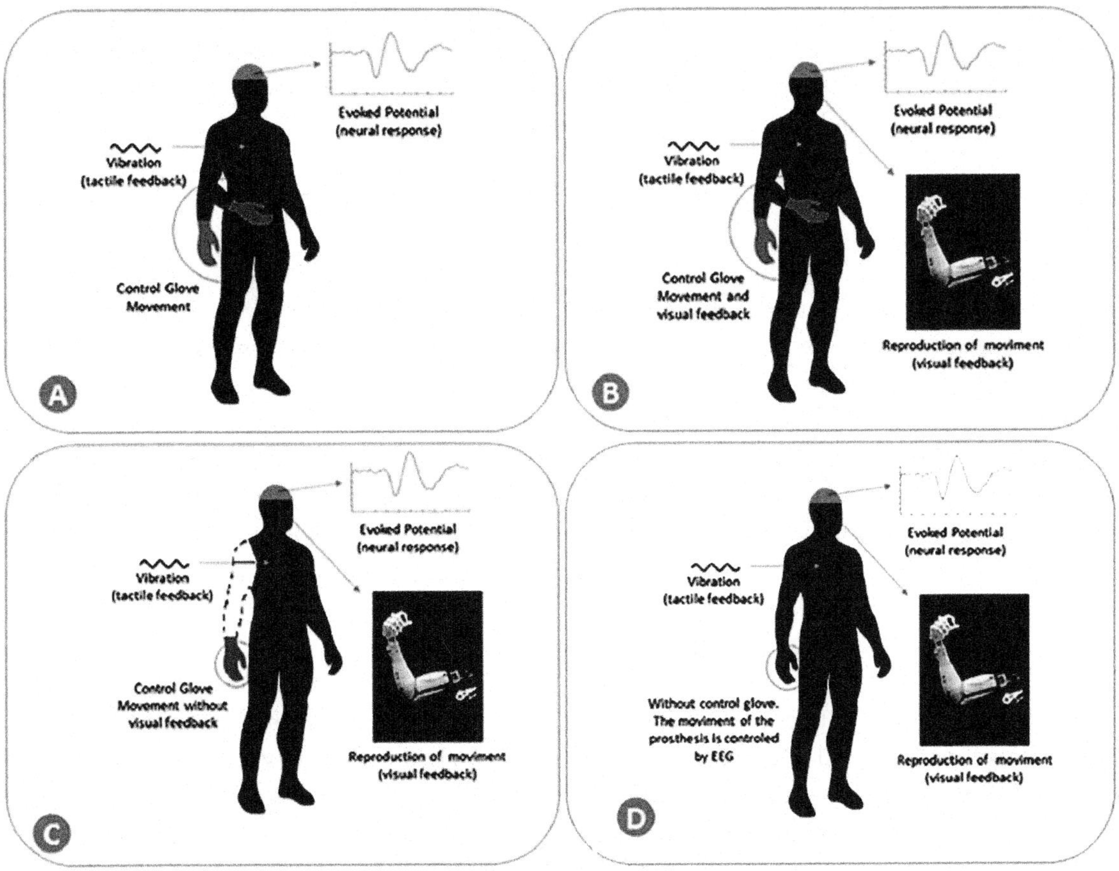

(described below), it is possible to track the individual's arm and mapping its movement into the robotic prosthetics movement. This step is important in order to induce an inferential causality about the controlling of the subject movements and the prosthetics movements. The subject will then associate that the prosthesis movements depend on his intention and effective arm moving (Figure 5 - B).

Integration of Tactile Feedback and Arm Movement

In this third step, the subject will control the robotic prosthesis by means of his own arm's movement and will also receive the vibro-tactile stimuli through the matrix of mini-motors along his/her body. This is a very important stage since in it is expected that the subject makes an association with the robotic arm movements and the tactile stimuli now. Here the subject will construct his/her cognitive mapping and it is expected a reformulation of his body schema, meaning he/she will start to incorporate the prosthesis, an external object, as part of his/her own body (Figure 5 - B).

Robotic Arm Control by means of the Real Arm without Visual Feedback

In this fourth step, it will be exactly the same as the previous one, but now the subject will not see his own arm. The subject arm will be hidden and only the artificial robotic arm will be visible, which will be placed close to the real one in order to give a more natural perspective. This stage represents an extension of the Rubber Hand experiment using a robotic arm and evaluating not only the tactile induction, but how the movement of the entire arm is relevant. Besides, we can evaluate the different parameters associated with the cognitive ownership induction, such as the distance between the artificial arm and subject's body, intensity and capacity of cognitive representation (Figure 5 - C).

Tactile Feedback in Different Scales of Time and Injury Threat

In this fifth step, there will be evaluated the intensity or the efficiency of the ownership induction realized by the previous steps. In order to evaluate the hypothesis of prospection and body's proprioception, it will be also modified the temporal scales of the tactile feedback and the robotic arm movement responses. By delaying and advancing the feedback responses in time we hope induce conflicting coordination of the arm, from the subject perspective. It will help to reveal how memory and prospection of our body movements are important in our own representation. Similar to the rubber hand experiment, some threat of injury might be also realized on the artificial arm in order to evaluate emotional links and the intensity of the ownership feeling of the prosthesis. By measuring the behavior and some physiological signals (such as ECG and EEG) we will be able to correlate these features (Figure 5 - D).

Brain Computer Interface Control by using EEG signals

In this last step, the idea is to control the prosthetic robotic arm by using only the subject's intention, evaluated by electroencephalographic (EEG) signals. That is, at this stage it will be implemented a BCI (Brain Computer Interface) system controlling some degrees of freedom associated with the robotic arm. Therefore, we will have all the other steps, including the hidden arm, but now the subject "only needs to think" to move his/her arm and the robotic arm will make the move he/she is intending to do. This link will strengthen the sense of ownership of the robotic arm. When the individual triggers his arm movement by "thinking" he should expect a feedback (tactile or visual), as we normally do, and this natural expectation induce an increase of the cognitive representation of the artificial arm (Figure 6).

Methodology of the Equipments and Devices

The system sets that are constructed to perform the control of the robotic arm can be divided in two parts. The first part consists of a glove, responsible to capture the movements of the fingers and wrist rotation. The second part consists of a covering which surrounds the rest of the user's arm, embedding sensors capable of capturing the coordinates associated with each movement performed by the rest of the upper

Figure 6. Diagram of the equipment set used in this experiment. Numbers 1 and 2 represent that there is a visual feedback between the subject and both arms, the robot and the real, respectively

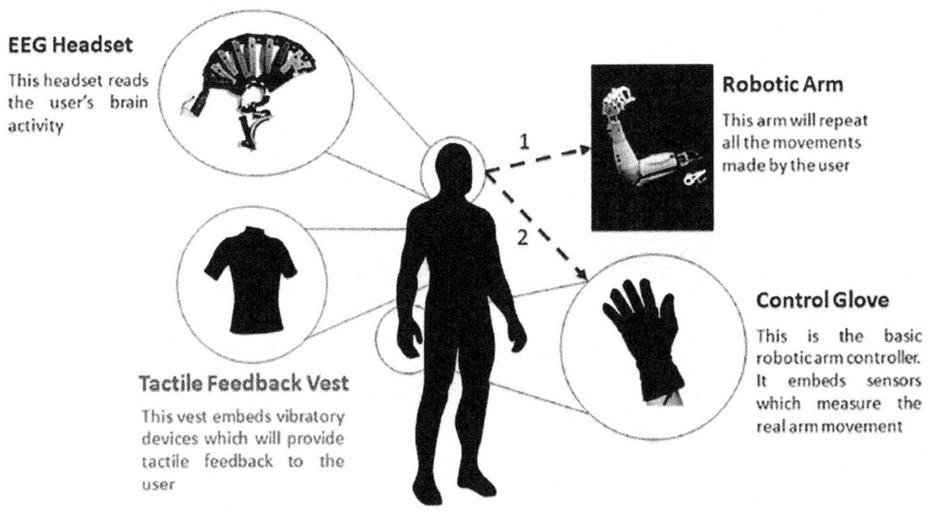

limb. In order to quantify the movement of fingers it will be used flexion sensors, which are variable resistors whose electrical resistance varies according to the degree of flexion. Each finger has a sensor, which is responsible for evaluating its position. For other degrees of freedom of the arm can be used Inertial Measurement Units (IMU) (Luinge & Velti, 2005; Dogan, 2006), which is an electronic device capable of capturing information about the position, velocity and acceleration of an object, Figure 6.

This type of device is a module boarding with a gyroscope and an accelerometer, able to read angles. Each IMU will be located in a center of rotation, which would make possible the reading of the wrist rotation angles, elbow and shoulder. All these sensors are responsible for providing the necessary information to a microcontroller, which sends control signals to the robotic arm. The block diagram depicting the system can be found in Figure 6 and 7.

Brain Computer Interface

In general, BCI systems can be defined as a set of techniques and methods, which seek to integrate different procedures for acquisition, processing (hardware and software), transfer and transduction of electrophysiological signals for 'action' of any artificial device, be it a robotic apparatus or software (Figure 8). For each task and sensory stimulus, a specific type of signature (synchronization or a neural code) is defined (Kanayama et al., 2009; Lenggenhager et al., 2011; Culpepper, 1999). This signature would carry the information about the possible behavioral responses, especially the motor ones. The central idea is to capture these neural signatures and interpret them similar to that performed by the brain. An algorithm (a mathematical model) should give an answer if a given pattern contains or not the execution information about a motor movement (for example, raise one arm). This interpretation is then sent to an external device (a robot, a wheelchair, or virtual simulation software) that will execute the movement itself or any other specified procedure.

Figure 7. A block diagram depicting the robotic arm control system using glove and accessory devices

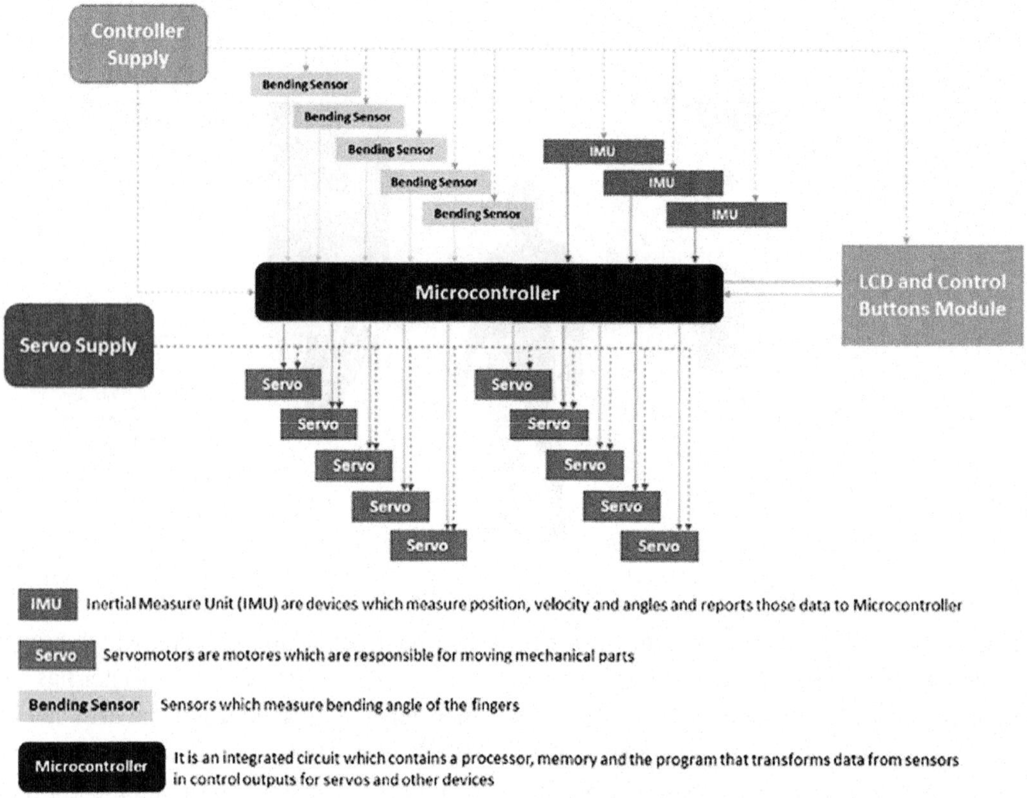

The common cycle of a BCI system begins with the acquisition of a neural signal associated with certain cognitive/motor task (Figure 6-7). The second step is filtering and signal amplification and application of any mathematical-computational tool that separates the best characteristics of the evaluated signal. The housekeeper task is the choice of some of these characteristics associated with specific neuronal patterns. The next step is the application of prediction techniques (linear or nonlinear) for modeling. Based on this prediction, the response associated with the individual's intention is determined. This last step of the cycle unfolds in two parts; the first is predicted itself and the other is the perception of this response by the individual in order to link his or her intention to the response (feedback) (Armel & Ramachandran, 2003; Neuper et al., 2005; Waldert et a., 2008; O'Doherty et al., 2011; Wessberg et al., 2000; Eliseyev et al., 2012; El Sadiik et al., 2011).

In this control method instead of receiving data directly from glove and arm-vest sensors, the microcontroller receives data coming from a computer where the EEG is connected. Basically, the computer receives the EEG readings and performs a processing of input signals received using a program that implements a mathematical model able to identify and classify specific brain signatures corresponding to each specific arm movement. Once identified, the computer produces an output which is received by the microcontroller, which sends control signals to the robotic arm. The methodology of this block diagram is illustrated in Figure 9.

Figure 8. A general scheme of a brain-computer interface (BCI) cycle with a tactile feedback. The cycle begins with the acquisition of some kind of neural signal associated with a particular cognitive/motor task. The second step is filtering and signal amplification and application of some mathematical-computational tool that separates the best characteristics of the evaluated signal. Then, the choice of some of these characteristics associated with specific neuronal pattern employed to task comes. The next step is the application of any prediction technique for modeling. Based on this prediction, the response associated to the individual's intention is determined. This last step of the cycle unfolds in two parts; the first is prediction itself and the other is the perception of this response by the individual in order to link his/her intention and produce a feedback.

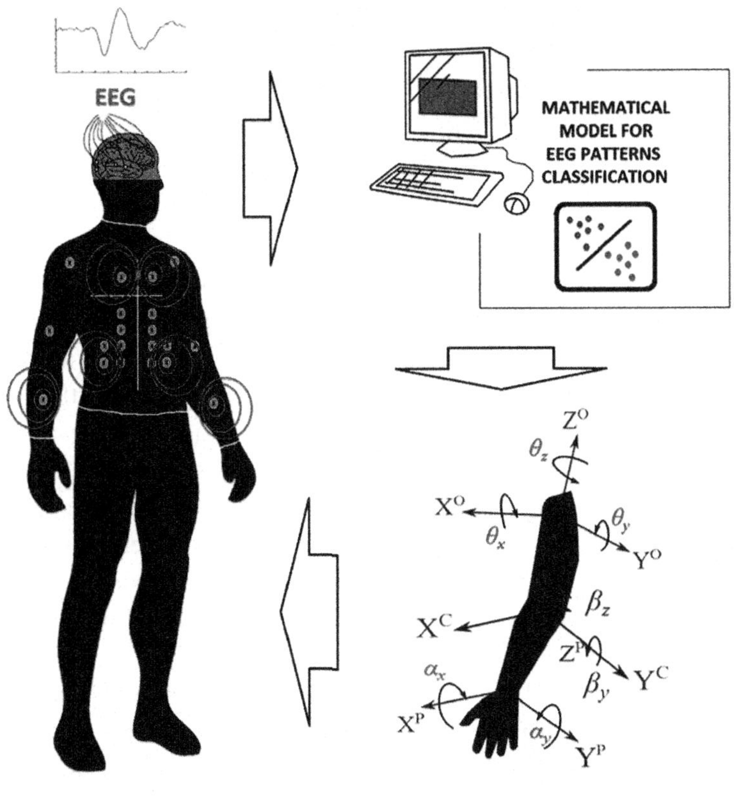

Vibro-Tactile Suit Device

The tactile feedback will be implemented by using a garment which will contain an array of vibrating devices, mini motors, spread across its surface in positions to facilitate a direct contact among these mini motors and the sensory points along the user's body, Figure 10. The vibration pattern is set in accordance with the movement of the arm. A set of specific arm movement will be pre-defined, say movement A = lift the arm without bending the elbow, to a height 'a' with a speed 'v'; movement B = lift the arm, bending the elbow, with angle 'w', height 'b' and speed 'v'. C = bends the arm to an angle 'w' and turns it into an angle 'c' with speed 'v'; and so on. For each pre-defined movement configuration, there will be a set of control signals that are responsible for modulating the vibrating motors, in intensity, frequency of vibration and position. A mathematical map is heuristically constructed in order to achieve optimal

Figure 9. A diagram block depicting the robotic arm control system using all devices and the EEG.

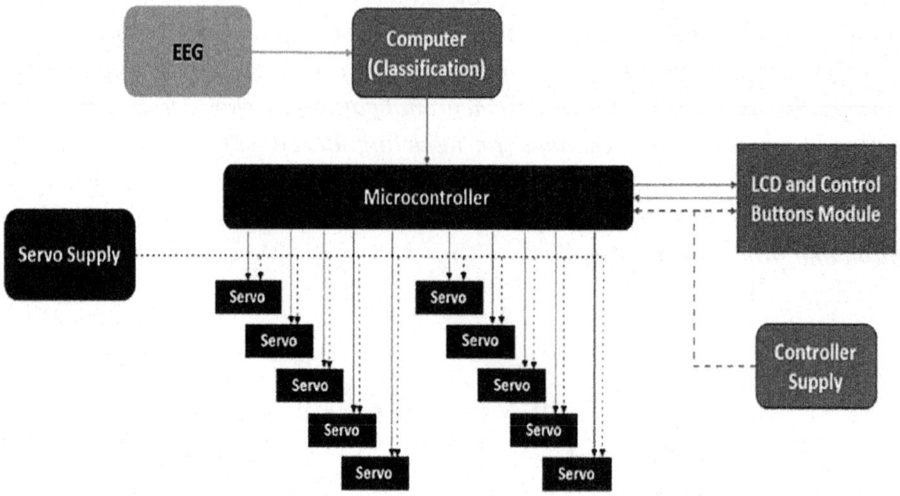

stimulation patterns associated with each person and each movement. That is, a function that takes as input the data from each arm position, at each instant of time, and its output signals that control the intensity, frequency and location of the vibration. Thus, there will be a drive for each vibration pattern with each of these three predefined attributes. By using the garment successive times, the process of association between the displayed movement and the standard of dress vibration will be strengthened. During the experiment is expected to appear a reevaluation of the individual's body schema resulting in optimization of the robotic prosthesis ownership feeling (El Sadiik et al., 2011; Dassonville, 1995), Figure 10.

The algorithms developed, in the BCI system, to predict the arm movements of the subjects are based on statistics of electrophysiological patterns generated from each motor movement and/or cognitive intent in previously combined protocol tasks. These tasks will be based on the set of arm movement, previously defined.

The premise under all BCI system is that in every action/intention there is a neural pattern and a specific electrophysiological condition, which can be captured by specific recording techniques, say, in this case, by an EEG, ECG and EMG. As each individual is unique in their own scale, condition, training, anatomical, and physiological setting, it is important to have a repertoire of subjects that can be considered for registration changes, behavior and even electrophysiological patterns. This variability will ensure a more general and robust methodology.

Alternative Protocol for Arm Movement Mapping by using Virtual 3D Image Capture Devices

In order to capture the movement of the arm without any physical equipment coupled directly onto the subject arm, we have been also using two equipments of image capture. One is the well know Microsoft Kinect® (Jana, 2012) and the other device is the Leap Motion® (Colgan, 2014). The Kinect® works with the combination of a RGB camera and a depth sensor, combining both they can map up to 48 joint points and up to six people. Its SDK provides the necessary mathematic to map the joints and movements and provides all the joints mapped in positions on the screen and in XYZ coordinates. The Leap

Figure 10. An illustration of the Tactile Feedback vest. Several vibrating devices are displaced over the vest in order to stimulate specific sensorial points of the individual. The quantity of red rings represents the vibration intensity and the yellow arrows represent the direction of this intensity

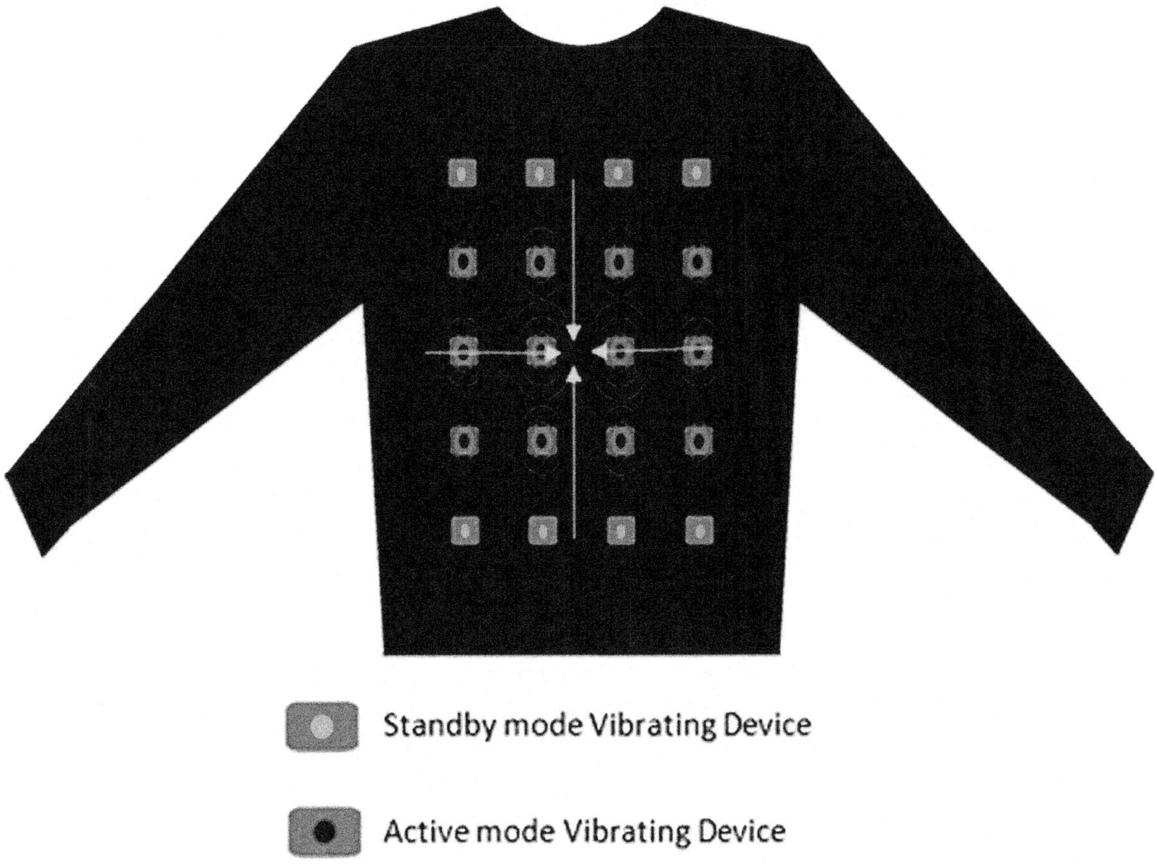

Motion consists of two cameras and three infrared LEDs pointed upward and can map 2 feet above the controller, by 2 feet wide on each side (150° angle), by 2 feet deep on each side (120° angle). Different from Kinect®, the Leap Motion® does not generate a depth map, in order to capture positions; it applies advanced algorithms to the raw sensor data.

Then, the data from these two devices are combined. The Kinect® has been used to map the whole body and to give an idea of the distance from the computer; and the Leap Motion® to give a more accurate data from the forearm and hand. The advantage of this technique over the gloves is that it could capture the arm without any physical contact, avoid possible spurious tactile feedbacks (Ofner & Muller-Putz, 2012).

Multidisciplinary Knowledge

The whole protocol exhibits different branches in different levels of knowledge. This structure requires an integration of different knowledge. And this integration is only possible with a multidisciplinary team, where each one has a deep domain in its branch, but at the same time each one is capable to understand

and participate on each step of the protocol as whole. Actually, the project itself only makes difference when it is integrated through its parts. If we look only onto the specific parts of the project there is no big technical innovation but by associating all of them in a unique interactive apparatus we are able to measure, access and modulate internal perspective of the participants.

It is involved in this project students and professors with different backgrounds, such as engineers of computation, biomedical engineers, physicists and neuroscientists in order to promote the integration of the knowledge under this project.

Ethics

In this stage of the project, the main aim is to build a preliminary protocol and device, to be after directed to subjects with special needs. Therefore, all stages of the project will be performed by voluntary subjects for all recording procedures and protocols. We declare we understand all risks and potential discomfort generated by the procedures, period of exposure of registering, tests and any possible embarrassment of the questionnaires. In this way, we designed a Consent and Informing formulary indicating each step, method and project objectives.

The algorithms developed to measure, predict actions and movement intentions of the subjects are based on statistics of signal patterns generated from each motor movement and/or cognitive intent in previously combined protocol tasks.

As each individual is unique in their own scale, condition, training, anatomical, and physiological setting, it is important to have a repertoire of subjects that can be considered for registration changes, behavior and even electrophysiological patterns. This variability will ensure a more general and robust methodology. However, by now, at this stage of development, there is no need for excessive or exhaustive number for fine adjustments in each step will come in the form of a new protocol, considering the circumstances and needs of each subject. At this stage, there should be not only tests in specific environments, but further study of the physiological and anatomical characteristics of each individual when exposed to interact with the proposed device.

Although at this stage, the subjects are all healthy and without any special needs, we believe that it will be sufficient to understand and adjust the key control parameters associated with the devices, both in number and variability. In general, in similar protocols, mainly those associated with BCI, each step follows the particular pattern of each individual without any concern about the generalization for the parameters and classifiers. It occurs because the variability patterns among subjects are extremely high, making difficult a generalization or producing unstable classification protocols. Here, however, we aim not only at the development of the equipment, but mainly at the study of variability to precise and define all parameters. Only after having these considerations we will be able to consider the physiological and anatomical needs, following the requirements of each condition and limitation of each patient.

FUTURE RESEARCH DIRECTIONS

Through this proposition we intend to formulate an experimental protocol in order to induce a cognitive ownership sensation of an artificial object, such as prosthesis of an upper limb, by means of a crossmodal stimulus. It also gives an extensive way of evaluation of how our body schema is associated with temporal resolution, mainly in memory and movement prospection.

In a near future the idea is to create a prototype for assistance and rehabilitation, able to yields a long-term functionality in a natural environment for people that has suffered a limb loss. Another important aspect of this system is its low cost and minimal special preparation of the subjects. Up today, there are already some analogues systems and protocols but in general they are not designed with crossmodal stimuli.

CONCLUSION

One of the greatest medical challenges today is to restore the movement or functional actions for individuals suffering traumatic injuries or degenerative diseases. In these cases, typically, the brain processes all motor commands, but the pathway responsible to transfer the information to the limbs (the spinal cord), or the proper limb, are committed or even absent, generating a consequent restriction or impossibility of the movement. A great perspective nowadays is the use of electrophysiological tools that allows the recording of brain activity patterns and its transfer to artificial models able to control a mechanical device, in order to restore the motor functionality. This type of apparatus allows a description of how neural activity is associated with specific behaviors. This research field is widely explored because it seeks the development of an integrated mechanical-electronic prosthesis device controlled by biological signals. The techniques that use these principles are usually called by Brain-Computer Interface systems.

The goal of a brain computer interface based system is providing a functional restoration to patients who have lost neurological functions, such as motor capability, or need the rehabilitation treatment. BCI systems essentially work by analyzing the electrical signals in/on the brain, which are produced during an intention to perform an action, such as arm movements. These signals are then processed using computer-based algorithms and delivered to external devices, such as a computer, a prosthetic arm, or a mobile robot. These devices then perform the function originally intended by the patient. Nevertheless, without a proper feedback the control of the prosthetic is not entire achieved. The subjects need to 'feel' the prosthetics as part of them. However, the problem is exactly how to generate or improve this ownership feeling of prosthetics in all individuals. Is there a general way that we could yield this property? Could we develop a systematic procedure in order to restore or modulate the human ability of proprioception?

Although there is a relative well knowing of the sensorial mapping about each part of our body, there are still several gaps about how the integration of all this representation is indeed accomplished; and the exact process under the cognitive ownership and integration feelings with an artificial object is still not totally understood.

The experimental protocol proposed here, introduces not only a clinical procedure of rehabilitation for people who may have lost a limb or have some motor impairment, but also proposes a fundamental experimental for investigation in real-time, in order to study the cognitive and behavioral relationships and how we represent our body and how we can reprogram this representation. The answers of these questions open new possibilities for new investigations about the way our brain-body network evolved along time and how it was configured to interact with ourselves and our environment, including other individuals.

REFERENCES

Alimardani, M., Nishio, S., & Ishiguro, H. (2013). Humanlike robot hands controlled by brain activity arouse illusion of ownership in operators. *Scientific Reports*, *3*, 2396. doi:10.1038/srep02396 PMID:23928891

Armel, K. C., & Ramachandran, V. S. (2003). Projecting sensations to external objects: Evidence from skin conductance response. *Proceedings. Biological Sciences*, *270*(1523), 1499–1506. doi:10.1098/rspb.2003.2364 PMID:12965016

Ball, T., Schulze-Bonhage, A., Aertsen, A., & Mehring, C. (2009). Differential representation of arm movement direction in relation to cortical anatomy and function. *Journal of Neural Engineering*, *6*(1), 016006. doi:10.1088/1741-2560/6/1/016006 PMID:19155551

Beeker, T. W., During, J., & Den Hertog, A. (1967). Artificial touch in a hand-prosthesis. *Medical & Biological Engineering*, *5*(1), 47–49. doi:10.1007/BF02478841 PMID:6037625

Biddis, E. A., & Chau, T. T. (2007). Upper limb prosthesis use and abandonment: A survey of the last 25 years. *Prosthetics and Orthotics International*, *31*(3), 236–257. doi:10.1080/03093640600994581 PMID:17979010

Borsook, D., Becerra, L., Fishman, S., Edwards, A., Jennings, C. L., Stojanovic, M., & Breiter, H. et al. (1998). Acute plasticity in the human somatosensory cortex following amputation. *Neuroreport*, *9*(6), 1013–1017. doi:10.1097/00001756-199804200-00011 PMID:9601659

Botvinick, M., & Cohen, J. (1998). Rubber hands 'feel' touch that eyes see. *Nature*, *391*(6669), 756. doi:10.1038/35784 PMID:9486643

Brasted, P. J., Bussey, T. J., Murray, E. A., & Wise, S. P. (2003). Role of the hippocampal system in associative learning beyond the spatial domain. *Brain*, *126*(5), 1202–1223. doi:10.1093/brain/awg103 PMID:12690059

Colgan, A. (2014). *How Does the Leap Motion Controller Work?* Retrieved from http://blog.leapmotion.com/hardware-to-software-how-does-the-leap-motion-controller-work/

Culpepper, J. (1999). *Discriminating Mental States Using EEG Represented by Power Spectral Density*. Claremont: Harvey Mudd College.

Dassonville, P. (1995). Haptic localization and the internal representation of the hand in space. *Experimental Brain Research*, *106*(3), 434–448. doi:10.1007/BF00231066 PMID:8983987

Dhillon, G. S., & Horch, K. W. (2005). Direct neural sensory feedback and control of a prosthetic arm. *IEEE Transactions on Neural Systems and Rehabilitation Engineering*, *13*(4), 468–472. doi:10.1109/TNSRE.2005.856072 PMID:16425828

Dogan, I. (2006). *Microcontroller Based Applied Digital Control*. Chichester, UK: Wiley.

Ehrsson, H. H., Holmes, N. P., & Passingham, R. E. (2005). Touching a Rubber Hand: Feeling of Body Ownership Is Associated with Activity in Multisensory Brain Areas. *The Journal of Neuroscience*, *25*(45), 10564–10573. doi:10.1523/JNEUROSCI.0800-05.2005 PMID:16280594

Ehrsson, H. H., Rosen, B., Stockselius, A., Ragno, C., Kohler, P., & Lundborg, G. (2008). Upper limb amputees can be induced to experience a rubber hand as their own. *Brain, 131*(12), 3443–3452. doi:10.1093/brain/awn297 PMID:19074189

El Sadiik, A., Orozco, M., Eid, M., & Cha, J. (2011). *Haptics Technologies: Bringing Touch to Multimedia*. New York: Springer. doi:10.1007/978-3-642-22658-8

Eliseyev, A., Moro, C., Faber, J., Wyss, A., Torres, N., Mestais, C., & Askenova, T. et al. (2012). L1-penalized N-way PLS for subset of electrodes selection in BCI experiments. *Journal of Neural Engineering*, *9*(4), 045010. doi:10.1088/1741-2560/9/4/045010 PMID:22832155

Funk, M., Shiffrar, M., & Brugger, P. (2005). Hand movement observation by individuals born without hands: Phantom limb experience constrains visual limb perception. *Experimental Brain Research*, *164*(3), 341–346. doi:10.1007/s00221-005-2255-4 PMID:15818499

Gallagher, S. (2006). *How the Body Shapes the Mind*. New York: Oxford University Press.

Gohil, V. J., Bhagwat, S. D., Raut, A. P., & Nirmal, P. R. (2013). Robotics arm control using haptic technology. *International Journal of Lastest Research in Science and Technology*, *2*(2), 98–102.

Graziano, M. S. A., & Botvinick, M. M. (2002). How the brain represents the body: insights from neurophysiology and psychology. In Prinz, W. and Hommel, B. (Eds.), Common Mechanisms in Perception and Action: Attention and Performance XIX, Oxford University Press, 136-157.

Hoffman, M., Marques, H. G., Arieta, A. H., Sumioka, H., Lungarella, M., Pfeifer, R. (2010). Body Schema in Robotics: A Review. *IEEE Transactions on Autonomous Mental Development*, *2*(4).

Jana, A. (2012). *Kinect for Windows SDK Programming Guide*. Retrieved from http://lirec.ict.pwr.wroc.pl/~jkedzier/urbi/Kinect%20for%20Windows%20SDK%20Programming%20Guide.pdf

Kahneman, D., & Tversky, A. (1979). Prospect Theory: An Analysis of Decision under Risk. *Econometrica*, *47*(2), 263–291. doi:10.2307/1914185

Kanayama, N., Sato, A., & Ohira, H. (2009). The role of gamma band oscillations and synchrony on rubber hand illusion and crossmodal integration. *Brain and Cognition*, *69*(1), 19–29. doi:10.1016/j.bandc.2008.05.001 PMID:18555572

Kim, K., Colgate, J. E., Santos-Munne, J. J., Makhlin, A., & Peshkin, M. A. (2010). On the Design of Miniature Haptic Devices for Upper Extremity Prosthetics. *IEEE/ASME Transactions on Mechatronics*, *15*(1), 27–39. doi:10.1109/TMECH.2009.2013944

Kuiken, T. A., Dumanian, G. A., Lipschutz, R. D., Miller, L. A., & Stubblefield, K. A. (2004). The use of targeted muscle reinnervation for improved myoelectric prosthesis control in a bilateral shoulder disarticulation amputee. *Prosthetics and Orthotics International*, *28*, 245–253. PMID:15658637

Kwok, R. (2013). Neurorposthetics: Once more, with feeling. *Nature*, *497*(7448), 176–178. doi:10.1038/497176a PMID:23657331

Lenggenhager, B., Halje, P., & Blanke, O. (2011). Alpha band oscillations correlate with illusory self-location induced by virtual reality. *The European Journal of Neuroscience*, *33*(10), 1935–1943. doi:10.1111/j.1460-9568.2011.07647.x PMID:21395867

Lewis, S. C. (2010). *The role of empathy in the rubber hand illusion*. Faculty of California State University.

Lisa, V. W., Anita, M. B., & Michelle, A. J. (2007). Reasons for Prosthetic Rejection by Children with Unilateral Congenital Transverse Forearm Total Deficiency. *Prosthetics and Orthotics International*, *19*(2), 51–54. doi:10.1097/JPO.0b013e3180421539

Luinge, H. J., & Veltink, P. H. (2005). Measuring orientation of human body segments using miniatures gyroscopes and accelerometers. *Medical & Biological Engineering & Computing*, *43*(2), 273–282. doi:10.1007/BF02345966 PMID:15865139

Marasco, P. D., Kim, K., Colgate, J. E., Peshkin, M. A., & Kuiken, T. A. (2011). Robotic touch shifts perception of embodiment to a prosthesis in targeted reinnervation amputees. *Brain*, *134*(3), 747–758. doi:10.1093/brain/awq361 PMID:21252109

Maravita, A., & Iriki, A. (2004). Tools for the body (schema). *Trends in Cognitive Sciences*, *8*(2), 79–86. doi:10.1016/j.tics.2003.12.008 PMID:15588812

McDaniel, M. A., & Einstein, G. O. (2007). *Prospective memory: An overview and synthesis of an emerging field*. London: Sage Publications Ltd.

McDaniel, M. A., Einstein, G. O., Graham, T., & Rall, E. (2004). Delaying execution of intentions: Overcoming the costs of interruptions. *Applied Cognitive Psychology*, *18*(5), 533–547. doi:10.1002/acp.1002

Mitz, A. R., Godschalk, M., & Wise, S. P. (1991). Learning-dependent neuronal activity in the premotor cortex: Activity during the acquisition of conditional motor associations. *The Journal of Neuroscience*, *11*, 1855–1872. PMID:2045890

Moseley, G. L., Gallace, A., & Spence, C. (2012). Bodily illusions in health and disease: Physiological and clinical perspectives and the concept of a cortical 'body matrix'. *Neuroscience and Biobehavioral Reviews*, *36*(1), 34–46. doi:10.1016/j.neubiorev.2011.03.013 PMID:21477616

Neuper, C., Scherer, R., Reiner, M., & Pfurtscheller, G. (2005). Imagery of motor actions: Differential effects of kinesthetic and visual-motor mode of imagery in single-trial EEG. *Brain Research. Cognitive Brain Research*, *25*(3), 668–677. doi:10.1016/j.cogbrainres.2005.08.014 PMID:16236487

O'Doherty, J. E., Lebedev, M. A., Ifft, P. J., Zhuang, S., Bleuler, H., & Nicolelis, M. A. (2011). Active tactile exploration using a brain-machine-brain interface. *Nature*, *479*(7372), 228–231. doi:10.1038/nature10489 PMID:21976021

Ofner, P., & Muller-Putz, G. R. (2012). Decoding of velocities and positions of 3D arm movement from EEG.*International Conference of the IEEE EMBS*. doi:10.1109/EMBC.2012.6347460

Ostlie, K., Franklin, R. J., Skjeldal, O. H., Skrondal, A., & Magnus, P. (2011). Musculoskeletal pain and overuse syndromes in adult acquired major upper-limb amputees. *Archives of Physical Medicine and Rehabilitation*, *92*(12), 1967–1973. doi:10.1016/j.apmr.2011.06.026 PMID:22133243

Ostlie, K., Lesjo, I. M., Franklin, R. J., Garfelt, B., Skjeldal, O. H., & Magnus, P. (2012). Prosthesis rejection in acquired major upper-limb amputees: A population-based survey. *Disability and Rehabilitation. Assistive Technology*, *7*(4), 294–303. doi:10.3109/17483107.2011.635405 PMID:22112174

Ostlie, K., Lesjo, I. M., Franklin, R. J., Garfelt, B., Skjeldal, O. H., & Magnus, P. (2012). Prosthesis use in adult acquired major upper-limb amputees: Patterns of wear, prosthetic skills and the actual use of prostheses in activities of daily life. *Disability and Rehabilitation. Assistive Technology*, *7*(6), 479–493. doi:10.3109/17483107.2011.653296 PMID:22315926

Paillard, J. (2013). Body schema and body image: a double dissociation in deafferented patients. In Motor control today and tomorrow. Academic Publishing House.

Paivio, A. (1971). *Imagery and verbal processes*. Holt, Rinehart, and Winston.

Prata, C. S., Masonb, R. A., & Justb, M. A. (2011). Individual differences in the neural basisi of causal inferencing. *Brain and Language*, *116*(1), 1–13. doi:10.1016/j.bandl.2010.08.004 PMID:21051084

Press, C., Heyes, C., Haggnard, P., & Eimer, M. (2008). Visuotactile learning and body representation: An ERP study with rubber hands and rubber objects. *Journal of Cognitive Neuroscience*, *20*(2), 312–323. doi:10.1162/jocn.2008.20022 PMID:18275337

Sampaio, E., Maris, S., & Bach-y-Rita, P. (2001). Brain plasticity: 'visual' acuitu of blind persons via the tongue. *Brain Research*, *908*(2), 204–207. doi:10.1016/S0006-8993(01)02667-1 PMID:11454331

Stephen, F. B., & Judith, A. B. (1985). Patterns of Acceptance and Rejection of Upper Limb Prostheses. *Orthotics and Prosthetics*, *39*(2), 40–47.

Tsakiris, M., Hesse, M. D., Boy, C., Haggard, P., & Fink, G. R. (2007). Neural Signatures of Body Ownership: A Sensory Network for Bodily Self-Consciousness. *Cereb Cortex*, *17*(10), 2235–2244. doi:10.1093/cercor/bhl131 PMID:17138596

Vignemont, F. (2003). The Body in the Mind. In *Proceedings of CUNY*.

Waldert, S., Preissl, H., Demandt, E., Braun, C., Birbaumer, N., Aertsen, A., & Mehring, C. (2008). Hand Movement Direction Decoded from MEG and EEG. *The Journal of Neuroscience*, *28*(4), 1000–1008. doi:10.1523/JNEUROSCI.5171-07.2008 PMID:18216207

Wessberg, J., Stambaugh, C. R., Kralik, J. D., Beck, P. D., Laubach, M., Chapin, J. K., & Nicolelis, M. A. et al. (2000). Real-time prediction of hand trajectory by ensembles of cortical neurons in primates. *Nature*, *408*(6810), 361–365. doi:10.1038/35042582 PMID:11099043

Yong, E. (2011, December7). Out-of-body experience: Master of illusion. *Nature*, *480*(7376), 168–170. doi:10.1038/480168a PMID:22158222

KEY TERMS AND DEFINITIONS

Adaptative Behavior: It is the behaviors that conduct effectively and progressively a successful of a subject in its interaction with the environment, considering as environment also other subjects.

Behavior: It is the observable responses, actions, or activities of someone.

Body Image: It is the way a person sees, thinks or feels about him/her body. It is the way someone imagines how others see him/her body aspect.

Body Schema: It is the, consciouss or unconsciouss, feelings that an individual has of his/her own body as an object displaced in time and space, independent and apart of from the external environement.

Brain-Computer Interface: It is the set of techniques that seeks to integrate different procedures for acquisition, processing (hardware and software) and direct transmission of neural signals to any artificial device, a robotic apparatus or software.

Conditioning: It is a way of learning in which a response is elicited by a neutral stimulus which previously had been repeatedly presented in conjunction with the stimulus that originally elicited the response.

Learning: It is the set of behavior changes that occurs as a result of practice by executing something.

Ownership Feeling: It is the sensation, or feeling, that something to the subject belongs to him. This feeling is increased by three factors: control of it, self-investment on it, and knowledge about it.

Phantom Limb: It is a limb that is felt to be present as part of the body after having been amputed. The often painful sensation of the presence of such a limb.

Prosthesis: It is an artificial device used to replace a missing body part, such as a limb, tooth, eye, or heart valve.

Representation (Cognitive): A mental representation, or a cognitive representation, is a hypothtetical internal mental or imaginary symbol that represents attributes of external reality, or else a mental process that makes use of such a symbol .

ENDNOTES

1 The term "representation" deserves a dedicated topic to exploit all possible implications and interpretation, mainly when used as cognitive feature. However, this discussion is out of the scope of this chapter and we have to consider the particular understanding of each reader.

Chapter 12
Cognitive Processes in Fashion Design:
Designing of Modelling Projects for the Visually Handicapped

Geraldo Coelho Lima Júnior
Universidade Anhembi Morumbi (UAM), Brazil

ABSTRACT

This chapter is concerned with the teaching and learning of modelling in fashion design courses. Following a series of observations, it was found that fashion design students, with normal sight, have difficulties in fully understanding how an item of clothing can be transposed to a modelling display bust, which represents the body of the wearer. The same obstacle affects visually-handicapped students. This study seeks to explore ways of overcoming this problem. It involves seeking to introduce features into teaching that can allow a comprehensive learning program to be taught and in particular, to concentrate on certain key factors - cognition, constancy and abstraction - with regard to the information on fashion projects that can be found in the surrounding learning environment.

INTRODUCTION

It was clear from the teaching practices of this author that a large number of students face obstacles in understanding modelling in fashion design when carried out in a creative way. This includes the question of how to devise a system to transpose the mould of items of clothing to form a three-dimensional structure. Problems regarding the teaching and learning system of modelling have been found in other research studies (Caro et al, 2014; Costa e Lima, 2006; Menezes e Spaine, 2010). The same problem of how to transpose the mould to a display bust in situations when the students were visually handicapped (either suffering from poor eyesight or blindness) was made apparent during the "Feeling and Dressing for Fashion" course which was run for this public.

In the light of this evidence, it was decided to investigate how this problem could be overcome. This involved attempting to introduce features in the learning which could enable the students to obtain infor-

DOI: 10.4018/978-1-5225-0510-5.ch012

mation about transposing a two-dimensional mould to a three-dimensional human body, as represented by the display bust. In this way, it was possible to recapture the whole period when they made the attempt or in other words, the learning could be shown to be complete.

The course referred to above is based on a higher education course in fashion design which was concerned with the learning activities for the creation of items of clothing and was aimed at visually handicapped people (VHs), such as students. The aim of this study is to describe the stages that must be followed by students in the area, whether they are students with normal sight or VHs, when undertaking all the detailed tasks of modelling and their application for creating an item of clothing. The purpose of this is to outline the necessary parameters for a methodology in which the teaching and learning of fashion design and the VHs are interwoven, and in this chapter, link this to a planned model for fashion design.

In seeking to employ a methodology which could assist in the key areas of the subject (apart from the research conducted in this area by the authors cited above) some important benchmarks were found in the work of Medeiros (2010), Souza (2007), Spaine (2010, 2013). These are research studies on the teaching and learning of planned modelling that underpin the ideas formulated here. In this way, an attempt is made to determine what resources will be employed, apart from the usual materials made use of in the classroom, for the teaching of modelling in fashion design to the students in question, while taking full account of their particular learning needs. According to Menezes and Spaine (2010), "the preparation of a mould includes a phase which involves studying ergonomic factors, anthropometry and a knowledge of the wearer´s body", and for this reason, an understanding is required of the forms, volumes and structures which shape the human body.

The mould for an item of clothing is first cut out of paper and is made up of all the parts of the garment; care is taken to ensure that it has the correct physical volume when it is manufactured in fabric and later worn. Each of the parts contains the formal elements needed to make the adjustment to the corresponding body of the wearer, with the aim of ensuring a match and meeting "the needs of comfort, durability and functionality" (Menezes & Spaine, 2010). However, the need to understand this is not always grasped by the students because they are unable to visualize the planned object – the mould and its parts – or obtain the right volume for the wearer´s body. As can be expected, this is a problem which students (whether able to see or not) have difficulty in tackling.

In view of this, this course combines knowledge related to the area of fashion design with research studies originating from different specialists in Neuro-science (Ackerman, 1992; Carter, 2012; Damásio, 2004, 2011, 2012; Martin, 2013;), particularly, factors related to cognition, constancy and abstraction with regard to information obtained from the surrounding environment and the objects found there (Gilead et al, 2014; Perlovsky e Ilin, 2012; Nee et al, 2014; Zeki, 2000, 2001). These are factors of great importance for learning in so far as the activities of each of them allow objects to form and materialize in the realm of fashion design projects. It is also important to enter into a dialogue with research studies aimed at the education of people with special needs who include those with visual impairments (Ballestero-Álvarez, 2003; Barros, 2013; Lima, 2006).

ABSTRACTION AND LEARNING

The brain is an organ that makes continuous contact with the environment. It is the means by which a person is able to assimilate information, reflect, understand and take action. Moreover, it intervenes in one´s feelings and the way that the outside world is perceived, as well as taking part in human activities and behaviour. Learning and the knowledge which can be obtained, comprise cerebral activities either through the simple recognition by a child of its own body or something which is accumulated throughout our lives and includes every kind of knowledge that is acquired.

However, some of these activities are concerned with the way the body itself functions and what is required for its maintenance and the existence of life. (Damásio, 2012). In view of this, the brain communicates both within and outside the human body, and this communication takes place through the interconnecting parts of the brain and the sensory motor system (Sporns, 2003).

The whole process of communication, whether of the brain with the body or the body with the surroundings, takes place by means of energy or electric impulses. These are currents that leave the brain in the direction of inner organs or the organs of the senses and then return directly to the brain or medulla. In the case of everybody, the brain forms what is called a mental image either of objects and the environment or of vivid moments which are understood to be "visual, auditory, tactile, olfactory or gustatory images" (DAMÁSIO, 2004, p. 205). In these cases, this includes the learning of specific materials such as those related to fashion design that are examined in this research study.

Thus it can be undserstood that there exists a complex system that is able to bring about a wide range of connections that involve distinct parts of the human brain. These connections evolve in the cerebral cortex which include

the physical structures responsible for most of what we call ''brainwork": cognition, mental imagery, the highly sophisticated processing of visual information, and the ability to produce and understand language. But underneath this layer reside many other specialized structures that are essential for movement, consciousness, sexuality, the action of our five senses, and more—all equally valuable to human existence. (ACKERMAN, 1992, p. 13).

The motor actions might be related to others of a sensitive character which means that there are different lobes in action at the same time. What features prominently in this research study is the limbic lobe, also referred to as the emotional brain (Ackerman, 1992; Martin, 2013), which is responsible "for many of the basic and emotional drives and associated with involuntary behaviour which is important for animal survival: pain and pleasure, fear, rage, sexual desires and even docility and affection" (ACKERMAN, 1992, p. 23).

Together with these functions, it is worth drawing attention to those that involve memories, thoughts and emotions. According to Martin (2013, p. 386-387):

The different sensing and motor systems are structurally and functionally independent regions that are only interconnected at the highest levels of independence [...] the neural circuits for reward, emotions and memory are closely integrated from the beginning.

Memory is a key factor of the study since it maintains a strict relationship with teaching and learning processes, as well as with the data on the links between the components of the limbic lobe which are interactors in these processes.

However, before memories are formed, the brain has to enter into contact with the environment and the objects that comprise it, whether this object is a landscape seen in the distance or something closer which can be touched. In either case, as Zeki (2001, p. 51, 52) makes clear, the brain processes this information in accordance with the laws of constancy and abstraction.

These laws (Zeki, 2001), establish a direct relationship with vision, since constancy and action are destined "to seek knowledge of the constant and essential properties of objects and surfaces when the information reaching it changes from moment to moment ".

It should be stressed that although the investigations of this author are aimed at the sense of vision, it can be assumed that the recognition of the properties of objects can take place through other sensory channels such as: a) touch for identifying surfaces and temperatures and the sensations these provoke, b) hearing through the detection of distinct sound frequencies emitted by an object or another individual; c) the palate through the recognition of tastes and textures - and this has a direct link with d) the olfactory system for the identification of smells. Thus the sensory mapping of an environment or object is shaped in this intertwined relationship between the senses which is made available and this is an example of what occurs to visually handicapped people.

With regard to abstraction Zeki (2001, p. 51, 52) states that:

The second Law is intimately linked to the first because abstraction is a critical step in the efficient acquisition of knowledge; without it, the brain would be enslaved to the particular. The capacity to abstract is also probably imposed on the brain by the limitations of the memory system because it does away with the need to recall every detail.

Abstraction is thus a natural practice since, as the author explains, it constitutes activities involved in the recognition of objects. In the view of Nee et al (2014, s/p), there are different degrees of abstraction which in turn are processed or apprehended in distinct regions of the prefrontal cortex, and "abstraction thus refers to the degree of processing/representation which is either linked to or separated from particular bodies". They stress the fact that the theories diverge with regard to the modus operandi of abstraction and draw attention to two of them – temporal and relational – each of which acts in its own way and gives distinct responses to the stimuli received by the brain. In these cases the kind of processing involved is linked to particular types of thoughts and intentions. "Temporal abstraction follows from the notion that more abstract goals (e.g., "buy a car") control responses for longer periods of time than more concrete subgoals (e.g., "go to dealership," "test drive," etc.)".

As a result, the temporal relation differs from the relational, to the extent that, in this case "The stimuli formed a pattern with one or more dimensions (e.g., shape, orientation, size) changing in an orderly manner across rows and down columns of the matrix" (Nee et al, 2014, s/p.). As pointed out earlier, there are two means of making contact with the objects, which are investigated in a distinct way by Zeki (2001). However, Nee et al (2014), highlight the fact that, despite differences, this factor does not imply that the two types of abstraction must be processed separately.

Perlovsky and Ilin (2012, s/p.), take up another position. Although these authors do not set out a clear neuro-scientific definition of abstraction, they state that the cognitive models are vague and uncoordinated. They add that "These crisp models of purpose are cultural models, formulated in language", or in other words, in the opinion of these authors, the representation of the objects is grounded on recognizable features of these objects. These are cultural traits, resulting from a convergence of factors which allow a configuration of objects which they can cohabit and which emerge from key reference points in the brain and memory, combined with their sensory perceptions.

Gilead et al (2013, s/p.) carried out experiments with 24 participants with a view to mapping cerebral regions for abstraction and concrete action and concluded that the former demonstrates heterogeneity with regard to specific areas. That is it can take place in different cerebral regions. According to these authors: "While the road from an abstract thought to concrete action appears paved and organized across the fronto-parietal cortex, the opposite direction (turning a particular object into an abstraction) goes through multiple sporadic, less-charted paths".

It is clear from what is pointed out in these research studies, that on the question of abstraction, at least three issues deserve attention: a) abstraction occurs in the presence of a stimulus originating from an object, for example, an image, sound, smell, taste or texture; b) there is a relationship between the perception of the object, memories relative to it and the ´concretization´ of the image of this object by the brain; c) cultural reference-points are important in so far as they act as a connecting-link between abstraction and ´concretization´. These points are important in the sphere of teaching and learning, in particular with regard to modelling in fashion design because the relationship between the mould and the body is to some extent, regarded as stemming from a recognition of the body of the wearer and the way that each part shaped on the mould is adapted to support this.

In the following section, two correlated issues will be discussed: learning processes and memory.

Learning and Memory

It is evident that several areas have been investigated in studies of neuro-science. The methods so far employed by neuro-educationalists, psychologists, pedagogical specialists and teachers have been aimed at broadening the knowledge taught by teachers in different subject-areas. With regard to teaching and learning, It is argued that many of the factors revealed, concern planning classes and the ways of applying and conveying knowledge to the students.

On the other hand, there are students who come across pedagogical methodologies which although grounded on an educational framework, disregard the need for reflection and fail to recognize the value of spontaneous memory for learning the content of their lessons. In reality, this enables students to pass their exams at the end of their studies but does not encourage them to consolidate what they have learnt. (Zaro et al, 2010; Santos, 2011; GrossiI, Lopes & Couto, 2014).

In this encounter between what is taught and what is learnt, there can be found billions of interconnected brain cells in different regions – neurons that allow someone to make sensory-motor responses, or in other words, "to modulate the act of signalling for a perceptive process" (BALDISSIN, 2014, s/p.).

For this reason, the study of neuro-science is becoming a crucial research field in this area. An awareness of cerebral processes, the mapping and forming of images and an understanding of human behaviour are tools which can be of benefit to the work of educators and in particular, to students. Understanding how the mind works, at the same time means learning to discern what takes place in the body. Damásio (2011, p. 139) believes that: "the living body is the central place. There is a need and drive to regulate life. The mapping in the brain is the driving-force and mechanism which transforms the simple rules of everyday life into a rule which can act as an intermediary for the mind and finally a rule for the conscious mind".

Both the teacher and the students must be involved in this scene so that they can make a joint contribution to learning. The stimuli are needed for the pupils because they can employ them when in the position of being both spectators of a class and actors within the learning process.

As can be foreseen, "studies in neuro-education are bound up with the emotions in all the senses" (TOKUHAMA-ESPINOSA, 2008, p. 63). In her research, this author underlines certain principles that are linked to teaching & learning, the first of these being the fact that "students learn better when they are highly motivated than when they lack any inner drive" (TOKUHAMA-ESPINOSA, 2008, p. 78). In other words, the presence of stimuli in the educational process triggers a reaction in the brain which is linked to sensory-motor patterns that carry out various functions including memory (Damásio, 2011).

On the question of education, it should be borne in mind that bringing some item of information to the memory has for a long time had the same meaning as learning by rote (Medeiros & Bezerra, 2013). Moreover, what is underlined here is precisely the need to explore another avenue – this entails understanding how the brain functions and the ways that the brain responds to more distinct stimuli emanating from the teaching/learning process. For this reason, the authors distinguish between two groups of memory and stress that the first of these is of crucial importance in the learning process. They are: "declarative or explicit memory which involves conscious knowledge and non-declarative or implicit memory which is related to incidental or non-conscious knowledge" (Medeiros & Bezerra, 2013, p. 183).

This activity implies that in the case of every human being, the information acquired should either be hoarded or rejected depending on its relevance to learning. Estrela and Ribeiro (2012, p. 144) explain that it "is impossible to imagine any human activity whether mental, motor or affective, without memory playing an active role. Without this, it would be necessary to learn how to carry out the same tasks every day, such as for example walking, speaking, reading, recognizing people and objects etc.".

Memory makes it possible to learn new knowledge and the relationships that allow supplementary material to be added to what has been taught and thus help to broaden the knowledge of a given subject.

The limbic system, discussed in an earlier section, and which is concerned with a varied set of cortical and sub-cortical regions that are essential for human behaviour, is closely linked to the processing and consolidation of memory, in particular the opkerations of the Hippocampus. Explicit memory and some cognitive aspects of memory take place in this region. As well as the Hippocampus, the Amygdala also has connections with memory in so far to it takes part in the acquisition, consolidation and storing of emotional memories, because it is in this region that the neural circuits involved in the behavioural emotions and expressions are processed (Martin, 2013; Ackerman, 1992).

The memory of facts, events and objects can be brought to the attention of the mind, while images are not reliable for the same facts, events and objects. In the opinion of Damásio (2011, p. 169), "what we generally call the memory of an object is a memory made up of sensory and motor activities arising from the interaction between the organism and object during a given period of time".

Several systems interact to bring about distinct processes of acquisition, assimilation of knowledge and subsequently, the consolidation of this information so that it can be committed to the memory for a long period of time A short-term memory can restore the whole context in which it is embedded because both sensory and motor factors are important for remembering a piece of information or an object. (Damásio, 2011).

The issues examined in this section can be combined – in other words, the teaching and learning processes, memory and abstraction. In undertaking this study, it should be understood that the relations between these factors have proved to be of great importance. This is due to the fact that learning and the contact with what is taught is filtered and reaches layers that act as reference-points during the period when abstraction takes place and as a result, the assimilation and remembering of new information can perhaps be consolidated. This procedure continues for the processing and representation of objects – or in other words what is learnt at every moment – in particular with regard to institutionalized teaching; it can enable a person to undergo each stage of learning and have a greater command over what is taught. In other words, it can be assumed that undertaking exercises based on memory or a set of benchmarks based on individual memory, can become a strategy for the consolidation of cerebral maps related to the knowledge conveyed in the classroom.

MATERIALS AND METHODS

At the beginning of this chapter, there was a description or a recurring learning problem among fashion design students in the subject-area of modelling. This was the question of how to transpose parts of a two-dimensional flat mould to the three-dimensional volume of the body of the wearer – and thus allow the item of clothing to be made. This difficulty was encountered by the author while giving some lessons in the "Feeling and Dressing for Fashion¨ Course in the classroom, both when supervising projects with students with normal sight and also with students who were (partially or wholly) visually handicapped.

It was realized that this obstacle was bound up with some key factors involved in teaching and learning with regard to abstraction. In other words, both a number of students with normal sight and all the student with visual impairments were helpless when it came to understanding how the two-dimensional mould that was formed with modelling paper or cardboard, could be adapted to the shape of the physical body. It was noted that this lack of understanding on the part of the students arose from the problem of how the prepared cut-out paper design could be adapted to the way the volume of the body was shaped and in particular, how the combined parts of a mould are able to conform to this structure.

The process of creating a collection of garments follows a sequence of different stages during which the creation of the moulds takes place. Each of these stages can be shown, together with a fashion design project, to provide the students with a preliminary understanding. The procedure can be undertaken in two ways:

1. A planned schedule that is aimed at building up a collection; or
2. The direct choice of a fabric.

A number of tasks continue the sequence in the following stages:

1. Illustrative sketchy drawings of the model;
2. Carrying out the technical design and specifications for the trial scheme;
3. Preparation of the mould;
4. Trialling or prototyping, when the garment is manufactured for the first time;
5. A clothes test when corrections are made to the trial garment;
6. Correction of any flaws in the trial garment or mould (if necessary);
7. Approval of the trial garment.

The illustration below shows the first stages of this process or in other words, until the time when the mould is formed, which is when the fashion design students first have problems in understanding the procedure (Figure 1).

This difficulty which was found in the teaching and learning of pupils with normal sight doing a course in fashion design, was also a persistent problem when teaching visually handicapped fashion design students.

After giving reflection to the difficulties of these pupils and conducting an analysis of the process of modelling a fashion design product, it became clear that there was a need to work out a means of assisting students to assimilate the information required for transposing the two-dimensional features of a mould to the three-dimensional nature of the wearer´s body.

Figure 1. Image illustrating the stages for forming a collection in fashion design. 1) choice of the fabric; 2) illustrative sketch of the model; 3) technical design of the model; 4) creation of the mould
Source: author's property

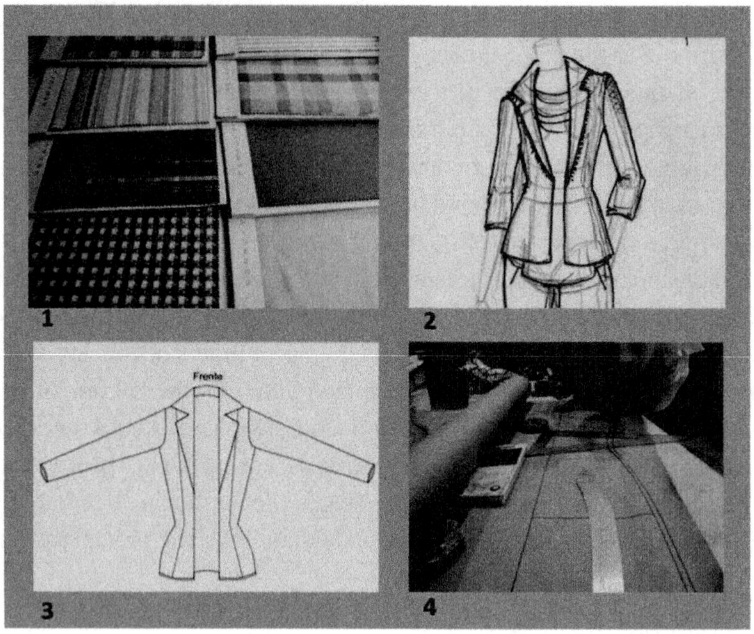

After making a series of observations of this group, it emerged that among the visually-handicapped students (VHs), there is a difficulty in understanding how to shape an image with flat objects, particularly when they are of such a large size that it is impossible to reach out and touch them. What this means is that it is hard to have an understanding of how to determine the whole object if the size, shape and lines of a flat object extend beyond the length of an arm. A similar situation can arise when one is dealing with a three-dimensional object. However, when recognizing the shape of a modelling display bust, this can be overcome since it is possible to grasp the physical volume because in this case, when the display bust acts as a representation of the body, it can be adjusted to the structure, shape and physical volume of the student and a suitable match can be made.

On the basis of these premises (and with a view to comparing matters related to the two-dimensional features of the mould with the three-dimensional body), it was decided to reverse the procedure traditionally followed in modelling classes. This meant that instead of beginning the lesson with the paper mould, the garment draped on the modelling display bust was used as a point of departure. An exercise carried out in an earlier stage of the procedure was resorted to (which dealt with the relationship between flat and volumetric objects, combined with the stimuli present at that time) and this was employed when planning this experiment, which in this case was much more complex (Figure 2).

Procedures

Three stages were planned for this process which was concerned with the "construction" and "deconstruction" of the modelling. These stages were closely aligned but each addressed specific objectives, which are described as follows. Stage 1 follows the path of construction to deconstruction. In this the

Figure 2. Two-dimensionality and three-dimensionality in the shapes and volumes of items of clothing. The picture shows teaching material used in the lesson with visually handicapped students, with flat moulds that increase the volume when wrapped round the body of the assembled dolls.
Source: author's property.

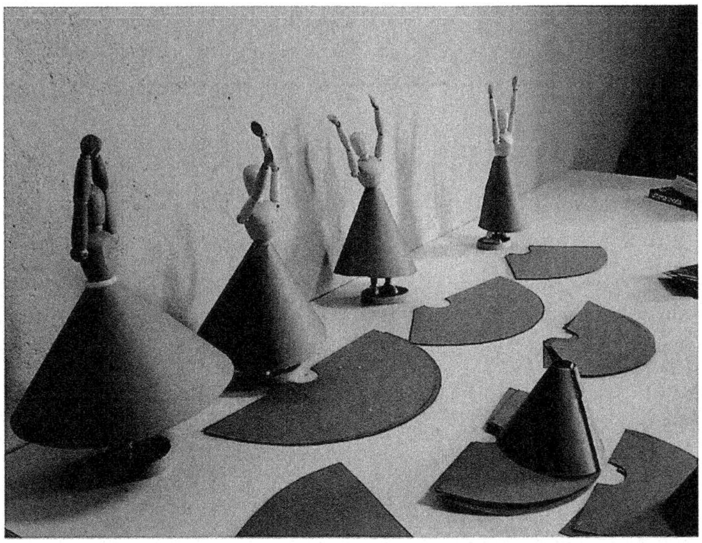

four students carried out the task of recognizing an item of clothing (a blazer) which was made of fabric and draped round a modelling display bust. The garment only showed parts that were fitted with pins but in general tyerms, it could be recognized as a blazer. After they had recognized the model, the students were requested to take out the pins bit by bit and undo the attachments between the different parts of the garment. The procedure began on the sides so that the structure remained supported on the shoulders of the display bust and thus allowed the handling to continue. When the sides were open, it was suggested that the corresponding moulds should be taken to the front and back and placed over the equivalent parts of the fabric. Through adopting this approach, the students began the process of ¨association by abstraction¨ of each part that they worked on.

It should be pointed out that the moulds were delivered ready made to the students and that at no time during the experiment did they design or cut the modelling that was used. This was not the object of the exercise and although what is in question is a course in fashion design, the learning does not usually involve the students in having to form the mould for the garments created for their collections, since this task is undertaken by a professional in the area who is the modeller.

In the next step, the students were requested to support the parts of the mould on the modelling table to give continuity to the process. This sequence entailed taking out all the pins and ensuring the parts of the mould in the fabric were completely separated so that these could be superimposed on the paper mould and the corresponding parts of the fabric and paper could be matched.

Stage 2 follows the same path as Stage 1, although with some minor alterations. The material garment (a short-sleeved blouse), which was draped on a modelling display bust, was taken to pieces and the material parts were directly superimposed on the corresponding parts of the paper mould. This stage of the process omitted the task of superimposing the paper mould on the material item that was still draped on the modelling display bust. However, after it had been taken apart, the garment was reconstructed on the bust and its parts pinned directly on the structure so that the item could remain recognizable after all the handling.

However, while the two stages discussed above were largely concerned with the principle of deconstruction, Stage 3 went in the opposite direction. The parts of a mould (for a long-sleeved shirt) were found on the modelling table cut in paper and their corresponding parts in material. This time the students were asked to take out the parts of the material which were then positioned on the paper, and attach them to the bust so as to form the structure of the garment by pinning together all the parts, beginning at the shoulders followed by the armpits and sides of the shirt.

The material that was used for the deconstruction and construction of the three garments was denim, which is a hard-wearing serge that is easily recognizable by the students. The use of a tough raw material was recommended as a means of making it easier for the student to handle (Figure 3).

The three stages were taught in a lesson but Stage 3 was repeated in three subsequent classes so as to fix what had been taught in the students´ memories. This set of basic moulds was supplemented with an exercise which required the students to superimpose the parts of the moulds on the manufactured garments so that they could remember what they had learnt both with regard to the parts above (the trunk) and below (the legs) (Figure 4).

RESULTS OBTAINED AND DISCUSSION

After the expkeriment had been conducted, it was confirmed that the methodology employed had proved effective to the extent that it consolidated a system of learning with regard to the transposing of sizes between the mould and body. Attention should be drawn to a number of points:

Figure 3. Piece of fabric resulting from flat modelling, smoothed out at the front. The mould of the fabric is superimposed on cardboard to form a relationship between the parts of this mould.
Source: author's property.

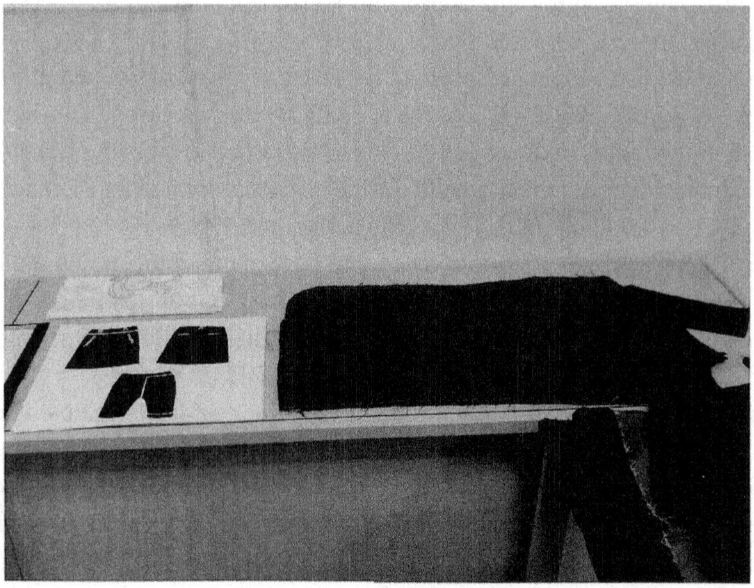

Figure 4. Recognition of the parts of an item of clothing. The handling of the garment on the modelling display bust allows key factoirs to be recognized such as points of support, the exact location of areas of the body and how these match with the item of clothing. Source: Lotus Art propertry. Used with permission.

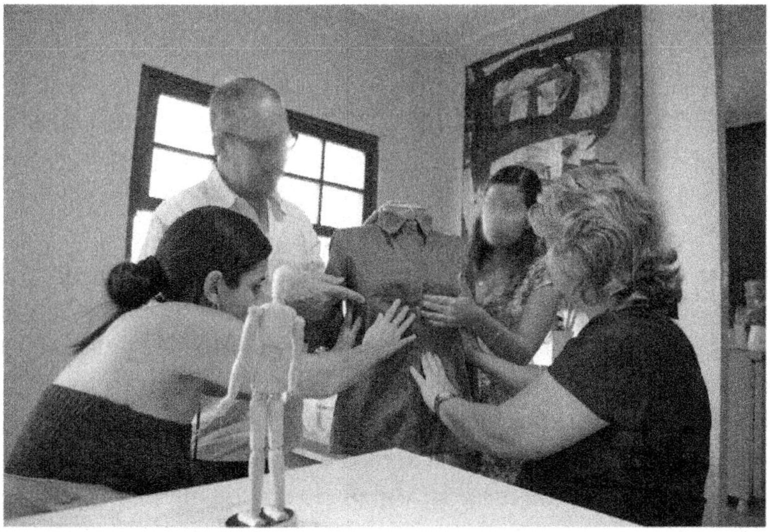

1. **A temporal equivalence must be established between the transmission and acquisition of information for a VH:** A visually handicapped person acquires information in a different way from people with normal sight because in their case, their vision reveals a picture made up of a series of identifiable objects seen in the distance. In the case of solid objects, the VHs must have the experience of touch which is a sense that reveals the details, texture, shape and volume of the object. With regard to visual images, an auditory description is necessary so that the VH can form the image of what is being seen. In both cases it can be said that the process of abstraction (Gilead et al, 2014; Nee et al, 2014; Perlovsky e Ilin, 2012; Zeki, 2001) is of crucial importance in so far to it enables the student to form an image and make a record through contacts with objects and the reference-points that can be found in his memory (Damásio, 2011). In view of this, whereas visual information is processed in an immediate way for the person with normal sight, in the case of those with visual impairments, there is a need for temporal space to allow this information to be identified and assimilated. It should be added that in attempting to cope with objects, the relations that they establish, and the patterns recorded in memory, which are closely bound up with emotions and feelings (Damásio, 2011), the body and brain display reactions and attitudes which in the case of learning, can be either beneficial or harmful to the student.
2. **VHs and people with normal sight interpret a mould of an item of clothing in different ways but the use of the methodology recommended can come to be effective for people with normal sight too:** The experiment carried out with visually handicapped people shows a system that can be applied in the same way to students with normal sight. The exercises undertaken stimulate the cerebral regions that are common to people with or without this impairment. It is not a question of whether or not someone has normal eyesight but rather, of how the stimuli that are triggered during the exercise are able to have the potential for neuro-plasticity and be important for everybody. As Nee et al (2014, s/p.) state "Humans have the remarkable ability to rapidly adjust behavior in

accordance with contextual demands. This flexibility of action is thought to depend upon the functions of the prefrontal cortex (PFC)". Employing deconstruction and construction as a learning method can be an effective tool to assist in the apprehension of kinds of knowledge that are not always formulated in concrete terms by fashion design students. It is believed that adapting to the teaching and learning process can broaden the interests of the students and thus make them capable of an induction of memory in the long term. As already discussed, the limbic cortex established cerebral connection that allow memory to be processed and consolidated in both the Hippocampus and Amygdala regions. Neural circuits can be found in these regions that are involved in emotions and expressions of behaviour (Martin, 2013; Estrelae Ribeiro, 2012; Damásio, 2011), and these are important for learning.

3. **The sequential procedure involved in dismantling and assembling the parts of an item of clothing is conducive to the formation of images. The perception of an object takes place at distinct times although these tend to overlap:** Damásio (2011) makes clear that images are formed in the brain as the result of relations that it establishes with the surrounding environment and also with the body itself. By means of the sense of touch – the most widely employed of the other senses in this experiment – it was possible to establish a link with, for example, equivalent factors between the moulds of paper and material. Apart from allowing a recognition of the differences of texture in the materials, the malleability of the fabric made it possible to adapt each part of the mould to the modelling display bust, which is a technical representation of the body of the wearer. According to Menezes and Spaine (2010), when thinking of a fashion product, it is necessary to take account of certain factors arising from a knowledge of the wearer´s body. This is to ensure that the model is suitable to its physical shape and to note the characteristics of the fabric and how it can be adapted to the body. In working on ideas for construction and deconstruction, each stage can be of value in ensuring that thoughts about future planning can be increasingly sophisticated. Work was also undertaken in two-dimensional and three-dimensional areas. The transposition between the two dimensions requires students, fashion-designers and professional modellers to solve problems which are linked to an understanding of two types of objects – flat and volumetric. That is they must create distinct images from a single shape that has particular configurations, depending on the purpose it serves. The mould, as a representation of a garment, also begins to act as the representation of the wearer´s body.
4. **It is easy to recognize and identify the mould from the textile material. The capacity of the fabric to adapt to the modelling display bust can assist one's understanding:** Students, and in general everybody else, are able to recognize an object more easily when they are familiar with its key features. Fabrics are the most widely-used raw material for the manufacture of items of clothing whether as fashion goods or clothes for professional purposes like uniforms. When working with denim – or jeans (which is a kind of serge or in other words, a strong hard-wearing textile product and a raw material that can be found in most people´s clothes) it was possible to reproduce all the details in the moulds of all the items used in the different stages of the experiment. Fabric acts as a possible means of tactile communication, initially as a substitute for paper and later because of its malleability, flexibility, resistance and stability. (Saltzman, 2004). With this material it was possible to trace all the details present in the modelling and to effect the transposition of the flat surface to the structure of the body represented by the modelling of the display bust. In separating the parts stitched together by pins, the fabric remained stable and allowed the mould to be superimposed on each of the component parts of the garments used in the experiment. Another key factor was to

check if the fabric of the bust was transposed to the flat mould because in this case, the handling of the raw material became simple owing to the qualities arising from its structure. Once they had recognized the lines, curves, and cuttings found among the two, the students showed more confidence in locating each item found in the paper with regard to the modelling display bust.

5. **As each new procedure was embarked on, the reassembly of the item of clothing proved to be easier or, it could be said, more natural to the students. The exercise carried out with other kinds of modelling achieved positive results which suggested that after the consolidation of this memory, the brain began to rely on this process being remembered and this could perhaps lead to other experiments:** The experiment which was divided into distinct and overlapping stages, resulted in a learning process that did not simply serve as a means of being given information. As well as being provided with a theoretical explanation with regard to preparing the mould, the students became involved in a practice that brought about a recognition of what was presented to them in theoretical terms. By touching, handling and investigating parts of the moulds, it is possible to identify the common areas between them such as the shoulders, the thorax (also called the bust in the case of female moulds), the waist and the hips. Hence when the students are requested to reassemble the moulds, these common areas begin to establish a way of matching the wearer´s body as represented by the modelling display bust. The reassembling is begun in the region of the shoulders because on the basis of previous exercises, there is an awareness that this is clearly the main point of support for the garment on the body (Saltzman, 2004). Furthermore, it can make it possible to adapt to the structure of the body in a way that allows the other areas to be harmonized. The assembly/re-assembly carried out in this encounter and repeated in subsequent lessons, is a possible means of consolidating the memory with regard to this kind of activity. Once this stage has been completed in Stage 1, the students are better prepared to undertake it in Stage 3.
6. **The exercise achieved satisfactory results even when it was suggested that there should be a direct assembly on the bust without any previous deconstruction, as carried out on the first occasion:** As mentioned in the previous section, the repetition of the exercise proved to be necessary for the retention or consolidation of memory (Martin, 2013; Damásio, 2011; Ackerman, 1992), with regard to the planned experiment. Once memorized by the students after successive repetitions, the initial linearity of the process could be broken. In other words, covering the intermediary stages did not prove to be anything more than a pretext for the final stage. This can now be attained directly. Another point that is worth highlighting in the method employed was to carry out an exercise as a kind of game or in other words, the practice was exercised as a ludic experience. In this case, what was sought, was to reject the generally prejudiced idea that making the transposition was difficult and to turn the activity into a kind of jigsaw puzzle. This was because when the plan for deconstruction and construction was attached to an attempt to unite all the parts of the moulds, the example could be given of a game which could only be completed when all the pieces are able to form a single shape or image.
7. **The recognition by the participants of the possibility of completing the exercises can be understood as a driving-force and hence a stimulus:** After the three stages of the experiment had been completed, the suggestion that the exercises should be carried out in subsequent lessons was greeted in a more relaxed way and with a certain degree of enthusiasm. This was because having managed to complete the stages in sequence, they felt encouraged to repeat Stage 3 at other times. It was found that what had originally been a cause of apprehension (or more specifically, when the students were able to overcome the problem of making the transposition from the two-dimensional

features of flat modelling to the three-dimensional aspects of the modelling display bust), they were instilled with confidence and enthusiasm and eager to carry out the task when requested.

FUTURE RESEARCH DIRECTIONS

After the method outlined here had been employed for visually handicapped students, there was found to be a need to validate it with groups of students with normal sight as well.

The example of modelling in research design pointed to other subject-areas which are taught in a way that requires students to learn classroom material by rote (Medeiros & Bezerra, 2013). This approach does not conform to contemporary academic standards and fails to give scope for a broader reflection of what is taught and hence fails to act as a stimulus for the students to develop their skills. The subject of modelling is closely bound up with the development of fashion products and in the sequence of stages outlined earlier, some other key questions emerged. These included the constraints experienced by fashion design students when seeking to recognize the fabrics or the way they are formed and the materials used in their manufacture. Moreover, there is a lack of due attention to the question of stimulating students in fashion design to make improvements in their training.

The use of materials is included in an earlier stage of modelling and as a result has an adverse effect on learning how to form textile structures since it is insufficient for an appropriate preparation of the moulds. In the light of this, the study of Textile Design is recommended as the next stage in devising a teaching method for fashion design courses that can lead to other means of sharing knowledge between teachers and students which take full account of neuro-education as an indispensable aid.

CONCLUSION

This study found that fashion design students with normal sight have difficulties in adapting the structure of a two-dimensional mould to the corresponding parts of the wearer´s body as represented by the modelling display bust. The same problem arose in the case of students who are visually handicapped – either partially or completely. This was addressed by conducting an experiment for students with impaired sight which entailed carrying out their modelling activities for fashion design in a ludic way. The purpose of this was to allow students to make connections with the materiality of the moulds and find a way of matching two-dimensional and three-dimensional features on the basis of an abstraction and consolidation of the objects involved. When the experiment was carried out it was subdivided into three stages in which the construction and deconstruction of the items of clothing was viewed as a means of studying the patches of fabric and the way the parts of the mould could be fitted on the body – or modelling display bust. It was concluded that procedures adopted in each of the stages of the experiment (and the repetition of Stage 3 at different times during the course) allowed the students to determine the features of each part of the mould. This enabled them to associate it with the modelling display bust and match the corresponding features between the parts of the mould and the wearer of the fashion garment in a satisfactory way. It should be understood that as well as seeking to underpin the theoretical framework established for flat modelling, the method employed is designed to broaden the scope of teaching/learning and devote particular resources to the apprehension and consolidation of memories related to the issues discussed here. For this reason, it was believed that combining the methodological aspects of teaching modelling with the fundamental principles of neuroscience would be of significant value to fashion design and neuro-education. Moreover, the methodolofical framework outlined here can be extended and adapted to other subject-areas and has a significant potential for broadening the line of inquiry into fields of design and neuroscience. This experiment provided evidence that as in the case of visually handicapped students, students with normal sight can achieve similar results. This is of great value in improving teaching and learning methods in this area and broadening the opportunities of the use of modelling for the development of fashion design products.

REFERENCES

Ackerman, S. (1992). *Discovering the Brain.* National Academies Press.

Baldissin, M. (2014). Learning and human perceptions according to anthroposophy and neurosciences. *Arte Médica Ampliada, 34*(1). Available at http://www.abmanacional.com.br/arquivo/f479d9701ee-676a3553251963d0515d0bb7c9d23-34-1-aprendizado-e-percepcoes.pdf

Ballestero-Alvarez, J. A. (2003). *Multisensorialidade no ensino de desenho a cegas* [Multisensoriality in the teaching of design to the blind]. São Paulo: USP. Available at http://www.teses.usp.br/teses/disponiveis/27/27131/tde-21032005-213811/pt-br.php

Barros, A. B. (2013). *Processo de inclusão no contexto da deficiência visual: dificuldades, desafios e perspectivas* [Social inclusion in the area of visual impairment: difficulties, challenges and perspectives]. (Master's Dissertation). São Luís, Brazil.

Caro, C. R., Rodrigues, J. R., & Pedro, E. S. (2014). *As técnicas de modelagem plana e moulage e suas aplicações na indústria do vestuário* [The techniques of modelling and moulage and their applications in the clothing industry]. Available at http://www.contexmod.net.br/index.php/segundo/article/view/141

Castilho, K., & Vicentini, C. G. (2008). O corte, a costura, o processo e o projeto de moda no re-design do corpo [The court and dressmaking – fashion projects in redesigning the body]. In Corpo e moda: por uma compreensão do contemporâneo [The body and fashion: through an understanding of the contemporary world]. Barueri, SP: Estação das Letras e Cores Editora.

Costa, F. Z. N., & Lima, I. M. M. R. D. (n.d.). *Metodologia de ensino de modelagem: uma experiência pedagógica no curso de moda* [A methodology for teaching modelling: a pedagogical experience in fashion courses]. Available at http://www.coloquiomoda.com.br/anais/anais/2-Coloquio-de-Moda_2006/artigos /46.pdf

Damásio. (2004). *António. Em busca de Espinosa: prazer e dor na ciência dos sentidos* [In search of Espinosa: pleasure and pain in the science of the feelings]. São Paulo: Companhia das Letras.

Damásio. (2011). *António R. E o cérebro criou o homem* [The brain created mankind]. São Paulo: Companhia das Letras.

Damásio. (2012). *O erro de Descartes: emoção, razão e o cérebro humano* [The mistake of Descartes: emotions, reason and the human brain]. São Paulo: Companhia das Letras.

Estrela, J. B. C., & Ribeiro, J. D. S. F. (2012). Análise das relações entre memória e aprendizagem na construção do saber [An analysis of the relationship between memory and learning when building knowledge]. *Caderno Intersaberes, 1*(1). Available at http://www.grupouninter.com.br/intersaberes/index.php/cadernointersaberes/issue/view/52

Fiorini, V. (2008). Design de Moda: abordagens conceituais e metodológicas [Fashion Design: conceptual approach and methodologies]. In Design de Moda: olhares diversos [Fashion Design: various viewpoints]. Barueri: Estação das Letras e Cores Editora.

Fornasier, C. B. R., Martins, R. F. D. F., & Demarchi, A. P. P. (2008). O ensino da disciplina de desenvolvimento de projetos como sistema de gestão de conhecimento [Teaching the subject-area of designing projects as a system of knowledge management]. In Design de Moda: olhares diversos [Fashion Design: various viewpoints]. Barueri: Estação das Letras e Cores Editora.

Gilead, M., Liberman, N., & Maril, A. (2014). *From mind to matter: neural correlates of abstract and concrete mindsets*. Available at http://www.ncbi.nlm.nih.gov/pmc/articles/PMC4014104/

Grossi, M. G. R., Lopes, A. M., & Couto, P. A. (n.d.). *A neurociência na formação de professores: um estudo da realidade brasileira* [Neuro-science in teacher-training: a study of the Brazilian situation]. Available at http://www.revistas.uneb.br/index.php/faeeba/article/view/821

Herculano-Houzel, S. (2009). *Brain Development: The Generation of Large Brains*. Elsevier.

Kheirbek, M. A., & Hen, R. (2015). *Mais neurônios, mais equilíbrio* [More neurons, more equilibrium]. São Paulo: Editora Segmento.

Kuhl, P. K. (2010, September9). Patricia K. Brain Mechanisms in Early Language Acquisition. Published in final edited form as. *Neuron, 67*(5), 713–727. doi:10.1016/j.neuron.2010.08.038 PMID:20826304

Lima, P. A. (2006). *Educação inclusiva e igualdade social* [Inclusive education and social equality]. São Paulo: Avercamp.

Lipovetsky, G. (1989). *O império do efêmero: a moda e seu destino da sociedade* [The empire of the ephemeral: fashion and its social goal]. São Paulo: Companhia das Letras.

Martin, J. H. (2013). *Neuroanatomia: texto e atlas* [Neuro-anatomy: text and atlas]. Porto Alegre: AMGH.

Medeiros, S. M. T. D. (2010). *Reflexões iniciais sobre a didática de ensino da disciplina de modelagem plana a partir do design e da ergonomia* [Preliminary reflections on the didactic principles of planned modelling based on design and ergonomics]. Available at http://www.coloquiomoda.com.br/anais/anais/6-Coloquio-de-Moda_2010/71756_Reflexoes_iniciais_sobre_a_didatica_de_ensino [Preliminary reflections on the didactics of teaching] .pdf

Medeiros, M., & Bezerra, E. D. L. (2013). *Contribuições das Neurociências à compreensão da aprendizagem significativa* [Contributions made by the neurosciences to the understanding of significant learning]. Available at http://www.revistadialogos.com.br/Dialogos_10/Mario_Bezerra_Neurociencia.pdf

Mendes, V. D., & Haye, A. D. L. (2009). A moda do século XX [Fashion in the 20th Century]. São Paulo: Martins Fontes.

Menezes, M. D. S., & Spaine, P. A. D. A. (2010). *Theoretical guidelines for the teaching of Modeling flat industrial clothing: from the point of view of industry and teaching and learning*. Available at http://oasisbr.ibict.br/vufind/Record/UEL_6f114ec715698f84e2c8609ef8572dde

Nee, D. E., Jahn, A., & Brown, J. W. (2014). *Prefrontal Cortex Organization: Dissociating Effects of Temporal Abstraction, Relational Abstraction, and Integration with fMRI*. Available at http://www.ncbi.nlm.nih.gov/pmc/articles/PMC4184367/

Nicolelis, M. (2011). *Muito além do nosso eu: a nova neurociência que une cérebros e máquinas* [Far beyond our ego; a new neuroscience that combines brains and machines]. São Paulo: Companhia das Letras.

Perlovsky, L., & Ilin, R. (2012). *Brain. Conscious and Unconscious Mechanisms of Cognition, Emotions, and Language*. Available at http://www.ncbi.nlm.nih.gov/pmc/articles/PMC4061812/

Purves, D., Augustine, G.J., & Fitzpatrick, D. (n.d.). *Neuroscience*. Available at http://www.ncbi.nlm.nih.gov/books/NBK11066/

Saltzman, A. (2004). *El cuerpo diseñado: sobre la forma em el proyecto de la vestimenta* [The designed body: on the form in a clothing project]. Buenos Aires: Paidós.

Santos, D. R. D. (n.d.). *Contribuições da neurociência à aprendizagem escolar na perspectiva da educação inclusiva* [Contributions of neuroscience to academic learning from the perspective of educational inclusion]. Available at http://www.psicopedagogia.com.br/new1_artigo.asp?entrID=1395#.VMvVIUfF-yU

Sorger, R., & Udale, J. (2009). Fundamentos do design de moda [Fundamental principles of fashion design]. Porto Alegre: Bookman.

Souza, W. G. D. (2007). *Modelagem no Design do Vestuário* [Modelling in Clothes Design]. Available at http://fido.palermo.edu/servicios_dyc/encuentro 2007/02_auspicios_publicaciones/actas_diseno/articulos_pdf/A6045.pdf

Spaine, P. A. D. A. (2010). *Modelagem plana do vestuário: diretrizes para a indústria do vestuário e o ensino-aprendizado*[Flat file modelling of clothing; guidelines for the clothing industry and teaching/learning]. Available at https://www.faac.unesp.br/Home/Pos-Graduacao/MestradoeDoutorado/Design/Dissertacoes/patricia-aparecida-de-almeida-spaine.pdf

Spaine. (2013). *Modelagem plana: contribuição para a evolução do vestuário* [Flat file modelling; a contribution to the evolution of clothing]. Available at http://www.cih.uem.br/anais/2013/?l=trabalhos&id=511

Sporns, O. (n.d.). *Network Analysis, Complexity, and Brain Function*. Available at http://csc.ucdavis.edu/~cmg/netdyn/Net_Brain_final.pdf

Tokuhama-Espinosa, T. N. (2008). *The scientifically substantiated art of teaching: a study in the development of standards in the new academic field of neuroeducation (mind, brain and education science)*. Doctoral Research Study. Capella University. Available at https://www.researchgate.net/publication/36710537_The_Scientifically_Substantiated_Art_of_Teaching_A_study_in_the_development_of_standards_in_the_new_academic_field_of_neuroeducation_mind_brain_and_education_science

Zaro, M. A., Rosat, R. M., Meireles, L. O. R., Spindola, M., Azevedo, A. M. P. D., Bonini-Rocha, A. C., & Timm, M. I. (2010). Emergência da Neuroeducação: a hora e a vez da neurociência para agregar valor à pesquisa educacional [The emergence of neuroeducation: the time and place for neuroscience to aggregate value for educational research]. *Ciências & Cognição, 15*. Available at http://www.cienciasecognicao.org/revista/index.php/cec/article/view/276

Zatorre, R.J., Fields, R.D., & Johansen-Berg, H. (2013, May 22). Plasticity in Gray and White: Neuroimaging changes in brain structure during learning. *Nat Neurosci.*

Zeki, S. (2000). *Abstraction and idealism: From Plato to Einstein: how do we acquire knowledge?* Available at http://www.vislab.ucl.ac.uk/pdf/platotoeinstein.pdf

Zeki. (2001). *Artistic Creativity and the Brain*. Available at http://www.sciencemag.org/content/293/5527/51.full

KEY TERMS AND DEFINITIONS

Piloted Item of Clothing: This concerns the first item manufactured for each model after all the details have been checked and any defects have been identified in the clothes test. (Menezes; Spaine, 2010).

Sketchy Drawings: Studies of models that are designed in a general way by free hand drawing and where the details are sketched of an item of clothing although without the need to outline the physical shape and proportions correctly (Sorger; Udale, 2009).

Technical Data Sheet: A document that identifies the item of clothing and provides descriptive information about the following: the raw materials and accessories, the amount of material required, technical design, stages in the execution of the model, the machinery used in the dress-making, the manufacturing time and the total costs (Sorger; Udale, 2009).

Technical Design: After the designs of the sketchy drawings and the models of the collection have been selected, the technical design is carried out. This follows the shape of the body and is a stage that precedes the building of the mould. (Sorger; Udale, 2009).

The Clothes Test: After this the manufactured item of clothing is tested by a model – male or female – to check any defects and make a comparison with the details described in the technical design. The clothes test continues until all the factors have been given assurance so that the quality of the product can be guaranteed (Menezes; Spaine, 2010).

Three-Dimensional Modelling (Moulage): The modelling carried out initially on the display which will later be transferred to the flat file modelling. This kind of modelling allows experiments to be conducted in forms and volumes and does not need to be based on basic moulds (Spaine, 2013).

Two-Dimensional Modelling: Also known as flat file modelling, this involves carrying out the mould directly on paper or cardboard from the basic moulds of items of clothing (Spaine, 2013).

Chapter 13
Cognitive Processes in the Reception of Interactive Short Film Script:
Mental Representations by Audiovisual Specialists

Patricia Bieging
University of Sao Paulo, Brazil

Raul Inácio Busarello
Federal University of Santa Catarina, Brazil

ABSTRACT

This chapter aims to present mental representations by seven audiovisual specialists on Sirena, the antagonist of a noir fictional story in a multilinear interactive short film script. To this end, a reception study was created with in-depth interviews with specialists after reading the script to discuss its several aspects. This approach is of qualitative character. The study was conducted between June and October 2015 in São Paulo, Brazil. It is possible to perceive that participants have interpreted the character differently in spite of its consistent build. We could verify that the mental representation of the antagonist was aided by the world-view and experience of each interviewed specialist, which has led to unique readings of the character.

INTRODUCTION

Human knowledge constitutes facts and rules stored in memory, formed by patterns observed in behavior (Clancey, 1993). Knowledge is formed in specific actions in which the situation orders the space-time between objects and agents conditioned by their physical and characteristic environmental limitations (Rohlfing; Rehm; Goecke, 2010). In this sense, action is seen as something broader than simple behavior. On the other hand, the context in which the action occurs depends on several factors, being defined by

DOI: 10.4018/978-1-5225-0510-5.ch013

the socio-cultural levels of the individuals and environments involved. Individual actions are limited by the context, which in turn is determined by the situation. Thus, there may be a myriad of contexts for a given situation, with possible overlaps. Vanzin (2005) sees the formation of knowledge occurring when situated, due to a process of social-interactive participation by the agent in a community of practice in which concepts and culture cannot be understood in isolation. In this sense, cognitive activities of knowledge acquisition and relation can only be understood when in relation to their context.

When facing a given situation, individuals construct a general representation to elaborate their behavioral plans. Therefore, cognitive phenomena occur through internal processing of information provided by the transformations in representation. Clancey (1993) understands reasoning (understanding, making meaning, speaking, conceiving, etc.) as the act of manipulating representations through stored models and schemata, in which meanings are a way to map primary information, concepts, facts, and rules. In this context, a mental representation is a way to understand situations (Vanzin, 2005). Representations are transitory and limited to the set of elements in a given situation occurring in a specific context and with specific aims. It is through mental representations, or mental imagery, that individuals form their world-view. For Clancey (1993), the act of representing is interactive, involving motor and sensory aspects - that is, it is necessary to take perception and movement into account in order to represent something. The process of knowledge formation occurs in every act of seeing, speaking, hearing, feeling, interacting, and so on. In this sense, people conceive an idea when interpreting something, not simply by recovering and assembling meanings, but creating meaning - that is, information is not determined by the informer, but constructed by the observer - because people as receivers conceive a unique form of knowledge when understanding something.

From a narrative point of view, Jiménez (1996) sees that it is inseparable from its social and cultural contexts, as discourse is constructed by the set of actions and propositions of a person. Relations between narratives and meanings result in a complexity of other autonomous microtexts, in which history and the way it is presented are an effect of the exchange between the narrator and the audience. The narrative is different as it is the action of retelling a given story with real or imaginary events. This story is made true by the receiver's mind, occurring in an environment of representation (Castro; Freitas, 2010). Narrative is thus not a static phenomenon, but a phenomenon emerging from the interpretation of readers and their interaction with a text (Tenenbaum; Tomizu, 2008). The act of narrating is the basic way to make meaning of experience. In this sense, people narrate to negotiate their experiences in the meaning systems that provide contexts.

With constant technological developments and the advent of new means of communication, storytelling devices have been adapted. Among these, interactive environments (Steiner; Tomkins, 2001) can be seen as the newest field of application to present narratives. This is possible because the narrative works in a flexible model, which enables the interpretation of a range of phenomena through rules involving a certain type of general knowledge. It thus constitutes a cultural pattern which allows for certain analogies to seem plausible and intelligible (Brockmeier; Harré, 2003). In this context, iCinema (interactive cinema) has been gaining strong interest of viewers as to their active participation in decision-making processes as the narrative develops. Interactivity (Manovich, 2014) is the key factor in iCinema, allowing for greater immersion and deeper experience of subjects in a story (Bieging; Aquino, 2014) and enabling interferences with the narrative flow and plot, reformulating it according to viewer choices. The multilinearity of structures collaborates with iCinema, inviting viewers to take important roles in developing the narrative. This growth of film possibilities can be perceived today in the broadening of film narratives, and in various moves in the medium, generating multiple sensations and intensifying emotions. The alternatives created from technological developments make way for perceptions and emotions that go beyond the concrete material. The new means of communication insert individuals within their productions, generating effects of closeness as they invite spectators to become immersed (Murray, 2003) based on the development of complex forms ingrained in relations.

Based on this context, the authors of this study constructed a multilinear script for an interactive short *film noir*, in order to observe in depth the cognitive relations of representation by viewers in relation to the aesthetic possibilities of this new narrative medium. In this sense, this chapter departs from a segment of this study, from the approach of Reception Studies, observing the mental representations by 7 participants of Sirena, antagonist character of the short film script created by the authors.

Firstly, the theoretical background on multilinear narrative features and its potentials are presented, in relation to the formation of knowledge. In sequence, the story and its multilinear interactive structure build is presented, with focus on relations between characters in the plot. The methodology of this study is described then. Later, participant representations of the character are analyzed based on the selected theoretical background. This chapter ends with a revisit to objectives, reflecting upon interviewee representations of the studied character.

Multilinear Narratives in Knowledge Formation

When people interact with the world, they depart from certain expectations of their surroundings (Azevedo, 2006). This interaction requires constant negotiation, which guarantees the possibility to alter and flex these expectations. Making meaning processes in a person's mind creates dynamic world models for people to deal with their environment. These models are ready-made expectation structures that are somehow supplied by their culture and (re)appropriated by people. In this sense, it is possible to argue that stories can provide world models, surfacing memories of what is "culturally expected, but dialectically celebrating canon 'transgression'. It solves problems as it finds new ones" (Azevedo, 2006, p. 48).

Although individual memory seems very particular for each person at first, it should be understood as a collective social construct (Halbwachs, 2006), considering their variations, transformations, and changes. When speaking, individuals are challenged by their past, their present, and their social or cultural references. Memory and representations are constructed during discourse. Therefore, the brain organizes, reinvents, and selects information, activating memory from each person's point of view, from their own place in the history of their time. Symbolic and imaginative aspects are especially implied in this constructive process (Thompson, 2002), as fragments of memories are rarely combined in equal or linear ways. For example, when individuals watch a given event in television, it can serve as reference for their consciousness, presenting and providing representations to be selected and stored with other fragments of memories (Lima, 2009). They can also serve to re-elaborate their understanding and perception of the world, and their construction of knowledge, (re)constructing states of mind. In addition, these audiovisual references can serve as models for new life styles and new social roles, with which one may relate to. Differently, in a digital environment, individuals may become protagonists departing from "the standardized narratives and representations of events, individuals, and social groups for the possibilities of self-representations and self-narratives" (Lima, 2009, p. 8).

There are varying dominions of reality whenever one is telling a story. The narrative usually approaches an event supposedly occurred or imagined, which constitutes the dominion of history with narrator, spectator, and the narrated events. For Azevedo (2006), the narrator has no dominion on the story itself, but on the events narrated - exerting this dominion when changing focus on its time and space. In this sense, perception by the reader/listener/spectator is always local and partial; however, they are aware that someone in another position will have different views of the same event.

Since the film is a record of physical reality in presenting characters, settings, and events, its images cannot pass unnoticed by spectators - particularly when we consider messages with strong connections to quotidian life and reality. That is when the appeal to sensitivity is established. It is through these aspects of life that spectators feel attached to the narrative and construct new knowledge, whether they are represented by images or even by its messages.

Grodal (2002) explains that movie spectators can experience sensations based on "perception, cognition, and dependence" when watching a scene. He also adds that this realism also depends on other factors, such as the mental connection that establishes a level of familiarity and recognition of the conveyed message. The conception of reality or realism in audiovisual messages is not only centered in the issue of representation itself, but in different ways as (depending on genre) the mind of a receiver can establish completely different parameters from film. The purest representation of reality may not provide immersive elements for spectators to live the story intensely.

For Grodal (2002), representations are directly simulated in the mind and confronted with reality, which can increase the feeling of realism. This perception of reality is often treated in relation to forms of expression, rather than to the production process. From the experience perspective, what transforms image into representation of reality is not knowing the subject in relation to the process of production or reception, but the realistic relevance generated independent of how the process was performed (Grodal, 2002).

A story offers the safety of being exterior to one's self, becoming an element in which individuals can project their feelings and dreams. This make-believe experience can be found in child's play, Shakespearean text, in cinema, and on the Web through a myriad of narratives. "Stories evoke our deepest wishes and fears because they are part of the magic" (Murray, 2003, p. 103). When a person finds oneself in a fictional world, they suspend disbelief and exert their creativity. By immersing, individuals focus their attention into the new world and surroundings, thus using their intelligence to reinforce the veracity of this experience, rather than question it.

For Courtés (1979), the narrative must be presented as a simple elementary structure, as it must consider that all of its components have meaning so it makes sense. In addition, the result of secondary developments of narration not found in a preliminary structure "constitute a subordinate structural layer: thus, the narration as a whole requires a hierarchical structure of content" (Courtés, 1979, p. 126). Barthes et al. (1976) sees the understanding of a narrative as not simply following a given story, but recognizing stages and projecting plot threads. The narrative as an object is the goal of this communication, which depends on two agents: the giver and the receiver of a narrative.

In this sense, it is the background of subjects in this process, with all their emotional and subjective baggage, that will provide a tone to the meanings established and the motivation in getting in touch with the contents of a story. Smith (2003) explains that emotions are constructed through personal relations in the social framework that each individual belongs to, making the community and its relations to be crucial in developing an emotional system controlled by the interests of its members.

Social interaction is a potential structure that enables the acquisition and development of knowledge. However, it does not limit individual aspects as they are necessary in group contexts. For Hansman and Wilson (2010), knowledge is acquired empirically and scientifically, easily obtained in any situation. Through this perspective, the authors believe that the cognition phenomenon favors equally the production of knowledge.

For Vanzin (2005), knowledge in cognition is developed during activities performed by individuals, especially when people create, interpret, and manifest their opinions about what they are doing. Clancey (1993) sees that each new activity is part of the coordinate course of perception, and thus the acquisition and interpretation of knowledge is inherently situated. The situated activity is ever-going interaction.

PRESENTATION OF THE SCRIPTED STORY

The object of this study is an Interactive Multilinear Fictional Script written by Raul Inácio Busarello and Patricia Bieging. It is a *film noir* about a detective that sees himself involved in the murder he is investigating. The story begins with the son of a deceased millionaire as the prime suspect of a young lady's murder. As the plot unravels, the detective begins suspecting of his stepmother and of his tutor, an attorney. The plot twists as we find out the murdered woman was pregnant. This incident reveals the millionaire's will, in which his son would only get his inheritance in case he had a male heir under age 21. From this point, the detective is used as a puppet in the hands of the lawyer and the stepmother, who conspire for the inheritance. As this is an interactive multilinear story, the events occur differently depending on spectator choices. The detective may become an accessory to murder by engaging in a romance with the stepmother, or by being bought by the attorney. The story has 3 different endings and a series of 7 points to make choices during the plot.

The story makes use of *film noir* tropes, which reflect in the construction of the characters and plot. In this sense, 6 characters are presented:

- **Detective:** A tall, skinny man with tired eyes and a charming look, of age around 35. He has a cynical view of the world and his work is his only reason to live. As the classic anti-hero, some of his actions are for his own benefit, although he usually acts according to the law;
- **Jessica**: An angelical-looking young lady of age 18. She is the murder victim and Igor's girlfriend;
- **Igor Magnus**: The only son of a deceased millionaire. He is an overprotected, reckless, and spoiled young man about to turn 21. The typical playboy who loves to party and get people's attention;
- **Alex Magnus**: The millionaire father of Igor. Competitive and ambitious, he loves to gamble. All of his relationships and business have a sense of adventure. He is cunning and his financial games are a tad sadistic. The only person he trusts is his attorney, Mr. Romano, who is Igor's legal guardian until he is 21. He was found dead after cardiac arrest in front of Sirena;
- **Mr. Romano**: A lawyer of approximately age 60. He is clever, well-read, and cautious, in spite of his naive appearance. He is the legal representative of the Alex Magnus estate, and the legal guardian of his son, Igor. He has a certain dispute with Sirena;
- **Sirena**: is The widow of millionaire Alex Magnus and Igor's stepmother - with whom she is not close to. She is a woman of about 40 years of age, who was once a famous model, now retired due to her aging. In spite of that, she is beautiful and sultry, intelligent, and ambitious. She cunningly uses her charms to reach her goals. Following the trope of the femme fatale that is typical of film noir as a genre, she somehow gets involved with the anti-hero of the plot.

The short film is structured in 16 scenes, with approximate linear runtime of 17 minutes. The narrative experience in the script are structured through the interactive process in which decisions are *branched*

(Galyean, 1995) in 7 interaction points inserted in the story. Each interactive node in the plot structure is a point in which decisions are processed, and from which a new *branch* will be available for interaction. The story is constructed in 3 acts (Field, 2009): *Set-up*, *Confrontation*, and *Resolution*, divided by *Plot Points*. The multilinear possibilities of the plot are explored within the classic model with *branch* after *essential events* of the story occur. The configuration of this interactive fictional world offers possibilities of choice not only from the main character thread, but in the entire plot as choices are not centered on the actions of a specific character, but in transforming the *story thread*.

In the *First Act*, the Detective questions Igor as a possible suspect in Jessica's murder. He is represented by Mr. Romano, that offers his help in the investigation, if necessary.

After that, the audiovisual receiver has the option to choose between 2 sequences. In both, the Detective has his first meeting with Sirena. In one option, she acts aggressively; she has nothing good to say of Igor, interrupting the questioning and asking the Detective to leave her, as she is late for an appointment. In the other option, Sirena seduces the Detective, and kisses him.

The *Inciting Incident* of the first act is marked by the news that Jessica was pregnant. The news gives motivation to the Detective, but exposes Igor's aggressive temper. As previously, viewers have the opportunity to choose between 2 sequences. They lead to the *Plot Point 1*, which is marked by the Detective receiving a file with documents explaining the conditions in Igor's inheritance. In the first option, the Detective meets Mr. Romano in his office. The attorney hands the documents to the Detective, who is then pensive. In this case, there are 2 more options to choose from: one continues in an interrogation room with Sirena, and the other has Mr. Romano toasting to the case success. In the second option, the Detective is in bed with Sirena, who hands him the documents. Again, 2 options are given to the receiver: he may access the scene in the interrogation room with Sirena, or watch the Detective visit Mr. Romano's office to receive a briefcase. Both toast to the case success.

In the beginning of Act II, Sirena is being questioned, and she reports she Igor was physically aggressive with her. When alone, she smiles slyly to the Detective. Once again, viewers are taken to a fork in the story, but this one has no choice to be made as the sequences are conditioned to choices previously made. In this case, if the receiver had selected the file handout by Mr. Romano, the next scene shows a perplex Detective asking Sirena if Igor was aware of his father's will. On the other hand, in case the viewer had selected receiving the file from Sirena, she pressures the Detective to get rid of both Igor and the attorney. To which, the Detective is appalled.

During the Act II, the story *Mid Point* shows the Detective receiving information that Igor's DNA was found in Jessica's body, providing enough evidence to incriminate him.

The Act II Plot Point is presented with 2 options for receivers to choose from. In both, Igor is being questioned while wearing inmate clothing. In the first option, the Detective is pleased the case is closed, and Igor is madly saying he did not kill Jessica, revealing that Sirena is pregnant with his child. Both the Detective and the attorney are surprised with the news. In the second sequence, the Detective is stressed out, threatening to indite Mr. Romano as accessory to murder. Igor claims he did not kill Jessica, and that Sirena is pregnant with his child. The Detective is shocked with the news.

This leads to the story *Act III* and the plot *Climax*. After the previous choices, the receiver gets 2 more options that rush the Detective into Sirena's apartment. In the first option, the Detective finds Sirena in tears while holding a pregnancy test. She tells him it's a girl. The Detective arrests her. In the second option, the Detective realizes that Sirena has escaped. Then, viewers can choose 2 more paths: One shows Sirena in an airport lobby, pregnant. Mr. Romano arrives and she smiles to him, telling him it's a boy. On the second option, Sirena is in the airport lobby, pregnant. The Detective arrives and arrests her. She smiles and says it's a boy.

The *Final Sequence* of the story is marked by 3 possible options. In the first option, the Detective collects all documents of the investigation, puts them into a box, and exits. In the second option, the Detective collects all documents in a box, and buries it. The third option is only available if the viewer chose the scene in which Mr. Romano hands the Detective a briefcase in the Act I Climax. In this ending, after the Detective collects all documents, Mr. Romano is seen burying the box.

METHOD

This study uses the aforementioned interactive multilinear fictional script as research object. It was written by Raul Inácio Busarello and Patricia Bieging in order to investigate the various cognitive aspects established towards the character of Sirena, aiming at understanding subjective factors in the mental representation of the participants, particularly in relation to the possibilities of this new interactive medium.

The field research was performed between June and October 2015, based on the method of Reception Studies by Martín-Barbero (2003). The understanding of Cultural Studies departs from the supposition that each receiver reads, perceives, and produces meanings differently from the same media text.

From this premise, 7 specialists with varying degrees of experience in the audiovisual market have been selected for the study. They were selected based on their production of fiction and non-fiction, with the main prerequisite on their professional and practical background. Their practical experience proved to be essential, as they needed to understand the technical structure of the script. It was extremely important to know how to read the script while imagining each composition, *device*, and interactive processes, as well as how scenes were constructed and sequenced.

It is possible to suppose that when individuals are in an environment with known people, the distance among them is already overcome and thus research relations are made easier, due to the existence of bonds that are favorable when "speaking frankly". This helps diminishing or excluding the possibility of the researcher to be considered socially above the research participants (Bourdieu et al., 1998). To this end, all of the 7 participants selected are in direct liaison with the research authors.

The sample of specialists (Struchiner; Ricciardi; Vetromille, 1998) to analyze the interactive multilinear fictional film script aimed at collecting advanced field data in relation to evaluating the object of study. This method was efficient as it gathered experienced professionals in condition to contribute significantly with the study. The participant sample size was not large, but it was significant as it involved select professionals in the main areas related to the study (Aedo; Catenazzi; Díaz, 1996).

The specialized participants are:

- **Ricardo**: A journalist with a graduate degree in Cultural Journalism. Screenwriter, winner of the Criando Histórias award in 2006.
- **Maicon**: A journalist with a graduate degree in Cultural Journalism. Director and presenter of a YouTube channel.
- **Willian**: An audiovisual producer with a graduate degree in Executive Production and Television Management. Director of photography, with wide experience in Brazil and internationally.
- **Walter**: A journalist, and a PhD in Brazilian Cinema. Editor and movie critic, collaborator in cultural marketing and audiovisuals.

- **Maria**: An audiovisual producer with a graduate degree in Cultural Journalism. Television producer, develops scripts and is a grid director in TV and radio.
- **Alex**: Scene director and movie producer. A publicist specialized in actor direction. Produced several films in countries such as Brazil, Italy, France, Switzerland, and Spain.
- **Jennifer:** A journalist and screenwriter. Has an MA in Language Sciences and a PhD in Applied Linguistics. Professor of graduate and undergraduate programs.

To reach the research objectives, in-depth interviews were performed with the specialists. Bauer and Gaskell (2008) explain that the in-depth interview technique is ideal when the research analysis focuses on an object or situation in which detailed individual experience is important; they also serve to evoke sensitivity in participants, or to test specific instruments.

The interviews sought to analyze the phenomenon configured in individual processes of mental representation in participants, from the characteristics and development of a character in the narrative. Above all, the symbolic mediation of participants was considered in their understanding of the story situations.

Data collection and analysis have a qualitative direction, with the objective of understanding and explaining social and subjective phenomena constructed by individuals through their interaction (Merriam, 1998).

JUDGMENT, IMAGINATION, AND RECOGNITION: RECEPTION

In this section, we present the mental representations of the 7 participants towards the character of Sirena, presented in the script as one of the antagonists in the story. The multilinear character of the story evidences the peculiar readings that each participant made of the script, which could have generated different interpretations of the narrative. However, character tropes and features were maintained in all versions.

The data analysis considers the entire subjectivity process leading to a meaning-making system that is inherent to each individual. Azevedo (2006) points out that although meanings are in people's minds, they are not static models - they are actually compatible with an interactive view of meaning that emerges from social interrelations which arereconstructed based on all dimensions of the context. At this point, Durand (1998) sees that a given symbol in a symbolic system is under constant threat of regionalism and (re)signification, becoming in each moment an image with the function of social recognition - that is, a convention. "For this reason, a symbolic image must always be revisited, similarly to how a song or a theatrical hero needs an interpreter" (Durand, 1988, p. 33).

From these premises, the main evidence related to the cognitive processes of participants were selected, paying special attention to mental processes and of world building in the understanding of Sirena as a character after reading the script written by the authors of this study. Factors such as perception, imagination, judgment, and internal ways to see the world were highlighted in the interactive process. In this sense, the first impression of participants in receiving the character of Sirena was triggered by emphasizing particularly the conversion of facts presented in the script face the perceptive uniqueness of the specialists volunteering for the study.

This approach starts with the testimonial by Ricardo, who is somewhat fascinated by Sirena and her actions in all versions of the story. For him, the character is a sultry woman, unreachable, and sure of herself. In spite of calculated actions, she uses seduction as a weapon in her favor, without being superficial. However, he does not consider her to be vulgar, or "slutty".

I think she's bold, daring, and audacious; I think this is the first time she meets the Detective... she seduces him. Going to bed or not, she already makes it implicit from the start. She has an affair with her stepson, which I didn't imagine, either. You think, 'wow, this woman's really very daring.' I think this audacity and seduction of hers... got her everything she wanted. (Ricardo)

Another participant, Maicon, considers Sirena to be "extremely seductive", comparing her to other "bombshells" in the history of cinema. For him Sirena is like Black Widow (character played by Scarlett Johansson in *The Avengers*), or like Julia Roberts as the protagonist of *Pretty Woman*. She is also compared to Clara, a character in the Brazilian soap opera *Passione* (TV Globo, May 2010 to January 2011). Maicon says Sirena "isn't a seductress like Sharon Stone in *Basic Instinct*. Sirena is more similar to the main character in *Disclosure* - so, she's a very strong woman who weakens other characters by seducing them. That's very interesting", explains Maicon. Similarly to Ricardo, Maicon was also captured by Sirena, similar to the sirens of Ancient Greek myth - a dangerous charm in her singing and beauty.

However, Alex saw Sirena as a show-off; in his choices, he sought after her most seductive side because he wanted to understand how far the character would go in her ability to involve and manipulate others. For him, it was important to "know how manipulative (or not) she was; how real or how fake she was" (Alex). For this participant, it was important to find out what was Sirena's attitude towards crime. According to him, "my focus wasn't on Romano or on her - I wanted to know who killed Jessica. I wanted to know what Sirena brought to the table in such a drastic moment as a crime. I wanted to see what she had to show me from her voluptuousness", completes Alex. Although having also been captured by Sirena and her charms, Alex tries to put himself in the place of the Detective, who seeks for clues to unravel the mysterious assassination at all costs. Alex seemed anxious and enthusiastic about Sirena at the same time during his reception of the script, enthusiastically hopping and saying "I knew it" or "I saw that coming". It is possible to say that the participant related to the Detective, but recognized something particularly strong about Sirena, who could envelop viewers in her voluptuousness.

On the other hand, Jennifer as an outspoken feminist related to the character, protecting Sirena's innocence since her first contact with the character. Her choices have revealed Sirena's aggressive side, as she believes that would be the attitude of someone who had been unjustly accused. "I put her in a defensive position [...] Aggressive is normal for her. I thought that if she were aggressive, she'd put herself in a position and would make her own defense", says Jennifer. The participant shows solidarity to the character because they are both women, thus becoming an accomplice to Sirena. She explains her choices pointing out that she had "that possibility. I had the choice to 'cuff her... but I didn't want to. It was my prerogative - as a woman, I didn't want her to be 'cuffed. It's a woman's standpoint. I have sorority towards her, as a woman." (Jennifer)

Similarly to Jennifer, Ricardo also rooted for Sirena, and was surprised to see she was involved with the murder in versions of the story. Ricardo reveals that he was sure the murderer was Igor, the millionaire's son and Sirena's stepson, from the beginning of the story. However, he was surprised to see Sirena as an accessory to murder. "Then you get the feeling that, 'oh, I got it'... or 'oh, I knew it'. But then I was surprised that Sirena was a part of it. I didn't expect that in any option. I thought that in a second reading she wasn't in on the crime, but then she was even more strongly a criminal", Ricardo explains. It is important to highlight that it is not clear whether Sirena took part in the murder of Jessica in any point of the story. Sirena is only accountable for getting pregnant with her stepson's child in order to keep his inheritance. It is thus possible to notice a hasty judgment by Ricardo in blaming her for the murder based on another action that he was shocked by, considering it to be of bad character.

Ricardo also did not choose to handcuff Sirena, revealing that he was rooting for the antagonist: "I wanted her to be punished. I think I personally adore villains in general... I thought she'd be guilty, but I liked her; and I think she has a well-defined personality." In addition, he shows a certain admiration for the character, as he relates to the Detective as a protagonist, and similarly to him, he let himself fall for the charms of the antagonist. "I thought she was nice because that kind of character is already attractive to me... Maybe because I relate to him, and since he fell for her charms, maybe that's what I think he would think", says Ricardo. He finishes his testimonial with, "She's a character that reflects on our wickedest desires. Gee, I wanted to do all the wrong stuff and not pay for it. I think this is part of the classic villain who does it all, and is so good at being bad, that gets no punishment."

Ricardo reflects the facts presented in the story script to his own real life, who seems moved for not being as relatively free as Sirena seems to be. Although life in fiction is a reflection of quotidian life, sometimes it triggers viewers to think about their own daily actions.

In this sense, Pelloso and Ferraz (2007, p. 119) explain that "the sense of production can be obtained from psychosocial practices [...] seen exactly as what people do in their quotidian life [...] it seems pertinent to consider literature as a subjectivity-producing practice which is, therefore, a producer of psychosocial realities." Thus, (re)signifying Sirena's actions lead Ricardo to negotiate the message, opposing her behavior to his own, in his relationships and the events occurred until then. It is possible to say that negotiating media meanings in general is a way of existing, "embedded in collective subjectivity, quotidian culture, and the most unconscious politics" (Escosteguy, 2001, p. 180). That is, people are incorporated and molded into social practices, values, and rules within their (local or global) community since birth. The mental process working in mechanisms such as converting teaching, learning, and (re) configuring the individual and subjective world will always define how each human being will interact, interpret, and (re)signify the world around them. This is what Ricardo and the other participants do in this study. To sum up, individuals are incorporated into signification systems that dictate ways of living within the small and large spheres of society.

When receiving an interactive multilinear fictional script, the patterns established within a society dominate each interpretation by the participants of this study in relation to Sirena as a character. Walter perceives Sirena as a *femme fatale*, but not as a standard of classical beauty - one who shows harmony and well-being since the first contact. For him, Sirena seems to be more a victim of circumstances than a villain. "Voluptuous... That lovely, shaken brunette that looks younger than she really is, that has suffered somehow in the past and thought her life was good at some point - and suddenly everything is taken from her", explains Walter. He sees the character as a woman who did not have an easy life, since her childhood. "She turned out terribly because she was left out of the will", completes Walter. That justifies her actions.

By relating to Sirena, Walter somehow allows her to justify her actions in the narrative. Walter helps Sirena when claiming her actions are coherent with a life of hardship that the character had before she could inherit. A clear process of negotiation with the meanings of the story can be seen, flexibilizing (Azevedo, 2006) possible consequences to be suffered by Sirena.

In the same line of thought, Tenenbaum and Tomizu (2008) claim that each narrative medium has its own set of conventions and specific techniques to develop cultural meanings, like the ones used by the authors to construct the story. A skillful screenwriter can thus use the specific conventions of the field, expecting the receiver to have a narrative experience imposed by the convention.

In Sirena's narrative, seduction is her strong ally - independent of her actions, she is protected by most receivers of the story. Maicon claims he has tried to put himself in Sirena's place when making his choices: "I wanted to do what she'd do; so I followed her steps exactly, to seduce and manipulate men, because that's fun." The participant says the antagonist has "a dangerous side people can't see", which makes her a seductive character that surprises whoever meets her. In spite of the admiration, her nature is unclearly represented by Maicon, who claims that,

I related to Sirena because she's seductive, she's a woman, she's determined, and she makes her strength and her feminism bend things to her favor. No, if I were a woman I wouldn't do it; but her getting pregnant with Igor to inherit is something I wouldn't do. So I relate to her nature, and that is reflected well. (Maicon)

In Maicon's standpoint, Sirena is practically a protagonist as she is the only woman in the story, and that calls his attention more than the other characters.

Sirena gains prominence as the plot develops by being strongly relatable to the research participants, be it by her background or by her strength in seeking her objectives no matter the cost. The connection mentioned by Maicon, Jennifer, or Ricardo demonstrate how representations (Grodal, 2002) of quotidian life are simulated in the mind of participants, and even by fragments scattered in the script. Perceiving the character and the justification of her actions is made stronger when put in perspectives that lead the simple reading of a script to transform the character in a representation of reality that may be actual life for anyone.

The narrative process is thus considered more than the simple creation of a logical structure of events, as it aims to allow for significant personal or subjective experiences. It is necessary to add a consciousness of the user experience of a sequence for a system to identify an emerging temporal sequence, or an action as narrative. Subjectivity is thus involved in the narrative cogwork, consituting a relation of meaning in itself (Nath, 2004).

The emotional and subjective background of participants with Sirena's story can be visualized particularly in a reading by Willian. At a first stage, Willian refers to every male character by name and calls Sirena a 'whore' throughout the entire in-depth interview. "I thought more of her character as a whore - that's what the character is: strong. I see her as a blonde or a redhead, wearing a red dress, with cigarette in hand", says Willian. This participant declares Sirena as "a total accomplice" to the murder. In addition, she thought it was interesting when the Detective and the antagonist got personally and sexually involved, justifying it as something professional: "She got involved more professionally so she could close the case. Later, it seems that this involvement of the Detective and her has different objectives. It looks like he'll kill the boy, find the inheritance, and disappear with the whore" (Willian).

Although Willian states that Sirena is a strong woman, this is intuitive to the strength given to the story protagonist, who also uses every possible means to reach his goals, even if less noble. However, what is noted from this participant's representation is that this impulse of not limiting actions for their own benefit is seen in both characters of Sirena and the Detective. However, Willian perceived the Detective as the hero in the story, while Sirena is seen pejoratively.

The interpretation by Willian can be explained when observing the accumulated knowledge from interacting with different people throughout his life. Cognition (Vanzin, 2005) is in this case established in (re)signifying the story and in participant perceptions face the character design in their own imagination. In Willian's imagination, what should present herself as seductive is transformed into a prostitute when Sirena begins to fight for her rights of the inheritance left by Alex Magnus. For this participant, the character's struggle for better ways of living does not dignify her to the same level as the Detective when he gets involved with Sirena to get more information on the case. The social relations of Willian are marked in his speech, especially when considering that a majority of the society is still patriarchal and relatively sexist. However, his impulses in the mental representation of the situation lead him initially to a position that reflects the discourse of a society (Smith, 2003) in which a controlling majority is still male, leaving women to spaces and behaviors considered feminine and "proper" of women.

It is important to highlight that Sirena is not put forward as a prostitute in any point of the story. Her *dramatis personae* is even mentioned by the authors as "a woman of about 40 years of age, who was once a famous model, now retired due to her aging. In spite of that, she is beautiful and sultry, intelligent, and ambitious." (Busarello; Bieging). However, Willian perceived her character through her interest in the inheritance of Alex Magnus. "She's a gold digger that was with the boy's father because of the money. I see her as the key of the film - very strong. And in the end you get surprised because she had an affair with her ex-husband's son, that's surprising", explains Willian. Once again, the script and the narrative do not declare in any moment that Sirena was married to Alex Magnus for money - *au contraire*, she suffers his death.

On a second reading, making different choices, Willian claims that his image of Sirena changed a little: "Even then, she had sexual relations with her ex-husband's son, she's still a whore in my second reading. Then again, her image improved in about 50% for me... but still, she's a tramp."

This projection of the events in the film in relation to Sirena brings to Willian the hard look that women are given every day when turned into objects based on their actions and choices, which are not always well-accepted by society. What can be seen is that the physical realism presented by the events in the script can be sensed to an extreme by Willian or by any other participant, varying according to causes internal or external (Kracauer, 1960). This state of mind, activated by a sense of judgment and specially by the subjective and mental processes of research participants can be changed by the reception of violent scenes, jump-scares, or any other action generated by the subjective gaze of a camera. They aspire to the events, transforming receivers in witnesses participating in the action. The disturbance and intimacy are present in fictional narratives, inviting the receiver to be immersed in the story, altering their state of mind in part, or even entirely.

Similarly to Willian, Maria did not see good intentions in Sirena, considering her vulgar. Maria says she imagined Sirena as "a beautiful woman, with a nice body, long hair, but pretty slutty. I imagined her as a pretty woman that talks and acts vulgarly. It's not a person that involves you much. You can see she's up to no good." The participant believes that Sirena and the attorney, Mr. Romano, were plotting against Igor, but she believes that Igor effectively committed the crime. According to her, "you see that he [Igor] committed the crime. It's a fake tone of line, right...? When he says he wanted the fortune." However, Maria makes clear she wanted to blame Sirena for the murder, saying this is very close to what happens in real life. "I think it's like in real life - sometimes you skip stages. You don't know all the context and you jump straight into what you want. I already wanted her [Sirena] to be guilty and I didn't even know how the crime was - there, this was fine, solved it. So, sometimes you jump to conclusions." Maria shows concern with her own attitude of judging people without trying to understand the whole: "We don't try to find out what it was, how it was, and to try to hear all of them to make an informed choice. I went straight to - she did it. I didn't even see the details of how it went, how he died, I skipped the part with the shovel."

The impulse in blaming Sirena came to Maria from her action of seducing her stepson, acting the way she did in the story. Maria explains that Sirena "was already a person with this instinct, of thinking about herself only, of thinking about her own personal behalf, and not of other people."

In the relation between participants and script story, it is possible to see that the social perception of each interviewee gains strength within various contexts of reading. In this sense, Grodal (2002) points out that *perception* is the primal component of experience when individuals meet the world, the objects, and life itself. This relation is established in both physical and social spheres, as everything that is seen and heard is especially determined by the complex and particular perception of each individual as something subjective and unique. According to the author, representations are directly simulated in our minds and confronted with reality, which may increase the sensation of realism, and guilt for Maria particularly when pre-judging by holding onto pre-established concepts when receiving the story.

In this sense, the concept of life, values, local and global rules - as well as mental processes and states of mind when receiving a story - are crucial in discourse and in judging events in various quotidian situations.

CONCLUSION

This chapter aimed at approaching the mental representations produced by 7 specialists in audiovisual production on the antagonist character in an interactive multilinear fictional *film noir*script. We depart from the idea that knowledge is generated in situations between individuals and their environments, and all of their previous experience are prepondering to form their current representations. In this case, when someone comes in touch with a narrative, mental representations on facts and elements in the story are originated by negotiating background experiences with what is being presented. Thus, the construction of meaning occurs within the individual, in a relation between their state of mind, the informational elements in the audiovisual product, and their environment.

It is possible to note that the character of Sirena was (re)signified differently depending on the participant. In spite of the multilinear structure in the narrative, the character design remains consistent, independently of the paths taken when viewing the scenes described in the script. In this sense, it is evident that each specialist has attributed a subjective impression of the character. From their world-view, some specialists related to Sirena to the point of trying to protect her and becoming themselves accomplices. On the other hand, she was incriminated by others even before collecting the facts in the plot.

It is evident that certain aspects of her character build printed in her actions have resulted in revealing her actions and her world-view as a character. Many of these aspects were more evident in the testimonials by some of the specialists, who have turned her into a victim when negotiating with their own experiences. However, she was accused of having participated in the murder even if that was not clearly stated in the script, only by judging her actions and considering them bad or uncommon. Many actions were also in a trope in which similar actions among different characters have yielded in varying judgments and representations, depending on the participant.

In a general sense, we can verify that the mental representations of each participant on the character are strongly sustained by their life experiences and judgments of value, leading to varying interpretations of the same character.

REFERENCES

Aedo, I., Catenazzi, N., & Díaz, P. (1996). The Evaluation of a Hypermedia Learning Environment: The CESAR Experience. *Journal of Educational Multimedia and Hypermedia*, *5*(1).

Azevedo, A. M. T. (2006). *Estrutura narrativa e espaços mentais*. Belo Horizonte: Faculdade de Letras da UFMG.

Barthes, R., Greimas, A. J., Bremond, C., & Eco, U. (1976). Análise estrutural da narrativa: pesquisas semiológicas. Tradução de Maria Zélia Cardoso Pinto. 4ª ed. Petrópolis: Vozes.

Bauer, M., & Gaskell, G. (2008). Pesquisa qualitativa com texto, imagem e som: um manual prático. 7. Ed. Rio de Janeiro: Vozes.

Bieging, P., & Aquino, V. (2014). A experiência estética no iCinema: sensação e interatividade a partir de narrativas multilineares. In Olhares do sensível: experiências e dimensões estéticas em comunicação (pp. 84-97). São Paulo: Pimenta Cultural.

Bourdieu, P. (1998). *A miséria do mundo*. Petrópolis: Editora Vozes.

Brockmeier, J., & Harré, R. (2003). Narrativa: Problemas e Promessas de um Paradigma Alternativo. *Psicologia: Reflexão e Crítica*, *16*(3), 525–535. doi:10.1590/S0102-79722003000300011

Castro, C., & Freitas, C. (2010). Narrativa Audiovisual para Multiplataforma - Um Estudo Preliminar. *Bibliocom*, *2*(7). Retrieved June 20, 2010 from http://www.intercom.org.br/bibliocom/sete/pdf/cosette-castro-cristiana-freitas.pdf

Clancey, W. (1993). Situated Action: A neuropsychological interpretation. *Cognitive Science*, 17–87.

Courtés, J. (1979). *Introdução à semiótica narrativa e discursiva. In Tradução de Norma Backes Tasca*. Coimbra: Livraria Almeida.

Durand, G. (1988). *A imaginação simbólica. In Tradução de Eliane Fittipaldi Pereira*. São Paulo: Cultrix.

Escosteguy, A. C. (2001). *Cartografias dos estudos culturais: uma versão latino-americana*. Belo Horizonte: Autêntica.

Field, S. (2009). *Roteiro: os fundamentos do roteirismo. Tradução Alice Leal*. Curitiba: Arte & Letra.

Galyean, T. A. (1995). *Narrative Guidance of Interactivity. In Program in Media Arts and Sciences*. Massachusetts Institute of Technology.

Grodal, T. (2002). The experience of realism audiovisual representation. In A. Jerslev (Ed.), *Realism and 'reality' in filme and media*. Copenhagen: University of Copenhagen.

Halbwachs, M. (2006). *A memória coletiva*. São Paulo: Centauro.

Hansman, C. A., & Wilson, A. L. (2010). *Situating Cognition: Knowledge and Power in Context*. Retrieved June 27, 2013 from http://www.eric.ed.gov/PDFS/ED478969.pdf

Jiménez, J. G. (1996). *Narrativa audiovisual*. 2ª. ed. Madri: Cátedra.

Kracauer, S. (1960). *Theory of film: The redemption of physical reality*. New York: Oxford University Press.

Lima, A. S. (2009). *Da cultura da mídia à cibercultura: as representações do eu nas tramas do ciberespaço*. Retrieved August 28, 2010 from http://www.ufg.br/this2/uploads/files/75/ciberespaco_representacoes_do_eu.pdf

Manovich, L. (2014). *¿Qué es el cine digital?* Retrieved January 20, 2014, from http://www.upv.es/labo luz/revista/

Martín-Barbero, J. (2003). Dos meios às mediações: Comunicação, cultura e hegemonia. 2 ed. Rio de Janeiro: Editora da UFRJ.

Merriam, S. B. (1998). *Qualitative Research and Case Study Applications in Education: Revised and Expanded from I Case Study Research in Education*. San Francisco: Jossey-Bass Publishers.

Murray, J. H. (2003). Hamlet no holodeck: o futuro da narrativa no ciberespaço. São Paulo: Itaú Cultural: UNESP.

Nath, S. (2004). *Narrativity in User Action: Emotion and Temporal Configurations of Narrative*. Croatia: University of Split.

Pelloso, R. R., & Ferraz, M. G. (2015). *Ética e moral como modos de produção de subjetividade*. Retrieved October 15, 2015 from http://www.scielo.br/pdf/ trans/v82n2/29418.pdf

Rohlfing, K. J., Rehm, M., & Goecke, K. U. (2010). *Situatedness: The Interplay between Context(s) and Situation*. Technology, Bielefeld University. Retrieved June 27, 2010 from http://mm-werkstatt.informatik.uni-augsburg.de/files/publications/68/situatedness.pdf

Smith, G. M. (2003). *Film Structure and the Emotion System*. New York: Cambridge Published in the United States of America by Cambridge University Press. doi:10.1017/CBO9780511497759

Struchiner, M., Ricciardi, R. M. V., & Vetromille, V. P. (1998). O painel de especialistas no processo de apreciação analítica de sistemas hipermídia para o ensino de graduação.*IV Congresso RIBIE*, Brasília.

Tenenbaum, J., & Tomizu, A. (2008). Narrative Meaning Creation in Interactive Storytelling. *International Journal of Computational*, *1*(2), 3–20.

Thompson, E. P. (2002). *A voz do passado, História Oral*. Rio de Janeiro: Editora Paz e Terra.

Vanzin, T. (2005). *TEHCo: modelo de ambientes hipermídia com tratametno de erros, apoiado na teoria da cognição situada. UFSC, 2005. Tese (Doutorado em Engenharia de Produção) Programa de Pós-Graduação em Engenharia de Produção*. Florianópolis, SC: Universidade Federal de Santa Catarina.

KEY TERMS AND DEFINITIONS

Act: Structure set-up by part of plot with a specific goal in story: set-up, confrontation and resolution.
Average Reversal Point: Incident that changes the story direction.

Branching: Interaction points in a multilinear story where the spectator must make decisions.
Climax: Plot's highlights tension.
iCinema: Interactive cinema, where the spectator decides the plot's course.
Inciting Incident: The event that informs about what is the story.
Multilinearity: Absence of traditional linearity in a story.
Plot: Relationship between the characters actions and the story incidents.
Story Wire: Order of incidents within a story.
Turning Point: Incident or event in story that marks the passage between acts.

Chapter 14
Understanding How the Mind Works:
The Neuroscience of Perception, Behavior, and Creativity

Claudia Feitosa-Santana
Federal University of ABC, Brazil & Albert Einstein Israelite Hospital, Brazil & School of The Art Institute of Chicago, USA

ABSTRACT

The understanding of the inner workings of the mind are relevant to enhance curriculum achievements, therefore optimizing the professional practice in general and of the arts and design in particular. The recent birth of neuroscience as a transdisciplinary field poses a challenge to the curriculum and is yet to be included as an integral part of its core. The lessons taught by #TheDress viral Internet phenomenon are here discussed with the intention of enlightening the urgency of a popularization of neuroscience knowledge, from daily life to the professional practice, as a tool to explain how context and experience influence our perception. Along the same lines, the section "The Roots of Human Behavior" addresses the fundamental concept of human behavior and how our emotions were built by our genes, helping us understand basic and complex human choices. Finally, the section "The Neuroscience of Creativity" discusses the neural basis of creativity and its relation to intelligence by dissecting what neuroscience already knows about the development of creativity and how the work environment could foster creativity. The discussion of these topics in this chapter aims to enlighten readers of the importance of neuroscience knowledge in the curriculum and how the arts and design practices can benefit society to become more tolerant.

INTRODUCTION

Although many leading concepts in modern neuroscience can be traced back to the speculation of ancient Greek philosophers (Crivellato & Ribatti, 2007), the neuroscience was not established as a unified discipline until 1971, date of the first meeting of the Society for Neuroscience. Neuroscience as a discipline is considered the rebirth of the mind study and one of the first truly transdisciplinary fields,

DOI: 10.4018/978-1-5225-0510-5.ch014

giving researchers a large conceptual umbrella under which they could posit hypotheses about the neural basis of thinking at all levels (Tokuhama-Espinosa, 2010).

The first section of this chapter presents an event that catapulted neuroscience to stardom as a pop phenomenon, when on the 26th of February 2015 the photo known as #TheDress became a viral Internet Meme with over 10 million people from around the world astonishingly arguing over the dichotomy in color perception caused by its ambiguity: #whiteandgold or #blueandblack. This image taught the general public that color perception can be as relative as human perception in general. The brain is equipped with a mechanism called perceptual constancy, responsible for bringing some stability to our already troubled lives. In the specific case of color, this mechanism is constantly compensating for changes in lighting in order to aid the color appearance of objects to remain stable. Without the color constancy, we would perceive objects changing color constantly because the light emitted by them actually changes according to the change in lighting. The full understanding of the reasons behind people perceiving #TheDress as either white-and-gold or blue-and-black and what they have in common is still a scientific mystery to be solved, but the most important lesson from this viral phenomenon is that neuroscience principles can be easily grasped by the general public and that color perception can be as subjective as a political opinion or a purchase decision.

The second section presents humans as a congeries of genes. These genes are responsible for humans greatly appreciating their own well being, and greatly dealing with their own pain. They are also responsible for mutual assistance among relatives. The closer the relationship the more likely they are to help a family member: a mother is always willing to sacrifice more for her son than her nephew, and the nephew more than the son of a neighbor, and so on. What lies behind all of this is the selfish gene. Humans don't think and act on their own need to increase their chances of being replicated. The selfish gene is a replicator and an almost immortal replicator lasting millions of years. The mortal lasting only a few years is the human. The selfish gene is responsible for humans fighting so much and almost never helping a stranger. But the selfish gene doesn't control everything and humans might be the only animals on the planet who are aware of being the result of selfish genes as well as having the capacity to transcend the selfishness of this very gene. Only humans can change the rules of the game so that they can become real humans.

The third section discusses the neuroscience of creativity and the ability of thinking outside the box, that is; thinking differently from the norm. Creativity is a combination of genetics and environment. However, neuroscience studies still cannot explain how exactly these two factors are combined in order to enhance the development of creativity. The increase in size of the human brain happened during two different periods of our evolution and may have generated a sophistication in human memory that provided humans with a better mental understanding of the environment and with greater creativity on how to socially behave. This section also discusses the neural basis of creativity and its relation to intelligence and also tackles the importance of enhancing the working environment in order to foster creativity.

CONTEXT IS EVERYTHING: LESSONS FROM #THEDRESS

Color vision scientists around the world went to sleep just to wake up on Friday the 27th of February, 2015 with several emails and messages questioning them about the color of #TheDress. At first glance, most scientists concluded that they were all seeing the same picture on different screens and that different screens producing the emission of different lights in turn ended up influencing the perception of the observer. However, the color difference was stark and even when people were looking at the same screen they were divided between gold-and-white and blue-and-black (Mahler, 2015; Rogers, 2015).

The general principle behind this dichotomy is "how we think we see". But to understand this sentence is necessary to understand at least some basic concepts of how the human mind works. Millions of people around the world were really passionate about the color of #TheDress and when someone perceived a different color it was like an affront, a threat to their own self-image (Rogers, 2016).

Color perception is as relative as human perception in general. A better way to understand this notion would be to use price as a measurement. For instance, the dress was priced at $77. Is this expensive or cheap? It depends. To a Hollywood celebrity is close to nothing but to a homeless person is quite a bit. So what is the relationship between price and color? The same way the notion of being expensive or cheap is related to a bank account, the color of the dress also depends on how each person's brain works. In the case of #TheDress, people are divided into two main categories, #whiteandgold or #blackandblue. The color of the original dress is black and blue (pictured below). Who is right? The answer is: everyone.

A brief testing was conducted by this author on February 27, 2015 with 24 observers using the same device: 12 individuals perceived gold-and-white and 12 others perceived blue-and-black. At the same time, across the town of Chicago, a student tested 29 co-workers: 11 perceived gold-and-white and 18 perceived blue-and-black. The conclusion pointed out the fact that the ratio of gold-and-white and blue-and-black could only be safely set after rigorously tested in a bigger sample. It also confirmed that the photo is actually controversial and people are divided between gold-and-white and blue-and-black. In general, color perception differences occur frequently but are so subtle that people usually don't pay much attention to them. However, #TheDress caused a stark color difference and that is why it turned viral.

The brain is equipped with a mechanism called perceptual constancy and that brings some stability to our already troubled lives. Among others, both the color constancy and the constancy of size or shape aim to ease the perceptual instability of our daily lives. In dealing with size, if a person is very close to you, their image projected onto your retina is different from the projected image when they are far away but your brain has no problem understanding that it is perceiving the same person with the same size, no more no less (Goldstein, 2011).

In the specific case of color, this mechanism is constantly compensating for changes in lighting in order to aid the color appearance of objects to remain stable. Without the color constancy, we would perceive objects changing color constantly because the light emitted by them actually changes according to the change in lighting - whether natural or artificial (Shevell, 2006).

The first known report on color constancy dates back to 1694 by Philippe De La Hire who addressed the fact that we don't realize that the colors are different under different lightings. Later on, a week before the French Revolution (1789) Gaspard Monge made a brilliant statement about color constancy, drawing attention to the phenomenon to the Royal Academy of Sciences in Paris. Clad in a red knit, he asked his colleagues to observe the clothing through red lenses. The result was an awestruck audience surprised that they had the feeling the mesh had a very whitish red tone - almost white (Feitosa-Santana et al., 2006; Mollon, 2006).

In the case of #TheDress, some brains assume that lighting is yellow and discount this lighting realizing the dress as blue-and-black while others assume that the lighting is blue, thus perceiving a gold-and-white dress.

Before the digital era, taking pictures required us to choose between Kodak and Fuji films – the former offering warmer tones and the latter cooler ones. Analog cameras were not equipped as our brain and were unable to discount the room lighting. So we had to do what we called white-balancing before taking a picture which is similar to what our brain does with color constancy in order to stabilize our perception of the colors.

But if we were aware of this phenomenon for so long, why is that the world is only addressing this mechanism now? It so happens that the color difference caused by #TheDress is brutal and in order to realize why, we need to understand how we build relationships between colors (color space) within our brain.

Just over a hundred years ago, Ewald Hering proposed that the experience of color results from the analysis of colors in opposite pairs: green opposing red and blue opposing yellow. Thus explaining why we are unable to perceive greenish-red/reddish-green or bluish-yellow/yellowish-blue. He also used examples of afterimages that we realize after setting a straight stare for about 30 seconds on the same image (Feitosa-Santana et al., 2006).

As explained by Hering, our perception of color is based on two chromatic channels working in opposition: the blue-and-yellow and green-and-red. And what does that mean, exactly? Blue is opposite to yellow; they do not co-exist and so we cannot see a bluish-yellow or yellowish-blue (figure 1). The

Figure 1. #TheDress

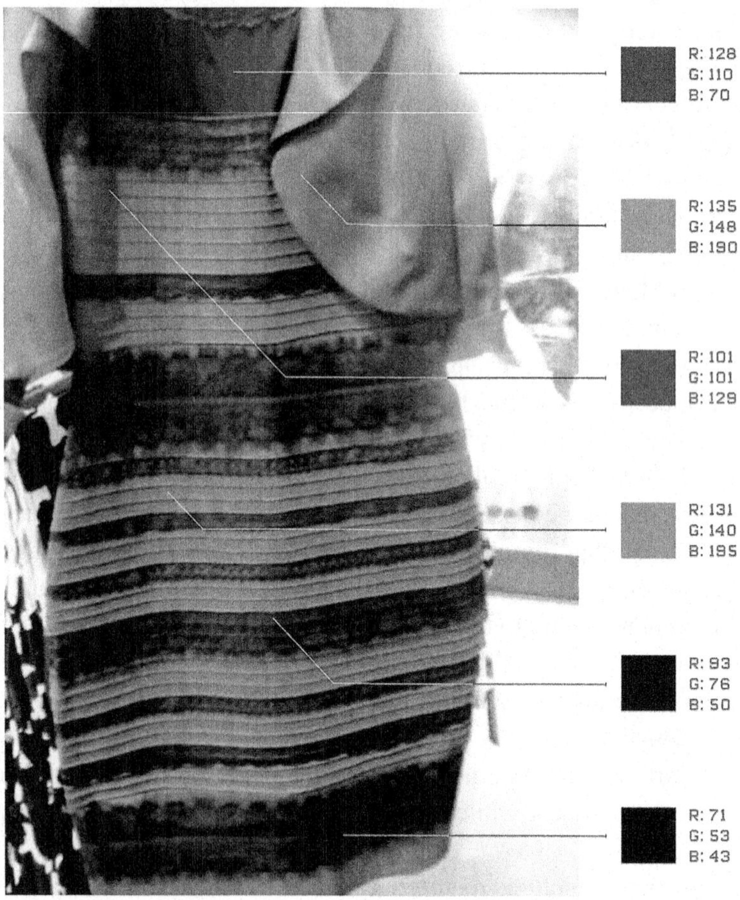

same goes for green and red. It is important to stress that our vision is based on light-colors but not pigment-colors – the latter being easily understood by observing the process of mixing pigment-colors used by printers and painters (Feitosa-Santana et al., 2006).

The original dress emits lights that usually cause the color perception of blue-and-black. The dress-image, however, came out very differently. The photo analysis (picture above) tells us that the dress should be perceived as blue-and-gold but the image is unstable and itself full of lights - requiring an arduous task from the brain in order to decide in milliseconds which is the predominant color of the lighting, so that it can then discount and generate the color perception: in this particular picture either blue lighting or yellow lighting.

If it were the case that our brains had to decide between green or blue lighting, the dress most likely would have not become a viral phenomenon, since blue and green are not opponents and perceptual differences would have passed unnoticed. Blue lighting creates a perception predominantly yellow, and green lighting generates a perception dominantly red. Yellow and red are two different colors but between each other there is a plethora of colors such as yellowish-red, yellowish-orange, orangish-red and many other similar descriptions that we are very used to, so it would not cause any real controversy.

In the case of blue and yellow lighting, which are opponents, the result is completely different depending on which of the two the brain decides to perceive. The blue lighting creates a perception predominantly yellow and soon people identify the color as gold or yellow. Discounting yellow lighting generates a perception predominantly blue. Between the blue and yellow there is no chromatic intermediary, creating the stark difference in colors of #TheDress.

The study conducted by this author and colleagues (Feitosa-Santana et al., 2016) in a sample of 52 subjects indicated that the distribution of color perception of #TheDress was approximately of 30% perceiving as blue-and-black, 40% as white-and-gold and 30% as blue-and-gold. The study conducted by Lafer-Sousa et al. (2015) reported different proportions for their 53 subjects: approximately 50% reporting blue-and-black, 40% reporting white-and-gold, and 10% reporting blue-and-gold. The difference in the color perception distribution between these studies could be explained by the fact that most of the subjects in the study conducted by Feitosa-Santana and co-authors were art students very interested in the discussion about #TheDress and may be influenced by the knowledge that the colored stripes of the dress-image were actually blue (and not white) and gold (or orange, but not black) when evaluated by the Photoshop RGB system, not considering the color constancy mechanism in action.

Still, it is unclear the relationship between the brains of those perceiving white-and-gold and those perceiving blue-and-black besides the fact that the former attains its observation in shadows and the latter under daylight (Brainard & Hurlbert, 2015; Gegenfurtner et al., 2015; Winkler et al., 2015). Lafer-Sousa and colleagues (2015) suggest that it all depends on people's daily schedules: morning people would perceive the dress as white-and-gold and evening people would perceive the dress as blue-and-black. However, Feitosa-Santana and colleagues (2016) did not find a correlation between the chronotype (morning/evening schedule) and the color of #TheDress.

Feitosa-Santana and colleagues (2016) found that the perception of the color of #TheDress was associated to the white perception and to color preferences. White-and-gold observers added more blue than the other groups to determine a white patch as well as they preferred more light colors than blue-and-black as well as blue-and-gold observers. Moreover, this study found that the preference for light green can be used as a predictor to the perception of the color of #TheDress, meaning that #whiteandgold observers had higher preferences for light green than others, blue-and-black or blue-and-gold.

The fully understanding of what people perceiving #TheDress in the same colors (white-and-gold or blue-and-black) have in common is still a scientific mystery to be solved, but the most important lesson from this viral phenomenon is being grasped by the general public. Color perception can be as subjective as a political opinion or a purchase decision. If the brain is unable to perceive the world as it is, what exactly does it do? It creates a world that can be useful to us and, therefore, becomes our reality. What we see or what we think is just a point of view, sometimes shared by many but sometimes shared by few or none. As renowned Brazilian Poet Carlos Drummond de Andrade once said "Each one opted in accordance with their whim, illusion or myopia". According to Wallisch (2015), just because we see something in a certain way it doesn't mean that everyone else will see it in the same way. Having this statement in mind helps us to built better relationships as well as a more tolerant society.

THE ROOTS OF HUMAN BEHAVIOR

Whether a denied favor or a broken deal, the reality is that usually the benefited person of the equation is a relative and the harmed one is not. In most cases, the excuse given for denying a favor or breaking a deal is directly related to the personal need of someone or that of a relative. Behind this attitude lies the selfish gene setting a scale of priorities according to the percentage of genes that they have in common with their relatives. The more genes in common the greater the benefit. The lesser degree of relation, the smaller the possibility of a benefit. In the calculation that defines the grantee, the winner is the one with the highest percentage of genes in common. The logic: to benefit the one with more genes in common increases the chances of preserving its own gene (Pinker, 1997).

This is a simplified summary of how the selfish gene, the central unit of the theory of evolution accepted by most scientists today, which has its roots in the association between the natural selection of Charles R. Darwin (1809-1882) and the genetics of Gregor J. Mendel (1822-1882) and was synthesized nearly half a century after the death of both. Since then it has been extensively studied in almost every branch of science. The selfish gene shows that animals were not adapted to preserve their species, but to preserve their genes. As Darwin noted, natural selection favors genes that replicate better, the selfish ones. Examples of the selfish gene in action are found in both the animal and plant kingdom, and with very little effort, you can identify them in the human society. From a simple broken deal to the killing of a stepson, there are many examples of the selfish gene speaking louder in *Homo sapiens* (Coyne, 2009; Dawkins, 2006; Trivers, 1971).

The term selfish gene was coined by Richard Dawkins, and despite some criticism, it is still widely used in discussing the evolution of species. Some authors prefer to call it the immortal gene or selfish genetic element and Dawkins regrets the fact that he didn't chose the term 'immortal gene' as strongly advised by Tom Maschler (Ridley, 2016). Anyway, it is worth noting that all authors are unanimous in saying that, although termed as the selfish gene, these genes have no conscience and therefore just do what needs to be done for the gene to be replicated. All authors, including the author of this article, agree that when the gene is selfish or, for example, the creator of something – it does not imply motivation or moral connotation (Dawkins, 2006).

The selfish gene is the one that needs to replicate, and the more replicators it gets, the greater the chances of success. To understand what is the selfish gene, it is necessary to differentiate body from gene. In the case of human beings, we must distinguish the person from the genes that compose them.

See: Robert and Elisabeth had one son, Bobby, who is not a clone of the parents but a package of genes composed of genetic material from both sides – half from the father and half from the mother. So what Bobby has in common with Robert and Elisabeth are genes. In this gene package, Bobby has one in two chances of having genes in common with both his father and his mother. Robert and Elisabeth didn't want Bobby as an only child, and because they don't live in China they could increase their family. Therefore, Bobby now has a sister, Beth, and again he has a one in two chances of having genes in common with his sister because they are children of the same father and same mother. Bobby and Beth grew. He had a daughter and she had a son. So Bobby has a one in four chances of having genes in common with his nephew, and Beth also has a one in four chances of having genes in common with her niece. So, the nephew and niece are cousins and have one in eight chances of having genes in common with each other. This probability in the percentage of genes in common creates a kinship scale as the number of genes in common increases. In this genetic relatedness there is no novelty, no mystery, and it is difficult to find someone who does not have this knowledge.

What is not new, but many still do not know, is that the scale of relatedness is directly related to the predisposition to cooperate (Hamilton, 1964). The more genes in common, the greater the chances are of cooperation. The less genes in common, the lower the chances of cooperation. It is within the motivation scale to make the largest investment in the selfish gene, and there are few biologists who do not accept this theory: genes are selfish because they need to ensure their own replication. The bodies do not replicate and therefore should not be selfish. Thus the selfish genes need to create neural connections of pleasure and pain to regulate the actions of the animal and thus increase their chances of replication. That's why animals appreciate their own welfare and suffer horrors with their own pain. On the issue of welfare, humans never get tired of self-benefit: shopping, traveling or any other action with no apparent purpose but to benefit oneself. "Never is enough" is the best expression for it. Thus, selfish genes need to create circuits in which humans can appreciate and feel good around each other. The more genes you have in common with someone the better you feel – increasing the chances of cooperation, and therefore, creating greater chances of replicating the selfish gene. Remember, the selfish gene has no conscience (Dawkins, 2006; Ridley, 2016).

Natural selection is not synonymous with preservation of the species and to think this way is a mistake. Darwin knew this, but was and still is often misunderstood. If the overarching goal were the species, the blood ties would be irrelevant. And what we see in the animal kingdom, including mankind, is just the opposite. In practice, it is easy to see that every mother is always more willing to sacrifice herself for her children than for her nephews, and she is more willing to help her nephews than the nephews of her husband and is less willing to help the nephews of her neighbor and so on (Pinker, 1997).

The truth is not nice and animals do not care about what happens to the species or ecosystem. In the film *March of the Penguins,* the penguin whose baby did not survive, suffers from his loss while simultaneously not caring one bit about the puppy beside him being dragged by a sea lion. This same penguin is afraid to dive for food because he does not know if there is a sea lion ready to attack him. So, what does he do? He waits for a much hungrier penguin to dive or he even tries to push the penguin neighbor in order to check whether the local water is free of sea lions and, therefore, safe to get food. Penguins, as much as men, are not concerned about their species or the ecosystem. Think in terms of humans; how many people do you know that skip their vacation to donate the money of their planned travel to benefit a stranger? How many couples do you know who chose to be foster parents instead of biological parents? By contrast, it is very common to find people who do not like their stepchildren, or worse, stepchildren who were obliged to leave home because they are victims of abuse by their stepparent. In general, animals behave in a selfish way depending on how the emotional circuits are designed, and selfish genes designed these circuits.

As stated earlier, it is necessary to separate the selfishness of the gene from the selfishness of the person, but because selfish genes built circuits in which people have more pleasure when they act in accordance to the wishes of the selfish gene, they end up being selfish too. Return to the exercise presented in the beginning of this paper. Most people are governed by their neural circuits, when in situations where they must choose between benefiting a relative or a non-relative. These neural circuits usually cause them to benefit the parent because of the well-being they feel from cooperating with the relative is much stronger when compared to the well-being they feel cooperating with a non-relative. For your friend, to benefit you at the expense of a relative would have much higher cost compared to the benefit given to a relative at your expense. The choice is often automatically and unconsciously based on cost-effectiveness (Hamilton, 1964; Pinker, 1997).

Now, imagine a change in scenery and your friend's dilemma is not to choose between his relative and you, but between his relative and his boss. If your friend benefits his relative, he runs the risk of being retaliated against by his boss. If your friend benefits his boss, he runs the risk of being retaliated against by the family. He probably will choose to benefit his boss and the chances of not being retaliated against by his family are enormous. In this dilemma, almost all human beings would benefit the boss. Why? When choosing a relative, the person is putting himself and his job in jeopardy, and he has more genes in common with himself than he has in common with a relative.

No one gets as sad about the pain of the neighbor as with his own pain. No one is as happy with the happiness of the neighbor as one's own happiness. But if that neighbor is a relative, the more genes in common, the more likely he is to have empathy and, therefore, the more likely he is to feel sadness or happiness for that family member with more intensity. And that is the essence of love, to feel pleasure with the pleasure of the other, and feel pain with the pain of the other. Today, knowing the selfish gene, it is possible to understand that cooperation between relatives or descendants of people with common genes is equivalent to helping themselves, and it is called kin selection. So if the love that exists between those who have genes in common is a mere reflection of kin selection, the question arises: love is nothing more than a neural circuit designed to work in favor of its creator, the gene selfish? Apparently, Yes!

But cooperation between people with common genes that corresponds to genes helping themselves is not the only type of cooperation in the animal kingdom, and especially in the human society. Cooperation between unrelated people exists, and friendship is an example (Trivers, 1971). The more you cooperate with your friend, the greater are your chances that they cooperate with you. This is mutual cooperation, also known as reciprocal altruism. From the point of view of the selfish gene, it may be good that you have the ability to cooperate with a non-relative. In practice, if you cooperate with your friend and your friend cooperates equally with you, you are even. But in the mathematics of friendship, no one wants to be abused and, therefore, you must be careful with the dodgers. The cheater is the one who wants to sell its cooperation for more than it's worth. Thus, he might be better benefited than those with whom he cooperates. A very clever dodger is one who can sell, successively, its cooperation for a little more than it is really worth. According to many evolutionists, cooperation between non-kin seems to have an important role in human evolution, and Robert Trivers (2011), in particular, suggests that humans are machines that have been adapted to cheat, to detect cheaters and avoid being seen as cheaters. In practice, it is easy to check, just look at the behavior of any child... It's no wonder that human society has its economy based on money, and also is full of laws, regulations, records, certificates, contracts, taxes, fines, etc. All examples that were made to prevent cheaters from having success.

Nobody wants to be cheated and there are impressive tactics developed through evolution to protect against a cheater. In much of the animal kingdom, females, unlike males, have much more certainty of having their genes in their offspring. Thus, males of several species have developed skills to ensure that they do not invest in offspring of other males and thus are not cheated. As an example, male mice secrete a chemical when a pregnant female smells him in order to check if he is indeed the father of the baby she is carrying. If she is not pregnant with him, the smell of the substance secreted causes abortion. Thus, it destroys their potential male stepson and the female is rendered to breed just for him. Think of the human race: the list of atrocities carried out by man to control the loyalty of "their" women is enormous. On the other hand, from the point of view of the female, most of them do not want a partner who is only interested in mating but do not want to collaborate in the creation of offspring. Therefore, in several species, the female requires a long courtship in order to check how much the male is really interested in investing in the future family. Recently, in modern society, another technique was developed to detect cheaters, man or woman, the DNA test.

In the eighties, the world followed a fairy tale that ended in tragedy. The story was a hoax, and many were the swindlers involved in it. Opinions are divided, and nearly all sided with Lady Di. According to the typical protocol, Prince Charles, who needed an heir to the English throne, could not marry Camilla and therefore sought the hand of his girlfriend, Diana, in marriage. Lady Di then discovered that she was not in a fairy tale and that her contribution to the marriage was only the "nobility" of her genes. Thus, she made no effort to hide from the world that she had been cheated. Human beings do not like cheaters. Lady Di was and still is admired by millions and millions of people worldwide. Recently, the couple's first child, William repeated the fairy tale and Kate was the chosen one. The royal family seems to have learned something and not encouraged, again, the realization of a marriage based only on genes. William and Kate were motivated to date for many years, trying to avoid another scandal that could, once again, upset the fans of the British royal family.

In the modern society, especially urban and virtual, there is a new type of cheating. Think of a friend who is a friend of all. They are the ones who please as many people as possible and rarely put themselves in conflict with others and they never take sides. They take sides only when they have no choice, and for sure their choice is based on who they see most likely to repay for their cooperation in the future. Whether consciously or unconsciously, their goal is always the same: self-benefit. Humans calculate the cost-benefit of any choice to cooperate in the future, always seeking chances to benefit from the cooperation of the chosen - short, medium or long term. Their motto: invest in their social network. You've just been introduced to one of the newest products of the selfish gene, the contemporary dodger. They are social chameleons, wearing the personality necessary to suit the environment, but you never really know who they are, much less what they think.

Genes have no morals, those who have morals are the animals - especially humans. Genes are not aware; those who are aware are people. The genes are potentially immortal while humans are not. Genes live for millions and millions and millions of years, people for just a few. Genes replicate, humans die. The gene does not learn, it just replicates. Humans learn, but take their knowledge to the grave. Not anymore. In oral society, humans lived an eternal cordless phone and knowledge could be lost easily.

But humans, unlike other animals, invented writing and have since avenged death. Therefore, with death avenged, the knowledge built by Darwin and Mendel is immortalized and enabled the development of the evolutionary theory presented here. Darwin descendants and some others have also contributed to the development of this theory. Mendel, in turn, had no children. His genes are gone, but his knowledge remained present. When his studies were rediscovered, many years after his death, they allowed the development of hundreds of studies that led to the understanding, treatment and often the cure of many genetic diseases (Heligman, 2009).

Humans are the only animals on earth that can break the totalitarianism of their genes. Genes created humans and they are the only ones that can turn against their creator. By rebelling, they can transcend their selfish nature and use free will to re-design the future of the human society in this planet.

THE NEUROSCIENCE OF CREATIVITY

Creativity is the ability to think differently from the norm, regardless of common sense. In other words, thinking "outside the box". What makes us creative? Like almost everything from personality traits to disease development, creativity is also a combination of genetic (nature/biology) and environment (nurture/culture), but the neuroscience studies still can not say how exactly these two factors are combined to the development of creativity.

Despite human lineage starting 6 million years ago, the first signs of human creativity are merely over 2 million years old – when *Homo habilis* was able to produce the first stone tools. Between *Homo habilis* and humans today, we have seen an increase in size of our brain that is associated with the improvement of our creativity, intellectual and technological productivity. The first increase in size of the head was noted in the *Homo erectus* (1.8 mi years ago) and coincides with the presence of creative thinking – the same that has adapted to climate differences, development of more sophisticated tools and long distance hunting abilities. The second significant increase in the brain has approximately 600 to 150 thousand years ago with a series of inventions like glue, protective mix of mosquitoes and fire, culminating in a giant leap of human creativity. Between 90 and 30 thousand years ago, it occurs the emergence of syntax in language along with the arts, sciences, and politics. Brain increase in these two moments may have generated a sophistication in human memory that enabled better mental representation (understanding) of the environment and greater skill in social behavior (Gabora & DiPaola, 2012).

Creative people in various fields like music, visual arts, poetry, etc. present creative processes with very similar standards such as the ease in creating metaphors. Besides the ease in creating metaphors, *Synesthetes* – people with synesthesia, also show 7 times more likelihood to pursue creative careers. Synesthesia is a consequence of cortical formation where 3 regions – temporal, parietal and occipital are not separated completely and the overlap of them is responsible for this perceptual phenomenon where two distinct senses are perceived simultaneously (such as seeing a color while at the same time listening to a certain musical note or tasting a flavor while at the same time feeling a certain texture of touch) (Ramachandram & Hubbard, 2003).

The study conducted by Jung and colleagues (2010) suggests that in neuroscience, creativity is correlated with increased cortical thickness in some regions of the brain and reduced cortical thickness in other regions of the brain. These findings indicate that creativity is a complex activity that requires excitement in some regions of the brain as well as inhibition in others – e.g., the improvisation arts require synaptic inhibition in a certain region of the brain that is associated with a power giant concentration. Lhommée and colleagues (2014) showed how the dopaminergic circuits (dependent dopamine) are important for stimulating creativity.

Creativity and intelligence require distributed neural participation in a complex combination of brain activities, although both work with partially different circuits, and it is suggested that creativity is not merely the result of a high intelligence but a circuit that has evolved parallel to intelligence (Jung et al., 2013). Insights are more likely to happen after a good night's sleep: those who sleep well are more likely to have a light bulb moment (Eureka!) than those who are sleep deprived (Wagner et al., 2004).

Humans avenged death when they invented writing, making the transmission of knowledge after one's passing a reality. This revenge is the basic difference between human culture and the culture of other species as we tend to develop culture faster. Chimpanzees teach their children how to exterminate ants but can't refine this technique adding something else because they are unable to pass on improvements in this behavior for future generations as humans do.

So, what exactly is needed to develop one's creativity? For children, a good diet, many hours of sleep and diverse stimulation (motor, visual and intellectual) because the ability of our brain must create new neural connections and our neural plasticity is much stronger when we are young. The creative process involves many simultaneous connections between different brain regions. For adults, in addition to a good diet and good night's sleep (and the amount of time varies from person to person), time, attention and a physical activity that results in well-being. In this context, time translates into *quality* and attention into *concentration* and *focus*. Some studies show that the only way to enable creative solutions in a short period of time is through attention (dedication or exclusive concentration towards the solution of the problem). Too much time without quality or little time without attention are harmful ways to stimulate creativity. Two great examples of time in the creative process are Darwin and Einstein, both worked years and years to synthesize their theories.

How does the work environment foster creativity? Most workplaces do not encourage creativity. One needs a flexible work environment, respecting the employee's personality so that in turn this can exercise their creativity, since contrary to popular belief, stressful environments work against creativity.

CONCLUSION

The primary goal of neuroscience is to understand how the human mind works in its various capabilities. Its main barrier is finding ways to dialogue with the general public and in the classroom. #TheDress Meme gave neuroscience the opportunity to become pop culture material and took scientific material to the daily discussions of millions around the globe. Despite a nostalgia constantly evoking the past, there are plenty of scientific evidence (Pinker, 2011) proving that humanity has never been better, and that we live in a much safer and compassionated society than our ancestors. Our biology may be still the same for thousands of years but our cultural changes are responsible for all this present positive change. Humans are the only animals on earth that can break the totalitarianism of their genes. And we hope to keep on going until we are done with that job.

REFERENCES

Axelrod, R. (2011). Lauching "The Evolution of Cooperation". *Journal of Theoretical Biology, 299*(2012), 21-24.

Baldassari, D., & Grossman, G. (2011). Centralized sanctioning and legitimate authority promote cooperation in humans. *PNAS Early Edition, 108*(27), 11023–11027. doi:10.1073/pnas.1105456108 PMID:21690401

Brainard, D. H., & Hurlbert, A. C. (2015). Colour vision: Understanding #TheDress. *Current Biology, 25*(13), R549–R568. doi:10.1016/j.cub.2015.05.020 PMID:26126278

Burt, A., & Trivers, R. (2011). *Genes in Conflits: the biology of selfish genetic elements.* Cambridge, London: The Belknap Press of Harvard University Press.

Coyne, J. A. (2009). *Why Evolution is True?* New York: Penguin Group.

Crivellato, E., & Ribatti, D. (2007). Soul, Mind, Brain: Greek philosophy and the birth of neuroscience. *Brain Research Bulletin, 71*(4), 327–336. doi:10.1016/j.brainresbull.2006.09.020 PMID:17208648

Dawkins, R. (2006). The Selfish Gene (30th ann. Ed.). New York: Oxford.

Feitosa-Santan, C., Lutze, M., Barrionuevo P., & Cao, D. (submitted). *The Perceived Color of #TheDress is Associated with Individual Color Preferences.*

Feitosa-Santana, C., Oiwa, N., Costa, M. F., Tiedemann, K. B., Silveira, L. C. L., & Ventura, D. F. (2006). Espaço de Cores. *Psicologia USP, 17*(4), 35–62. doi:10.1590/S0103-65642006000400003

Gabora, L., & DiPaola, S. (2012). How did humans become so creative? *A computational approach. International Conference on Computational Creativity.*

Gegenfutner, K. R., Bloj, M., & Toscani, M. (2015). The many colours of 'the dress'. *Current Biology, 25*, R1–R2. PMID:25562290

Goldstein, S. E. (2011). *Cognitive Psychology: Connecting mind, research, and everyday life experience* (3rd ed.). Belmont: Wadsworth Cengage Learning.

Grossberg, S. (2014in press). The visual world as illusion: The ones we know and the ones we don't. In A. Shapiro & D. Todorovic (Eds.), *Oxford Compendium of Visual Illusions.* Oxford, UK: Oxford University Press.

Hamilton, W. (1964a). The genetical evolution of social behaviour I. *Journal of Theoretical Biology, 7*(1), 1–16. doi:10.1016/0022-5193(64)90038-4 PMID:5875341

Hamilton, W. (1964b). The genetical evolution of social behaviour II. *Journal of Theoretical Biology, 7*(1), 17–52. doi:10.1016/0022-5193(64)90039-6 PMID:5875340

Heligman, D. (2009). *Charles and Emma – The Darwin's Leap of Faith.* New York: Harry Holt and Company, LLC.

Jung, R. E., Mead, B. S., Carrasco, J., & Flores, R. A. (2013). The structure of creative cognition in the human Brain. *Frontiers in Human Neuroscience*, *7*(330), 1–13. PMID:23847503

Jung, R. E., Segall, J. M., Bockholt, H. J., Flores, R. A., Smith, S. M., Chavez, R. S., & Haier, R. J. (2010). Neuroanatomy of Creativity. *Human Brain Mapping*, *31*(3), 398–409. PMID:19722171

Lafer-Sousa, R., Hermann, K., & Conway, B. R. (2015). Striking individual differences in color perception uncovered by 'the dress' photograph. *Current Biology*, *25*(13), pR545–R546. doi:10.1016/j.cub.2015.04.053 PMID:25981795

Lhommée, E., Batir, A., Quesada, J.-L., Ardouin, C., Fraix, V., Seigneuret, E., & Krack, P. et al. (2014). Dopamine and the biology of creativity: Lessons from Parkinson's disease. *Frontiers in Neurology*, *5*, 55. PMID:24795692

Mahler, J. (2015, February 27). The White and Gold (No, Blue and Black!) Dress that Melted the Internet. *New York Times*.

Mollon, J. (2006). Monge: The Verriest Lecture, Lyon, July 2005. *Visual Neuroscience*, *23*(3-4), 297–309. doi:10.1017/S0952523806233479 PMID:16961961

Pinker, S. (1997). *How the Mind Works*. New York: W. W. Norton & Company.

Pinker, S. (2011). *The Better Angels of Our Nature: Why Violence Has Declined*. New York: Viking Books.

Raihani, N. J., & Bshary, R. (2011). Resolving the iterated prisoner's dilemma: Theory and reality. *Journal of Evolutionary Biology*, *24*(8), 1628–1639. doi:10.1111/j.1420-9101.2011.02307.x PMID:21599777

Ramachandran, V. S., & Hubbard, E. M. (2003). The phenomenology of synaesthesia. *Journal of Consciousness Studies*, *10*, 49–57.

Rogers, A. (2015, February 26). The Science of why no one agrees on the color of this dress. *Wired*.

Rogers, A. (2016, February 26). A year ago, a Dress murdered the idea of objective color. *Wired*.

Shevell, S. K. (2003). Color appearance. In S. K. Shevell (Ed.), *The Science of Color* (2nd ed.; pp. 149–190). Oxford, UK: Elsevier. doi:10.1016/B978-044451251-2/50005-2

Tokuhama-Espinosa, T. (2010). *Mind, Brain, and Education Science: The new brain-based learning*. New York: W. W. Norton & Company.

Trivers, R. (1971). The evolution of reciprocal altruism. *The Quarterly Review of Biology*, *46*(1), 35–57. doi:10.1086/406755

Trivers, R. (2011). *The Folly of Fools: The Logic of Deceit and Self-Deception in Human Life*. New York: Basic Books.

Trivers, R. L. (1972). Parental investment and sexual selection. In B. Campbell (Ed.), *Sexual selection and the descent of man, 1871-1971* (pp. 136–179). Chicago: Aldine.

Wagner, U., Gals, S., Halder, H., Verleger, R., & Born, J. (2004). Sleep inspires insight. *Nature*, *427*(6972), 352–355. doi:10.1038/nature02223 PMID:14737168

Wallisch, P. (2015, February 27). Lessons from the Dress – Vision is fundamentally ambiguous but usually not this ambiguous. *Slate*.

Winkler, A. D., Spillmann, L., Werner, J. S., & Webster, M. A. (2015). Asymmetries in blue–yellow color perception and in the color of 'the dress'. *Current Biology*, *25*(13), R547–R548. doi:10.1016/j.cub.2015.05.004 PMID:25981792

KEY TERMS AND DEFINITIONS

#TheDress: A photo that became a viral meme on 26 February 2015 also associated with the hashtags #whiteandgold and #blackandblue.

Color Constancy: One type of perceptual constancy in which our brain compensates for changes in lighting in order to aid the color appearance of objects to remain stable.

Color Perception: A perceptual phenomenon determined by neural processes in the brain.

Context: The frame of reference, circumstances, factors or conditions involved in a scenario or event, background.

Creativity: The ability to think differently from the norm, regardless of common sense. In other words, thinking "outside the box".

Illusion: The deception of the senses, changing the actual appearance for a fake one.

Selfish Gene: The immortal gene that is the core view of evolution theory.

Compilation of References

Abbagnano, N. (2007). *Dicionário de filosofia.* Martins Fontes.

Ackerman, S. (1992). *Discovering the Brain.* National Academies Press.

Aedo, I., Catenazzi, N., & Díaz, P. (1996). The Evaluation of a Hypermedia Learning Environment: The CESAR Experience. *Journal of Educational Multimedia and Hypermedia, 5*(1).

Aharon, I., Etcoff, N., Ariely, D., Chabris, C., O'Connor, E., & Breiter, H. (2001). Beautiful faces have variable reward value: FMRI and behavioral evidence. *Neuron, 32*(3), 537–551. doi:10.1016/S0896-6273(01)00491-3 PMID:11709163

Aicher, O. (2001). *Analógico y Digital.* Barcelona: Gustavo Gili.

Alberti, L. B. (1989). *Da Pintura.* Campinas, Brazil: Editora da UNICAMP.

Alcantara, C. P. & Machado, J. H. C., Jr. (2010) Arte e Design: para a construção de uma relação mais afetiva entre homem e objeto. *Proceedings of: Congresso Brasileiro de Pesquisa e Desenvolvimento em Design.*

Aleksander, I. (2005). *The world in my mind, my mind in the world: key mechanisms of consciousness in people, animals and machines.* Exeter: Imprint Academic.

Alimardani, M., Nishio, S., & Ishiguro, H. (2013). Humanlike robot hands controlled by brain activity arouse illusion of ownership in operators. *Scientific Reports, 3*, 2396. doi:10.1038/srep02396 PMID:23928891

Almeida, D. Di M. de. (2001). Corpo, Tecnologia, Cultura. In Corpo e cultura. São Paulo, SP: Xamã.

Almeida, R. R. de. (2010). Emoção na Pesquisa de Público-alvo: Estudo de caso em disciplina de projeto de produto. *Proceedings of: Congresso Brasileiro ee Pesquisa e Desenvolvimento em Design.*

Aristotle, . (1984). *The Complete Works of Aristotle, in two volumes* (J. Barnes, Ed.). Princeton, NJ: Princeton University Press.

Armel, K. C., & Ramachandran, V. S. (2003). Projecting sensations to external objects: Evidence from skin conductance response. *Proceedings. Biological Sciences, 270*(1523), 1499–1506. doi:10.1098/rspb.2003.2364 PMID:12965016

Arnheim, R. (2002). *Arte e percepção visual: uma psicologia da visão criadora.* São Paulo: Pioneira.

Ashton, J. (2011). *Two problems with Neuroesthetic Theory of Interpretation.* Emory College of Arts and Sciences: nonsite.org. Available from http://nonsite.org/issues/issue-2/two-problems-with-a-neuro-aesthetic-theory-of-interpretation

Aumont, J. (1995). *A imagem.* Campinas: Papirus.

Au-Young, S. M., Shen, H., & Yang, C. R. (1999). Medial prefrontal cortical output neurons to the ventral tegmental area (VTA) and their responses to burst-patterned stimulation of the VTA: Neuroanatomical and in vivo electrophysiological analyses. *Synapse (New York, N.Y.), 34*(4), 245–255. doi:10.1002/(SICI)1098-2396(19991215)34:4<245::AID-SYN1>3.0.CO;2-D PMID:10529719

Axelrod, R. (2011). Lauching "The Evolution of Cooperation". *Journal of Theoretical Biology, 299*(2012), 21-24.

Azevedo, A. M. T. (2006). *Estrutura narrativa e espaços mentais.* Belo Horizonte: Faculdade de Letras da UFMG.

Bachelar, G. (1996). *A Formação do Espirito Científico: Contribuição para uma psicanálise do conhecimento.* Rio de Janeiro, Brazil: Contraponto.

Baldassari, D., & Grossman, G. (2011). Centralized sanctioning and legitimate authority promote cooperation in humans. *PNAS Early Edition, 108*(27), 11023–11027. doi:10.1073/pnas.1105456108 PMID:21690401

Baldissin, M. (2014). Learning and human perceptions according to anthroposophy and neurosciences. *Arte Médica Ampliada, 34*(1). Available at http://www.abmanacional.com.br/arquivo/f479d9701ee-676a3553251963d0515d0bb7c9d23-34-1-aprendizado-e-percepcoes.pdf

Ball, P. (2013). *Neuroaesthetics is Killing Your Soul: Can Brain Scans Ever Tell Us Why We Like Art?* Nature Publishing Group. Available from http://www.nature.com/news/neuroaesthetics-is-killing-your-soul-1.12640

Ballestero-Alvarez, J. A. (2003). *Multisensorialidade no ensino de desenho a cegas* [Multisensoriality in the teaching of design to the blind]. São Paulo: USP. Available at http://www.teses.usp.br/teses/disponiveis/27/27131/tde-21032005-213811/pt-br.php

Ball, T., Schulze-Bonhage, A., Aertsen, A., & Mehring, C. (2009). Differential representation of arm movement direction in relation to cortical anatomy and function. *Journal of Neural Engineering, 6*(1), 016006. doi:10.1088/1741-2560/6/1/016006 PMID:19155551

Barbour, J. B. (2000). *The End of Time: The Next Revolution in Our Understanding of the Universe.* London, UK: Phoenix-The Orion Publishing Group Ltd.

Barros, A. B. (2013). *Processo de inclusão no contexto da deficiência visual: dificuldades, desafios e perspectivas* [Social inclusion in the area of visual impairment: difficulties, challenges and perspectives]. (Master's Dissertation). São Luís, Brazil.

Barthes, R., Greimas, A. J., Bremond, C., & Eco, U. (1976). Análise estrutural da narrativa: pesquisas semiológicas. Tradução de Maria Zélia Cardoso Pinto. 4ª ed. Petrópolis: Vozes.

Basili, V. R., Lindvall, M., & Costa, P. (2001). Implementing the experience factory concepts as a set of experience bases. In *Proceedings of the 2001 International Conference on Software Engineering & Knowledge Engineering*.

Baudelaire, C. (1995). The Painter of Modern Life and Other Essays. London: Phaidon Press.

Bauer, M., & Gaskell, G. (2008). Pesquisa qualitativa com texto, imagem e som: um manual prático. 7. Ed. Rio de Janeiro: Vozes.

Bayles, D., & Orland, T. (2007). *Art and Fear. Observations on the perils (and rewards) of artmaking*. Eugene, OR: Image Continuum Press Edition.

Beeker, T. W., During, J., & Den Hertog, A. (1967). Artificial touch in a hand-prosthesis. *Medical & Biological Engineering*, *5*(1), 47–49. doi:10.1007/BF02478841 PMID:6037625

Bergeron, V. (2011). *What Should We Expect from the New Aesthetic Sciences?* Aesthetics: American Society for Aesthetics. Available from http://www.aesthetics-online.org/articles/index.php?articles_id=55

Berg, M., & Mol, A. (Eds.). (1998). *Differences in Medicine: Unravelling Practices, Techniques and Bodies*. Durham, NC: Duke University Press.

Bergson, H. (1990). Matter and Memory. New York, NY: Zone Book.

Bergson, H. (2010). *Mind-Energy: Lectures and Essays*. Kessinger Publishing.

Bertolino, A., Breier, A., Callicott, J. H., Adler, C., Mattay, V. S., Shapiro, M., & Weinberger, D. R. et al. (2000). The relationship between dorsolateral prefrontal neuronal N-acetylaspartate and evoked release of striatal dopamine in schizophrenia. *Neuropsychopharmacology*, *22*(2), 125–132. doi:10.1016/S0893-133X(99)00096-2 PMID:10649825

Best, S., & Kellner, D. (1997). *The post modern turn*. New York, NY: The Guilford Press.

Biddis, E. A., & Chau, T. T. (2007). Upper limb prosthesis use and abandonment: A survey of the last 25 years. *Prosthetics and Orthotics International*, *31*(3), 236–257. doi:10.1080/03093640600994581 PMID:17979010

Bieging, P., & Aquino, V. (2014). A experiência estética no iCinema: sensação e interatividade a partir de narrativas multilineares. In Olhares do sensível: experiências e dimensões estéticas em comunicação (pp. 84-97). São Paulo: Pimenta Cultural.

Bollnow, O. F. (2008). *O homem e o espaço*. Curitiba: Editora UFPR.

Bomfim, G. A. (1999). Coordenadas cronológicas e cosmológicas como espaço das transformações formais. Formas do Design. Rio de Janeiro: 2AB.

Bonsiepe, G. (2011). *Design, Cultura e Sociedade*. São Paulo: Blucher.

Borges, P., Machado, R. J., & Ribeiro, P. (2012).Mapping RUP Roles to Small Software Development Teams. In *Proceedings of International Conference on Software and System Process*, (pp. 190-199). doi:10.1007/978-3-642-27213-4_5

Borsook, D., Becerra, L., Fishman, S., Edwards, A., Jennings, C. L., Stojanovic, M., & Breiter, H. et al. (1998). Acute plasticity in the human somatosensory cortex following amputation. *Neuroreport*, *9*(6), 1013–1017. doi:10.1097/00001756-199804200-00011 PMID:9601659

Boselie, F. (1984). Complex and simple proportions and the aesthetic attractivity of visual patterns. *Perception*, *13*(2), 91–96. doi:10.1068/p130091 PMID:6504684

Bosi, I. A. (2006). Fenomenologia do olhar. In A. Novaes (Ed.), *O olhar* (p. 66). São Paulo: Companhia das Letras.

Botvinick, M., & Cohen, J. (1998). Rubber hands 'feel' touch that eyes see. *Nature*, *391*(6669), 756. doi:10.1038/35784 PMID:9486643

Bourdieu, P. (1998). *A miséria do mundo*. Petrópolis: Editora Vozes.

Bowker, G. (2005). *Memory Practices in the Sciences*. Cambridge, MA: MIT Press.

Boyd, B. (2009). *On the origin of stories: Evolution, cognition and fiction*. Cambridge, MA: The Belknap Press of Harvard University Press.

Brainard, D. H., & Hurlbert, A. C. (2015). Colour vision: Understanding #TheDress. *Current Biology*, *25*(13), R549–R568. doi:10.1016/j.cub.2015.05.020 PMID:26126278

Brasted, P. J., Bussey, T. J., Murray, E. A., & Wise, S. P. (2003). Role of the hippocampal system in associative learning beyond the spatial domain. *Brain*, *126*(5), 1202–1223. doi:10.1093/brain/awg103 PMID:12690059

Brattico, E., Bogert, B., & Jacobsen, T. (2013). Toward a neural chronometry for the aesthetic experience of music. *Frontiers in Psychology*, *4*, 206. doi:10.3389/fpsyg.2013.00206 PMID:23641223

Brockmeier, J., & Harré, R. (2003). Narrativa: Problemas e Promessas de um Paradigma Alternativo. *Psicologia: Reflexão e Crítica*, *16*(3), 525–535. doi:10.1590/S0102-79722003000300011

Brown, S., & Dissanayake, E. (2009). The arts are more than aesthetics: Neuroaesthetics as narrow aesthetics. In S. Martin & V. Oshin (Eds.), *Neuroaesthetics*. New York, NY: Baywood Publishing Company Inc.

Brown, S., & Dissanayake, E. (2009). The arts are more than aesthetics: neuroaesthetics as narrow aesthetics. In M. Skov & O. Vartanian (Eds.), *Neuroaesthetics* (pp. 43–57). Amityville, NY: Baywood.

Bryson, N. (2003). Introduction. In W. Neidich (Ed.), *Blow-Up: Photography, cinema and the brain*. New York: D.A.P., Distributed Art Publishers.

Bulatov, D. (Ed.). (2004). *Biomediale. Contemporary Society and Genomic Culture*. Kaliningrad: The National Centre for Contemporary Art.

Bullinger, H.-J., Warschat, J., Schumacher, O., Slama, A., & Ohlhausen, P. (2005). Ontology-Based Project Management for Acceleration of Innovation Project. *Lecture Notes in Computer Science*, *3379*, 280–288. doi:10.1007/978-3-540-31842-2_28

Burt, A., & Trivers, R. (2011). *Genes in Conflits: the biology of selfish genetic elements*. Cambridge, London: The Belknap Press of Harvard University Press.

Busatto Filho, G. (2000). Imagens do funcionamento cerebral durante tarefas cognitivas e emocionais: aplicações da técnica de ressonância Magnética funcional em psiquiatria. *Revista de Psiquiatria Clínica*, *27*(3). Retrieved from www. Hcnet.usp.br/Ipq/revista/vol27/n3/art164.html

Cairns-Smith, A. G. (1996). *Evolving the mind on the nature of matter and the origin of consciousness*. Cambridge, UK: Cambridge University Press.

Campbell, R., Heywood, C. A., Cowey, A., Regard, M., & Landis, T. (1990). Sensitivity to eye gaze in prosopagnosic patients and monkeys with superior temporal sulcus ablation. *Neuropsychology*, *28*(11), 1123–1142. doi:10.1016/0028-3932(90)90050-X PMID:2290489

Cara, M. (2010). *Do desenho industrial ao design no Brasil: uma bibliografia crítica para a disciplina*. São Paulo: Blucher.

Cardoso, R. (2013). *Design para um mundo complexo*. São Paulo: Cosac Naify.

Cares, C., Franch, X., & Mayol, E. (2006). Perspectives about paradigms in software engineering. In *Proc. 2nd International workshop on Philisophiocal Foundations on Information Systems Engineering*.

Carey, J. (2010). *What Good Are the Arts?* Oxford, UK: Oxford University Press.

Caro, C. R., Rodrigues, J. R., & Pedro, E. S. (2014). *As técnicas de modelagem plana e moulage e suas aplicações na indústria do vestuário* [The techniques of modelling and moulage and their applications in the clothing industry]. Available at http://www.contexmod.net.br/index.php/segundo/article/view/141

Castilho, K., & Vicentini, C. G. (2008). O corte, a costura, o processo e o projeto de moda no re-design do corpo [The court and dressmaking – fashion projects in redesigning the body]. In Corpo e moda: por uma compreensão do contemporâneo [The body and fashion: through an understanding of the contemporary world]. Barueri, SP: Estação das Letras e Cores Editora.

Castro, C., & Freitas, C. (2010). Narrativa Audiovisual para Multiplataforma - Um Estudo Preliminar. *Bibliocom*, *2*(7). Retrieved June 20, 2010 from http://www.intercom.org.br/bibliocom/sete/pdf/cosette-castro-cristiana-freitas.pdf

Catts, O., & Zurr, I. (2002). The Aesthetics of Care?. SymbioticA, School of Anatomy and Human Biology, UWA, Perth.

Cézanne, P. (n.d.). *Paul Cézanne: French Draftsman and Painter*. Retrieved from http://www.theartstory.org/artist-cezanne-paul.htm

Chalmers, D. (1996). *The conscious mind: in search of a fundamental theory*. Oxford, UK: Oxford University Press.

Chandrasekaran, B., Josephson, J. R., & Benjamins, V. R. (1998). Ontology of tasks and methods. In *Proceedings of the Workshop on Applications of Ontologies and Problem-Solving Methods, held in conjunction with ECAI'98*.

Changeaux, J.-P. (2013). *O verdadeiro, o belo e o bem*. Rio de Janeiro: Civilização Brasileira.

Changeux, J. P. (1985). *Neuronal man*. Princeton, NJ: Princeton University Press.

Changeux, J. P. (2013). *O verdadeiro, o belo e o bem: Uma nova abordagem neuronal*. Rio de Janeiro: Civilização Brasileira.

Chatterjee, A. (2010). Neuroaesthetics: A coming of age story. *Journal of Cognitive Neuroscience*, *23*(1), 53–62. doi:10.1162/jocn.2010.21457 PMID:20175677

Chatterjee, A. (2013). The illogic of beauty. In *The Aesthetic Brain: How We Evolved to Desire Beauty and Enjoy Art*. Oxford, UK: Oxford University Press. doi:10.1093/acprof:oso/9780199811809.003.0010

Chatterjee, A., Thomas, A., Smith, S. E., & Aguirre, G. K. (2009). The neural response to facial attractiveness. *Neuropsychology*, *23*(2), 135–143. doi:10.1037/a0014430 PMID:19254086

Chib, V. S., Yun, K., Takahashi, H., & Shimojo, S. (2013). Noninvasive remote activation of the ventral midbrain by transcranial direct current stimulation of prefrontal cortex. *Translational Psychiatry*, *11*(3), e268. doi:10.1038/tp.2013.44 PMID:23756377

Cinzia, D. D., & Vittorio, G. (2009). Neuroaesthetics: A review. *Current Opinion in Neurobiology*, *19*(6), 682–687. doi:10.1016/j.conb.2009.09.001 PMID:19828312

Clancey, W. (1993). Situated Action: A neuropsychological interpretation. *Cognitive Science*, 17–87.

Clark, V. P., & Raj, P. (2014). Neuroenhancement: Enhancing brain and mind in health and in disease. NeuroImage, 85, 889-894.

Cloutier, J., Heatherton, T. F., Whalen, P. J., & Kelley, W. M. (2008). Are attractive people rewarding? Sex differences in the neural substrates of facial attractiveness. *Journal of Cognitive Neuroscience*, *20*(6), 941–951. doi:10.1162/jocn.2008.20062 PMID:18211242

Coelho, L. A. (2006). *Textos Selecionados de Design* (G. C. Lima, Ed.). PPDESDI UERJ.

Colgan, A. (2014). *How Does the Leap Motion Controller Work?* Retrieved from http://blog.leapmotion.com/hardware-to-software-how-does-the-leap-motion-controller-work/

Conway, B. R., & Rehding, A. (2013). Neuroaesthetics and the Trouble with Beauty. *PLoS Biology*, *11*(3), e1001504. doi:10.1371/journal.pbio.1001504 PMID:23526878

Corso, H. V., Sperb, T. M., Jou, G. I., & Salles, J. F. (2013). Metacognição e funções executivas: relações entre conceitos e implicações para a aprendizagem. *Psicologia: Teoria e Pesquisa*, *29*(1), 21-29. Retrieved from http://www.scielo.br/pdf/ptp/v29n1/04.pdf

Costa, F. Z. N., & Lima, I. M. M. R. D. (n.d.). *Metodologia de ensino de modelagem: uma experiência pedagógica no curso de moda* [A methodology for teaching modelling: a pedagogical experience in fashion courses]. Available at http://www.coloquiomoda.com.br/anais/anais/2-Coloquio-de-Moda_2006/artigos /46.pdf

Costa, C. Z. (2010). *Além das formas: introdução ao pensamento contemporâneo no design, nas artes e na arquitetura*. São Paulo: Anablume.

Courtés, J. (1979). *Introdução à semiótica narrativa e discursiva. In Tradução de Norma Backes Tasca*. Coimbra: Livraria Almeida.

Couto, R. M., & Oliveira, A. J. (1999). *Formas do Design*. Rio de Janeiro: 2AB.

Coyne, J. A. (2009). *Why Evolution is True?* New York: Penguin Group.

Crary, J. (1992). *Techniques of the Observer: On Vision and Modernity in the Nineteenth Century*. Cambridge, MA: MIT Press.

Crivellato, E., & Ribatti, D. (2007). Soul, Mind, Brain: Greek philosophy and the birth of neuroscience. *Brain Research Bulletin*, *71*(4), 327–336. doi:10.1016/j.brainresbull.2006.09.020 PMID:17208648

Culpepper, J. (1999). *Discriminating Mental States Using EEG Represented by Power Spectral Density*. Claremont: Harvey Mudd College.

Cunningham, M., Barbee, A., & Philhower, C. (2002). Dimensions of facial physical attractiveness: The intersection of biology and culture. In G. Rhodes & L. Zebrowitz (Eds.), *Facial attractiveness. Evolutionary, cognitive, and social perspectives* (pp. 193–238). Westport, CT: Ablex.

Damásio. (2004). *António. Em busca de Espinosa: prazer e dor na ciência dos sentidos* [In search of Espinosa: pleasure and pain in the science of the feelings]. São Paulo: Companhia das Letras.

Damásio. (2011). *Antônio R. E o cérebro criou o homem* [The brain created mankind]. São Paulo: Companhia das Letras.

Damásio. (2012). *O erro de Descartes: emoção, razão e o cérebro humano* [The mistake of Descartes: emotions, reason and the human brain]. São Paulo: Companhia das Letras.

Damásio, A. (1994). *Descartes' error: emotion, reason, and the human brain*. New York, NY: The Hearst Corporation.

Damásio, A. (1996). *O erro de Descartes*. São Paulo: Cia das Letras.

Damásio, A. (1996). *O Erro de Descartes: Emoção, Razão e o Cérebro Humano*. São Paulo: Companhia das Letras.

Damásio, A. (2000). *O mistério da consciência*. São Paulo: Cia das Letras.

Damásio, A. (2000). *O Mistério da consciência: do corpo e das emoções ao conhecimento*. São Paulo: Companhia das Letras.

Damásio, A. (2004). *Em Busca de Espinosa: prazer e dor na ciência dos sentimentos*. São Paulo: Companhia da Letras.

Damásio, A. (2004). *Em busca de Espinosa: prazer e dor na ciência dos sentimentos*. São Paulo: Companhia das Letras.

Damásio, A. (2010). *Self comes to mind: constructing the conscious brain*. New York, NY: Pantheon Books.

Damásio, A. (2011). *E o cérebro criou o homem*. São Paulo: Companhia das Letras.

Damasio, A. R. (2004). *Em busca de Espinosa: Prazer e dor na ciência dos sentimentos*. São Paulo: Companhia das Letras.

Danto, A. C. (2014). *After the End of Art: Contemporary Art and the Pale of History*. Princeton, NJ: Princeton University Press.

Dassonville, P. (1995). Haptic localization and the internal representation of the hand in space. *Experimental Brain Research, 106*(3), 434–448. doi:10.1007/BF00231066 PMID:8983987

Davis, S. T., & Jahnke, J. C. (1991). Unity and the golden section: Rules for aesthetic choice? *The American Journal of Psychology, 104*(2), 257–257. doi:10.2307/1423158

Dawkins, R. (2006). The Selfish Gene (30th ann. Ed.). New York: Oxford.

Delacour, J. (2001). *Conscience & cerveau: La nouvelle frontière des neurosciences*. Bruxelas: Éditions De Boeck Université.

Dennet, D. C. (1988). *Consciousness in Modern Science*. Oxford, UK: Oxford University Press.

Dennett, D. C. (1994). The role of language in intelligence. In J. Khalfa (Ed.), *What is Intelligence? The Darwin College Lectures*. Cambridge, UK: Cambridge University Press.

Descartes. (1984). The Principles of Philosophy. In *The Philosophical Writings of Descartes*. Cambridge, UK: Cambridge University Press.

Descartes, R. (1973). Discurso do método. In *Coleção Os Pensadores* (J. Guisnburg & P. B. Júnior, Trans.). Abril Cultural.

Desmet, P. M. A. & Hekkert, P. (2009). Special Issue Editorial: Design & Emotion. *International Journal of Design, 3*(2), 1–6.

Dewey, J. (2010). *Arte como experiência*. São Paulo: Martins Fontes.

Dhillon, G. S., & Horch, K. W. (2005). Direct neural sensory feedback and control of a prosthetic arm. *IEEE Transactions on Neural Systems and Rehabilitation Engineering, 13*(4), 468–472. doi:10.1109/TNSRE.2005.856072 PMID:16425828

Di Dio, C., Macaluso, E., & Rizzolatti, G. (2007). The golden beauty: Brain response to classical and renaissance sculptures. *PLoS ONE, 2*(11), e1201. doi:10.1371/journal.pone.0001201 PMID:18030335

Dogan, I. (2006). *Microcontroller Based Applied Digital Control*. Chichester, UK: Wiley.

Donald, M. (1991). *Origins of the modern mind: Three stages in the evolution of culture and cognition*. Cambridge, MA: Harvard University Press.

Dondis, D. A. (1991). *Sintaxe da linguagem visual*. São Paulo: Martins Fontes.

Dumit, J. (2004). *Picturing Personhood: Brain Scans and Biomedical Identity*. Princeton, NJ: Princeton University Press.

Durand, G. (1988). *A imaginação simbólica. In Tradução de Eliane Fittipaldi Pereira*. São Paulo: Cultrix.

Dutton, D. (2010). *The Art Instinct: Beauty, Pleasure, and Human Evolution*. Oxford, UK: Oxford University Press.

Eden, A. H., & Turner, R. (2007). Problems in the Ontology of Computer Programs. In Applied Ontology (vol. 2, pp. 13–36). Amsterdam: IOS Press.

Edward de Bono Foundation. (2014). *Edward de Bono Foundation*. Retrieved April 9, 2014, from http://edwarddebonofoundation.com/

Ehrsson, H. H., Holmes, N. P., & Passingham, R. E. (2005). Touching a Rubber Hand: Feeling of Body Ownership Is Associated with Activity in Multisensory Brain Areas. *The Journal of Neuroscience*, *25*(45), 10564–10573. doi:10.1523/JNEUROSCI.0800-05.2005 PMID:16280594

Ehrsson, H. H., Rosen, B., Stockselius, A., Ragno, C., Kohler, P., & Lundborg, G. (2008). Upper limb amputees can be induced to experience a rubber hand as their own. *Brain*, *131*(12), 3443–3452. doi:10.1093/brain/awn297 PMID:19074189

Eisner, E. W. (1981). On the Differences between Scientific and Artistic Approaches to Qualitative Research. *Educational Researcher*, *10*(4), 5–9. doi:10.3102/0013189X010004005

El Sadiik, A., Orozco, M., Eid, M., & Cha, J. (2011). *Haptics Technologies: Bringing Touch to Multimedia*. New York: Springer. doi:10.1007/978-3-642-22658-8

Eliseyev, A., Moro, C., Faber, J., Wyss, A., Torres, N., Mestais, C., & Askenova, T. et al. (2012). L1-penalized N-way PLS for subset of electrodes selection in BCI experiments. *Journal of Neural Engineering*, *9*(4), 045010. doi:10.1088/1741-2560/9/4/045010 PMID:22832155

Engelmann, A. (2012). A Psicologia da Gestalt e a ciência empírica contemporânea. *Psicologia: Teoria e Pesquisa*, *18*(1). 10.1590/S0102-37722002000100002

Escosteguy, A. C. (2001). *Cartografias dos estudos culturais: uma versão latino-americana*. Belo Horizonte: Autêntica.

Estrela, J. B. C., & Ribeiro, J. D. S. F. (2012). Análise das relações entre memória e aprendizagem na construção do saber [An analysis of the relationship between memory and learning when building knowledge]. *Caderno Intersaberes*, *1*(1). Available at http://www.grupouninter.com.br/intersaberes/index.php/cadernointersaberes/issue/view/52

Fabbrini, R. (2006). O fim das vanguardas. In *Cadernos de Pós-graduação da UNICAMP* (vol. 8). Retrieved November 18, 2015, from: http://www.iar.unicamp.br/dap/vanguarda/artigos_pdf/ricardo_fabrini.pdf

Fabris, A. (1990). *Candido Portinari*. São Paulo: Perspectiva.

Fechner, G. T. (1876). *Vorschule der Aesthetic*. Berlin, Germany: Breitkopf & Härtel.

Feitosa-Santan, C., Lutze, M., Barrionuevo P., & Cao, D. (submitted). *The Perceived Color of #TheDress is Associated with Individual Color Preferences*.

Feitosa-Santana, C., Oiwa, N., Costa, M. F., Tiedemann, K. B., Silveira, L. C. L., & Ventura, D. F. (2006). Espaço de Cores. *Psicologia USP*, *17*(4), 35–62. doi:10.1590/S0103-65642006000400003

Feldmann, J. A. (2006). *From molecule to metaphor: a neural theory of language*. Cambridge, MA: MIT Press.

Field, S. (2009). *Roteiro: os fundamentos do roteirismo. Tradução Alice Leal*. Curitiba: Arte & Letra.

Fiorini, V. (2008). Design de Moda: abordagens conceituais e metodológicas [Fashion Design: conceptual approach and methodologies]. In Design de Moda: olhares diversos [Fashion Design: various viewpoints]. Barueri: Estação das Letras e Cores Editora.

Flusser, V. (2007). *O mundo codificado: por uma filosofia do design e da comunicação*. São Paulo: Cosac Naify.

Fogliano, F., & de Carmargo, D. C. F. (2010). Linguagem e materialidade na experiência fotográfica. In *Proceedings of Intercom - Sociedade Brasileira de Estudos Interdisciplinares da Comunicação XXXIII Congresso Brasileiro de Ciências da Comunicação* (pp. 1-14). Caxias do Sul, RS.

Fogliano, F. (2002). Fotografia e Complexidade. In L. Leão (Ed.), *Interlab: Labirintos do pensamento contemporâneo* (pp. 157–162). São Paulo: Editora Iluminuras.

Fogliano, F. (2012). A experiência estética: consciência, linguagem e narrativa. In C. Rocha, M. B. de Medeiros, & S. Venturelli (Eds.), *Art - Arte e Tecnologia // Modus Operandi Universal* (pp. 72–80). Brasília: Programa de Pós-Graduação em Arte da Universidade de Brasília.

Fogliano, F. (2013). Novos Paradigmas para a relação entre Imagem e Interação. In *Proceedings of #12. Art, Encontro Internacional de Arte e Tecnologia*. Brasília: Universidade Federal de Brasília.

Fogliano, F. (2014). Identidade e interação: Considerações sobre o retrato e suas funções. In *Proceedings of Simpósio Internacional de Inovação em Mídias Interativas - SIIMI* (pp. 55–66). Goiânia: Universidade Federal de Goiás.

Fornasier, C. B. R., Martins, R. F. D. F., & Demarchi, A. P. P. (2008). O ensino da disciplina de desenvolvimento de projetos como sistema de gestão de conhecimento [Teaching the subject-area of designing projects as a system of knowledge management]. In Design de Moda: olhares diversos [Fashion Design: various viewpoints]. Barueri: Estação das Letras e Cores Editora.

Foucault, M. (1995). *Discipline and Punish: The Birth of the Prison*. New York, NY: Vintage.

Foucault, M. (2003). *The Birth of the Clinic: Archaeology of Medical Perception*. Taylor and Francis e-Library.

Freedberg, D., & Gallese, V. (2007). Motion, Emotion and Empathy in Aesthetic Experience. *Trends in Cognitive Sciences*, *11*(5), 197–203. doi:10.1016/j.tics.2007.02.003 PMID:17347026

Freeman, W. J. (2003). O cérebro solitário. In *O livro de ouro da mente*. Rio de Janeiro: Ediouro.

Freire, K. (2009). Reflexões sobre o conceito de design de experiências. *Strategic Design Research Journal*, *1*(2), 37-44. Retrieved November 20, 2010 from http://www.unisinos.br/sdrj/pdf/63.pdf

Freitas, A. S. (2015). *Rencontre des arts: correspondances entre œuvres sonores et visuelles au xx*[e] *siècle*. Paris: Harmattan.

Fukuyama, F. (2002). *Our Post human Future: Consequences of the Biotechnology Revolution*. New York, NY: Farrar, Straus, and Giroux.

Funk, M., Shiffrar, M., & Brugger, P. (2005). Hand movement observation by individuals born without hands: Phantom limb experience constrains visual limb perception. *Experimental Brain Research*, *164*(3), 341–346. doi:10.1007/s00221-005-2255-4 PMID:15818499

Gabora, L., & DiPaola, S. (2012). How did humans become so creative? *A computational approach. International Conference on Computational Creativity*.

Gallagher, S. (2006). *How the Body Shapes the Mind*. New York: Oxford University Press.

Gallese, V., Keysers, C., & Rizzolatti, G. (2004). A unifying view of the basis of social cognition. *Trends in Cognitive Sciences*, *8*(9), 396–403. doi:10.1016/j.tics.2004.07.002 PMID:15350240

Galyean, T. A. (1995). *Narrative Guidance of Interactivity. In Program in Media Arts and Sciences*. Massachusetts Institute of Technology.

Gauch, H. G. Jr. (2003). *Scientific method in practice*. Cambridge, UK: Cambridge University Press.

Gazzaniga, M. S. (1998). *The mind's past*. Berkeley, CA: University of California Press.

Gegenfutner, K. R., Bloj, M., & Toscani, M. (2015). The many colours of 'the dress'. *Current Biology*, *25*, R1–R2. PMID:25562290

George, M. S., & Aston-Jones, G. (2009). Noninvasive techniques for probing neurocircuitry and treating illness: Vagus nerve stimulation (VNS), transcranial magnetic stimulation (TMS) and transcranial direct current stimulation (tDCS). *Neuropsychopharmacology*, *35*(1), 301–316. doi:10.1038/npp.2009.87 PMID:19693003

Gibbons, J. (2007). *Contemporary Art and Memory: Images of Recollection and Remembrance*. London: IB Tauris.

Gibbs, R. W. Jr. (2007). *Embodiment and cognitive science*. Cambridge, UK: Cambridge University Press.

Gilead, M., Liberman, N., & Maril, A. (2014). *From mind to matter: neural correlates of abstract and concrete mindsets*. Available at http://www.ncbi.nlm.nih.gov/pmc/articles/PMC4014104/

Goguen, J. A. (2000). What is Art? Journal of Consciousness Studies, 7(8-9), 7-15.

Gohil, V. J., Bhagwat, S. D., Raut, A. P., & Nirmal, P. R. (2013). Robotics arm control using haptic technology. *International Journal of Lastest Research in Science and Technology*, *2*(2), 98–102.

Goldstein, R. (2005). *Incompleteness: the proof and paradox of Kurt Godel*. New York, NY: W. W. Norton & Company, Inc.

Goldstein, S. E. (2011). *Cognitive Psychology: Connecting mind, research, and everyday life experience* (3rd ed.). Belmont: Wadsworth Cengage Learning.

Gombrich, E. H. (2000). Concerning "The Science of Art": Commentary on Ramachandran and Hirstein. Journal of Consciousness Studies, 7(8-9), 17-20.

Gombrich, E. H. (1951). *The History of Art*. New York, NY: Phaidon Publishers Inc.

Graziano, M. S. A., & Botvinick, M. M. (2002). How the brain represents the body: insights from neurophysiology and psychology. In Prinz, W. and Hommel, B. (Eds.), Common Mechanisms in Perception and Action: Attention and Performance XIX, Oxford University Press, 136-157.

Grodal, T. (2002). The experience of realism audiovisual representation. In A. Jerslev (Ed.), *Realism and 'reality' in filme and media*. Copenhagen: University of Copenhagen.

Grossberg, S. (2014in press). The visual world as illusion: The ones we know and the ones we don't. In A. Shapiro & D. Todorovic (Eds.), *Oxford Compendium of Visual Illusions*. Oxford, UK: Oxford University Press.

Grossi, M. G. R., Lopes, A. M., & Couto, P. A. (n.d.). *A neurociência na formação de professores: um estudo da realidade brasileira* [Neuro-science in teacher-training: a study of the Brazilian situation]. Available at http://www.revistas.uneb.br/index.php/faeeba/article/view/821

Guarino, N. (2009). *The Ontological Level: Revisiting 30 years of knowledge representation. conceptual modelling: foundations and applications, Essays in Honor of John Mylopoulos*. Springer-Verlag.

Guarino, N., Oberle, D., & Staab, S. (2009) What is an Ontology? In S. Staab & R. Studer (Eds.), Handbook on Ontologies, Second Edition. International handbooks on information systems. Springer Verlag. doi:10.1007/978-3-540-92673-3_0

Halbwachs, M. (2006). *A memória coletiva*. São Paulo: Centauro.

Hamilton, W. (1964a). The genetical evolution of social behaviour I. *Journal of Theoretical Biology*, *7*(1), 1–16. doi:10.1016/0022-5193(64)90038-4 PMID:5875341

Hamilton, W. (1964b). The genetical evolution of social behaviour II. *Journal of Theoretical Biology*, *7*(1), 17–52. doi:10.1016/0022-5193(64)90039-6 PMID:5875340

Hansman, C. A., & Wilson, A. L. (2010). *Situating Cognition: Knowledge and Power in Context*. Retrieved June 27, 2013 from http://www.eric.ed.gov/PDFS/ED478969.pdf

Heidegger, M. (2006). *Ensaios e conferências*. Petrópolis: Editora Vozes.

Heligman, D. (2009). *Charles and Emma – The Darwin's Leap of Faith*. New York: Harry Holt and Company, LLC.

Henninger, S. (2003). Tool Support for Experience-based Software Development Methodologies. *Advances in Computers*, *59*, 29–82. doi:10.1016/S0065-2458(03)59002-7

Henrywood, R. H., & Agarwal, A. (2013). The aeroacoustics of a steam kettle. *Physics of Fluids, 25*(10). Retrieved from http://scitation.aip.org/content/aip/journal/pof2/25/10/10.1063/1.4821782

Herculano-Houzel, S. (2009). *Brain Development: The Generation of Large Brains*. Elsevier.

Hirstein, W. (2012). Neuroaesthetics: Responding to the Critics. *Psychology Today*. Available from https://www.psychologytoday.com/blog/mindmelding/201212/neuroaesthetics-responding-the-critics

Hoffman, M., Marques, H. G., Arieta, A. H., Sumioka, H., Lungarella, M., Pfeifer, R. (2010). Body Schema in Robotics: A Review. *IEEE Transactions on Autonomous Mental Development*, *2*(4).

Höge, H. (1995). Fechner's experimental aesthetics and the golden section hypothesis today. *Empirical Studies of the Arts*, *13*(2), 131–131. doi:10.2190/UHTQ-CFVD-CAU2-WY1C

Huyssen, A. (2000). Present Pasts: Media, politics, amnésia. *Public Culture, 12*(1), 21-38.

Hyman, J. (2010). Art and Neuroscience. In *Beyond Mimesis and Convention*. Dordrecht: Springer Netherlands. doi:10.1007/978-90-481-3851-7_11

Iida, I. & Mühlenberg, P. (2006). O bom e o bonito em design. *Proceedings of: Congresso Brasileiro ee Pesquisa e Desenvolvimento em Design*.

Ikeda, M., Seta, K., Kakusho, O., & Mizoguchi, R. (1998). Task ontology: Ontology for building conceptual problem solving models. In *Proceedings of ECAI98 Workshop on applications of ontologies and problem-solving model*.

Iqbal, R. (2013). An Analysis of Ontology Engineering methodologies: A literature review. *Research Journal of Applied Sciences, Engineering and Technology*, *6*(6), 2993–3000.

Irwin, T., Kossoff, G., Tonkinwise, C., & Scuppelli, P. (2015). *Transition Design Overview*. Carnegie Mellon School of Design. Retrieved August 25, 2015, from https://www.academia.edu/13122242

Ishai, A. (2007). Sex, beauty and the orbitofrontal cortex. *International Journal of Psychophysiology*, *63*(2), 181–185. doi:10.1016/j.ijpsycho.2006.03.010 PMID:16759727

Ishizu, T., & Zeki, S. (2011). Toward a brain-based theory of beauty. PLoS ONE, 6(7). doi:10.1371/journal.pone.0021852

Jacobsen, T., Schubotz, R. I., Hofel, L., & Cramon, D. Y. (2006). Brain correlates of aesthetic judgment of beauty. *NeuroImage*, *29*(1), 276–285. doi:10.1016/j.neuroimage.2005.07.010 PMID:16087351

Jacobson, I., Ng, P.-W., McMahon, P., Spence, I., & Lidman, S. (2012). The Essence of Software Engineering: The SEMAT Kernel. Queue, 10(10).

Jacobson, I., Ng, P.-W., & Spence, I. (2007). Enough of process-let's do practices. *Journal of Object Technology*, *6*(6), 41–67. doi:10.5381/jot.2007.6.6.c5

Jana, A. (2012). *Kinect for Windows SDK Programming Guide*. Retrieved from http://lirec.ict.pwr.wroc.pl/~jkedzier/urbi/Kinect%20for%20Windows%20SDK%20Programming%20Guide.pdf

Jaros, V. (2011). Concept Development and Experimentation in project STRUCTURE. Cybernetic Letters.Brno: Expertia, o.p.s., 2, 36-41.

Jeffery, D. R., & Scott, L. (2002). Has twenty-five Years of Empirical Software Engineering Made a Difference. In *Proceedings of the 2nd Asia-Pacific Software Engineering Conference*. doi:10.1109/APSEC.2002.1183076

Jiménez, J. G. (1996). *Narrativa audiovisual*. 2ª. ed. Madri: Cátedra.

Johnson, M. (2007). *The meaning of the body: Aesthetics of human understanding*. Chicago: The University of Chicago Press. doi:10.7208/chicago/9780226026992.001.0001

Jonhson, G. A. (2010). The retrieval of beautiful: thinking though Merleau- Ponty's aesthetics. Evanston, IL: Northwestern University Press.

Judson, H. F., & Maddox, J. (1979). *The Eight Day of Creation: Makers of the Revolution in Biology*. London: Jonathan Cape.

Julier, G. (2006). From Visual Culture to Design Culture. *Design Issues*, *22*(1), 64–76. doi:10.1162/074793606775247817

Jung, R. E., Mead, B. S., Carrasco, J., & Flores, R. A. (2013). The structure of creative cognition in the human Brain. *Frontiers in Human Neuroscience*, *7*(330), 1–13. PMID:23847503

Jung, R. E., Segall, J. M., Bockholt, H. J., Flores, R. A., Smith, S. M., Chavez, R. S., & Haier, R. J. (2010). Neuroanatomy of Creativity. *Human Brain Mapping*, *31*(3), 398–409. PMID:19722171

Kac, E. (Ed.). (2007). *Signs of Life – Bio Art and Beyond*. Cambridge, MA: MIT Press.

Kahneman, D., & Tversky, A. (1979). Prospect Theory: An Analysis of Decision under Risk. *Econometrica*, *47*(2), 263–291. doi:10.2307/1914185

Kampe, K., Frith, C., Dolan, R., & Frith, U. (2001). Reward value of attractiveness and gaze. *Nature*, *413*(5856), 589. doi:10.1038/35098149 PMID:11595937

Kanayama, N., Sato, A., & Ohira, H. (2009). The role of gamma band oscillations and synchrony on rubber hand illusion and crossmodal integration. *Brain and Cognition*, *69*(1), 19–29. doi:10.1016/j.bandc.2008.05.001 PMID:18555572

Kandel, E. (2012). *The Age of Insight. The quest to understand the unconscious in art, mind, and brain, from Vienna 1900 to the present*. Published by Random House.

Kandel, E. R. (2007). *In search of the memory: the emergene of a new science of mind*. New York: W. W. Norton & Company.

Kandel, E. R., Schwartz, J. H., & Jessell, T. M. (Eds.). (2000). *The constructive nature of visual processing. In Principles of neural science* (Vol. 4, pp. 556–576). New York: McGraw-Hill.

Kanwisher, N., McDermott, J., & Chun. (1997). The Fusiform Face Area: A Module in Human Extrastriate Cortex Specialized for Face Perception. *The Journal of Neuroscience*, *17*(11), 4302-4311.

Kaptelinin, V., & Nardi, B. A. (2006). *Acting with technology: Activity theory and interaction design*. Cambridge, MA: MIT Press.

Karray, F., Alemzadeh, M., Saleh, J. A. & Arab, M. N. (2008). Human-Computer Interaction: Overview on State of the Art. *Smart Sensing and Intelligent Systems*, *1*(1), 138-159.

Kawabata, H. & Zeki, S. (2004). Neural correlates of beauty. *Journal of Neurophysiology*, *91*(4), 1699-705.

Kawabata, H., & Zeki, S. (2004). Neural correlates of beauty. *Journal of Neurophysiology*, *91*(4), 1699–1705. doi:10.1152/jn.00696.2003 PMID:15010496

Kevles, B. H. (1997). *Naked to the Bone: Medical Imaging in the Twentieth Century*. New Brunswick, NJ: Rutgers University Press.

Keysers, C., Wicker, B., Gazzola, V., Anton, J. L., Fogassi, L., & Gallese, V. (2004). A touching sight: SII/PV activation during the observation and experi- ence of touch. *Neuron*, *42*(2), 335–346. doi:10.1016/S0896-6273(04)00156-4 PMID:15091347

Kheirbek, M. A., & Hen, R. (2015). *Mais neurônios, mais equilíbrio* [More neurons, more equilibrium]. São Paulo: Editora Segmento.

Kim, K., Colgate, J. E., Santos-Munne, J. J., Makhlin, A., & Peshkin, M. A. (2010). On the Design of Miniature Haptic Devices for Upper Extremity Prosthetics. *IEEE/ASME Transactions on Mechatronics*, *15*(1), 27–39. doi:10.1109/TMECH.2009.2013944

Knoblich, G., & Sebanz, N. (2006). The social nature of perception and action. *Current Directions in Psychological Science*, *15*(3), 99–10. doi:10.1111/j.0963-7214.2006.00415.x

Koffka, K. (1957). *Princípios da Psicologia da Gestalt*. São Paulo: Cultrix.

Köhler, W. (1980). Psicologia da Gestalt. Belo Horizonte: Itatiaia.

Kossovith, L. (1996). Permanência e Renovação nas Artes. *Discurso*, *26*, 83–92.

Kossovith, L. (2006). *Tradição Clássica. Desígnio: revista de história de arquitetura e do urbanismo. nº5*. São Paulo, Brazil: Ed. Annablume/FAU-USP.

Kracauer, S. (1960). *Theory of film: The redemption of physical reality*. New York: Oxford University Press.

Kranz, F., & Ishai, A. (2006). Face perception is modulated by sexual preference. *Current Biology*, *16*(1), 63–68. doi:10.1016/j.cub.2005.10.070 PMID:16401423

Kringelbach, M. L. (2005). The human orbitofrontal cortex: Linking reward to hedonic experience. *Nature Reviews. Neuroscience*, *6*(9), 691–702. doi:10.1038/nrn1747 PMID:16136173

Kuhl, P. K. (2010, September9). Patricia K. Brain Mechanisms in Early Language Acquisition. Published in final edited form as. *Neuron*, *67*(5), 713–727. doi:10.1016/j.neuron.2010.08.038 PMID:20826304

Kuhn, T. S. (1996). *The Structure of Scientific Revolutions*. Chicago, DC: The University of Chicago Press. doi:10.7208/chicago/9780226458106.001.0001

Kuiken, T. A., Dumanian, G. A., Lipschutz, R. D., Miller, L. A., & Stubblefield, K. A. (2004). The use of targeted muscle reinnervation for improved myoelectric prosthesis control in a bilateral shoulder disarticulation amputee. *Prosthetics and Orthotics International*, *28*, 245–253. PMID:15658637

Kumar, M., & Garg, N. (2010). Aesthetic principles and cognitive emotion appraisals: How much of the beauty lies in the eye of the beholder? *Journal of Consumer Psychology*, *20*(4), 485–494. doi:10.1016/j.jcps.2010.06.015

Kwok, R. (2013). Neurorposthetics: Once more, with feeling. *Nature*, *497*(7448), 176–178. doi:10.1038/497176a PMID:23657331

L., M. M. (2013). *Representational Strategies on Alzheimer's Disease: A PracticeBased Arts Research in a Neuroscience Laboratory*. (Unpublished doctoral dissertation). University of Brighton and University for the Creative Arts, Farnham, UK.

Lafer-Sousa, R., Hermann, K., & Conway, B. R. (2015). Striking individual differences in color perception uncovered by 'the dress' photograph. *Current Biology*, *25*(13), pR545–R546. doi:10.1016/j.cub.2015.04.053 PMID:25981795

Lakoff, G., & Johnson, M. (1999). *Philosophy in the flesh: The embodied mind and its challenge to western thought*. New York, NY: Basic Books.

Landim, P. C. (2009) *Design/Empresa/Sociedade*. (Unpublished associate professor dissertation). São Paulo State University.

Latour, B. (2004). Why has critique run out of steam? from matters of fact to matters of concern. *Critical Inquiry*, *30*(2). Retrieved from http://criticalinquiry.uchicago.edu/past_issues/issue/winter_2004_v30_n2/

Latour, B. (2008). *A cautious Prometheus? A few steps toward a philosophy of design (with special attention to Peter Sloterdijk)*. Keynote lecture for the Networks of Design meeting of the Design History Society. Retrieved from http://www.bruno-latour.fr/sites/default/files/112-DESIGN-CORNWALL-GB.pdf

Latour, B., & Woolgar, S. (1986). *Laboratory Life: The Construction of Scientific Facts*. Princeton, NJ: Princeton University Press.

Latour, B. (1987). *Science in Action*. Cambridge, MA: Harvard University Press.

Latour, B. (1987). *Science in Action: How to Follow Scientists and Engineers through Society*. Cambridge, MA: Harvard University Press.

Le Bot, M. (2008). Arte/Design. *ARS Revista do Programa de Pós-graduaçnao em Artes Visuais*, *6*(11), 7–21.

Leder, H. (2013). Next steps in neuroaesthetics: Which processes and processing stages to study? *Psychology of Aesthetics, Creativity, and the Arts*, *7*(1), 27–37. doi:10.1037/a0031585

Leder, H., Belke, B., Oeberst, A., & Augustin, D. (2004). A model of aesthetic appreciation and aesthetic judgments. *British Journal of Psychology*, *95*(4), 489–508. doi:10.1348/0007126042369811 PMID:15527534

Ledoux, J. (2001). *O cérebro emocional: os misteriosos alicerces da vida emocional*. Rio de Janeiro: Objetiva.

Lenggenhager, B., Halje, P., & Blanke, O. (2011). Alpha band oscillations correlate with illusory self-location induced by virtual reality. *The European Journal of Neuroscience*, *33*(10), 1935–1943. doi:10.1111/j.1460-9568.2011.07647.x PMID:21395867

Leote, R., Oliveira, H. C., & Baraúna, D. (2014). *Teoria neurológica da experiência estética de Ramachandran e Hirstein: apresentação, comentários e proposta de aplicação*. Paper presented at the *5º* Encontro Internacional de Grupos de Pesquisa: Realidades Mistas & Convergências entre Arte, Ciência e Tecnologia. Escola de Artes e Comunicações ECA-USP, São Paulo.

Leote, R. (2012). Multisensorialidade e Sinestesia: Poéticas Possíveis?*Proceedings of 6th International Conference on Digital Arts, ARTECH*.

Lesage, D. (2009, Spring). "Who's Afraid of Artistic Research? On measuring artistic research output" (PDF). Art&Research - A Journal of Ideas. *Contexts and Methods*, *2*, 2.

Lewis, C. I. (1929). *Mind and the World Order: Outline of a Theory of Knowledge*. New York: Charles Scribner's Sons.

Lewis, S. C. (2010). *The role of empathy in the rubber hand illusion*. Faculty of California State University.

Lhommée, E., Batir, A., Quesada, J.-L., Ardouin, C., Fraix, V., Seigneuret, E., & Krack, P. et al. (2014). Dopamine and the biology of creativity: Lessons from Parkinson's disease. *Frontiers in Neurology*, *5*, 55. PMID:24795692

Lima, A. S. (2009). *Da cultura da mídia à cibercultura: as representações do eu nas tramas do ciberespaço*. Retrieved August 28, 2010 from http://www.ufg.br/this2/uploads/files/75/ciberespaco_representacoes_do_eu.pdf

Lima, P. A. (2006). *Educação inclusiva e igualdade social* [Inclusive education and social equality]. São Paulo: Avercamp.

Lipovetsky, G. (1989). *O império do efêmero: a moda e seu destino da sociedade* [The empire of the ephemeral: fashion and its social goal]. São Paulo: Companhia das Letras.

Lisa, V. W., Anita, M. B., & Michelle, A. J. (2007). Reasons for Prosthetic Rejection by Children with Unilateral Congenital Transverse Forearm Total Deficiency. *Prosthetics and Orthotics International, 19*(2), 51–54. doi:10.1097/JPO.0b013e3180421539

Liu, T. Y., Chen, Y. S., Su, T. P., Hsieh, J. C., & Chen, L. F. (2014). *Abnormal Early Gamma Responses to Emotional Faces Differentiate Unipolar from Bipolar Disorder Patients*. BioMed Research Internacional.

Livingstone, M. (2002). Vision and art: The biology of seeing. New York: HNA Ed.

Livio, M. (2003). *The Golden Ratio: The Story of Phi, the World's Most Astonishing Number*. New York, NY: Broadway Publishing.

Luinge, H. J., & Veltink, P. H. (2005). Measuring orientation of human body segments using miniatures gyroscopes and accelerometers. *Medical & Biological Engineering & Computing, 43*(2), 273–282. doi:10.1007/BF02345966 PMID:15865139

Lundy-Ekman, L. (2004). *Neurociência: Fundamentos para reabilitação*. Rio de Janeiro: Elsevier.

Lynch, M., & Woolgar, S. (1990). Introduction: Sociological Orientations to Representational Practice in Science. In Lynch & Woolgar (Eds.), *Representation in Scientific Practice* (pp. 1-18). Cambridge, MA: MIT Press.

Mahler, J. (2015, February 27). The White and Gold (No, Blue and Black!) Dress that Melted the Internet. *New York Times*.

Manovich, L. (2014). *¿Qué es el cine digital?* Retrieved January 20, 2014, from http://www.upv.es/labo luz/revista/

Manzini, E. (2008). *Design para inovação social e sustentabilidade: comunidades criativas, organizações colaborativas e novas redes projetuais*. Rio de Janeiro: E-papers.

Marasco, P. D., Kim, K., Colgate, J. E., Peshkin, M. A., & Kuiken, T. A. (2011). Robotic touch shifts perception of embodiment to a prosthesis in targeted reinnervation amputees. *Brain, 134*(3), 747–758. doi:10.1093/brain/awq361 PMID:21252109

Maravita, A., & Iriki, A. (2004). Tools for the body (schema). *Trends in Cognitive Sciences, 8*(2), 79–86. doi:10.1016/j.tics.2003.12.008 PMID:15588812

Martin, J. H. (2013). *Neuroanatomia: texto e atlas* [Neuro-anatomy: text and atlas]. Porto Alegre: AMGH.

Martín-Barbero, J. (2003). Dos meios às mediações: Comunicação, cultura e hegemonia. 2 ed. Rio de Janeiro: Editora da UFRJ.

Martín-Loeches, M., Hernández-Tamames, J. A., Martín, A., & Urrutia, M. (2014). Beauty and ugliness in the bodies and faces of others: An fMRI study of person esthetic judgement. *Neuroscience, 26*(277), 486–497. doi:10.1016/j.neuroscience.2014.07.040 PMID:25086316

Massey, I. (2009). *The neural imagination: Aesthetic and neuroscientific approaches to the arts*. Austin, TX: University of Texas Press.

McDaniel, M. A., & Einstein, G. O. (2007). *Prospective memory: An overview and synthesis of an emerging field*. London: Sage Publications Ltd.

McDaniel, M. A., Einstein, G. O., Graham, T., & Rall, E. (2004). Delaying execution of intentions: Overcoming the costs of interruptions. *Applied Cognitive Psychology*, *18*(5), 533–547. doi:10.1002/acp.1002

MCM-0056-2010 NATO Concept Development and Experimentation (CD&E) Process. (2010). Brussels: NATO HQ.

Medeiros, M., & Bezerra, E. D. L. (2013). *Contribuições das Neurociências à compreensão da aprendizagem significativa* [Contributions made by the neurosciences to the understanding of significant learning]. Available at http://www.revistadialogos.com.br/Dialogos_10/Mario_Bezerra_Neurociencia.pdf

Medeiros, S. M. T. D. (2010). *Reflexões iniciais sobre a didática de ensino da disciplina de modelagem plana a partir do design e da ergonomia* [Preliminary reflections on the didactic principles of planned modelling based on design and ergonomics]. Available at http://www.coloquiomoda.com.br/anais/anais/6-Coloquio-de-Moda_2010/71756_Reflexoes_iniciais_sobre_a_didatica_de_ensino [Preliminary reflections on the didactics of teaching] .pdf

Mendes, V. D., & Haye, A. D. L. (2009). A moda do século XX [Fashion in the 20th Century]. São Paulo: Martins Fontes.

Menezes, M. D. S., & Spaine, P. A. D. A. (2010). *Theoretical guidelines for the teaching of Modeling flat industrial clothing: from the point of view of industry and teaching and learning*. Available at http://oasisbr.ibict.br/vufind/Record/UEL_6f114ec715698f84e2c8609ef8572dde

Menezes, M. (2003). The Artificial Natural: Manipulating Butterfly Wing Patterns for Artistic Purpuses. *Leonardo*, *36*(1), 29–32. doi:10.1162/002409403321152257

Menezes, M. (2011). *"(Use)less is More! Is the artist a master of the useless?" in Useless, by Max Bruinsma and Federico Duarte for Experimenta Design 2011*. Lisbon: Babel.

Merleau-Ponty, M. (2004). Conversas, 1948- Maurice Merleau-Ponty. São Paulo: Martins Fontes.

Merleau-Ponty, M. (2002). *Causeries. 1948, établies et annotées par S. Ménasé*. Paris: Seuil.

Merleau-Ponty, M. (2012). *Phenomenology of Perception* (D. Landes, Trans.). London: Routledge.

Merriam, S. B. (1998). *Qualitative Research and Case Study Applications in Education: Revised and Expanded from I Case Study Research in Education*. San Francisco: Jossey-Bass Publishers.

Miller, A. (2014). *Colliding Worlds*. London: W.W. Northon and Company.

Miller, G. A. (2003, March). The Cognitive Revolution: A Historical Perspective. *Trends in Cognitive Sciences*, *7*(3), 141–144. doi:10.1016/S1364-6613(03)00029-9 PMID:12639696

Mitz, A. R., Godschalk, M., & Wise, S. P. (1991). Learning-dependent neuronal activity in the premotor cortex: Activity during the acquisition of conditional motor associations. *The Journal of Neuroscience*, *11*, 1855–1872. PMID:2045890

Mogi, K. (2013) Cognitive factors correlating with the metacognition of the phenomenal properties of experience. *Scientific Reports*, *3*(3354). Retrieved from http://www.nature.com/srep/2013/131127/srep03354/full/srep03354.html

Moles, A. (1981). *Teoria dos Objetos*. Rio de Janeiro: Tempo Brasileiro.

Mollon, J. (2006). Monge: The Verriest Lecture, Lyon, July 2005. *Visual Neuroscience*, *23*(3-4), 297–309. doi:10.1017/S0952523806233479 PMID:16961961

Mont'Alvão, C., & Damazio, V. (Eds.). (2008). Design, ergonomia e emoção. Rio de Janeiro, Mauad X: FAPERJ.

Moraes, D. (1999). *Limites do design*. São Paulo: Studio Nobel.

More, M., & Vita-More, N. (Eds.). (2013). The Transhumanist Reader. Wiley-Blackwell doi:10.1002/9781118555927

Moseley, G. L., Gallace, A., & Spence, C. (2012). Bodily illusions in health and disease: Physiological and clinical perspectives and the concept of a cortical 'body matrix'. *Neuroscience and Biobehavioral Reviews*, *36*(1), 34–46. doi:10.1016/j.neubiorev.2011.03.013 PMID:21477616

Moura, C. A. R. (2005). *Nietzsche: civilização e cultura*. São Paulo, Brazil: Martins Fontes.

Muñoz, F., & Martín-Loeches, M. (2015). Electrophysiological brain dynamics during the esthetic judgment of human bodies and faces. *Brain Research*, *1594*, 154–164. doi:10.1016/j.brainres.2014.10.061 PMID:25451119

Murray, J. H. (2003). Hamlet no holodeck: o futuro da narrativa no ciberespaço. São Paulo: Itaú Cultural: UNESP.

Nadal, M., & Pearce, M. T. (2011). The Copenhagen Neuroaesthetics conference: Prospects and pitfalls for an emerging field. *Brain and Cognition*, *76*(1), 172–183. doi:10.1016/j.bandc.2011.01.009 PMID:21334125

Nadal, M., & Skov, M. (2013). Introduction to the special issue: Toward an interdisciplinary neuroaesthetics. *Psychology of Aesthetics, Creativity, and the Arts*, *7*(1), 1–12. doi:10.1037/a0031842

Nakamura, K., Kawashima, R., Nagumo, S., Ito, K., Sugiura, M., Kato, T., & Kojima, S. et al. (1998). Neuroanatomical correlates of the assessment of facial attractiveness. *Neuroreport*, *9*(4), 753–757. doi:10.1097/00001756-199803090-00035 PMID:9559951

Nalbantian, S. (2008). Neuroaesthetics: Neuroscientific theory and illustration from the arts. *Interdisciplinary Science Reviews*, *33*(4), 357–368. doi:10.1179/174327908X392906

Nath, S. (2004). *Narrativity in User Action: Emotion and Temporal Configurations of Narrative*. Croatia: University of Split.

Nee, D. E., Jahn, A., & Brown, J. W. (2014). *Prefrontal Cortex Organization: Dissociating Effects of Temporal Abstraction, Relational Abstraction, and Integration with fMRI*. Available at http://www.ncbi.nlm.nih.gov/pmc/articles/PMC4184367/

Neuper, C., Scherer, R., Reiner, M., & Pfurtscheller, G. (2005). Imagery of motor actions: Differential effects of kinesthetic and visual-motor mode of imagery in single-trial EEG. *Brain Research. Cognitive Brain Research*, *25*(3), 668–677. doi:10.1016/j.cogbrainres.2005.08.014 PMID:16236487

Neves, J. D. (2010). Sobre o Design e Consumo de Experiências à Luz do Consumo de Bens. *Proceedings of: Congresso Brasileiro de Pesquisa e Desenvolvimento em Design.*

Nicolelis, M. (2011). *Muito além do nosso eu: a nova neurociência que une cérebros e máquinas* [Far beyond our ego; a new neuroscience that combines brains and machines]. São Paulo: Companhia das Letras.

Niemeyer, L. (2000). *Design no Brasil: origens e instalação* (3rd ed.). Rio de Janeiro: 2AB.

Norman, D. A. (2008). *Design emocional: por que adoramos (ou detestamos) os objetos do dia-a-dia.* Rio de Janeiro: Rocco.

Novaes, A. (1994). *Artepensamento.* São Paulo, Brazil: Cia das Letras.

Nunes, B. (Ed.). (2008). *Memória: Funcionamento, Perturbações e Treino.* Lisbon, Portugal: Lidel.

O'Doherty, J. E., Lebedev, M. A., Ifft, P. J., Zhuang, S., Bleuler, H., & Nicolelis, M. A. (2011). Active tactile exploration using a brain-machine-brain interface. *Nature*, *479*(7372), 228–231. doi:10.1038/nature10489 PMID:21976021

O'Doherty, J., Winston, J., Critchley, H., Perrett, D., Burt, D. M., & Dolan, R. J. (2003). Beauty in a smile: The role of medial orbitofrontal cortex in facial attractiveness. *Neuropsychologia*, *41*(2), 147–155. doi:10.1016/S0028-3932(02)00145-8 PMID:12459213

Ofner, P., & Muller-Putz, G. R. (2012). Decoding of velocities and positions of 3D arm movement from EEG.*International Conference of the IEEE EMBS.* doi:10.1109/EMBC.2012.6347460

Onians, J. (2007). *Neuroarthistory: From Aristotle and Pliny to Baxandall and Zeki.* New Haven, CT: Yale University Press.

Ostlie, K., Franklin, R. J., Skjeldal, O. H., Skrondal, A., & Magnus, P. (2011). Musculoskeletal pain and overuse syndromes in adult acquired major upper-limb amputees. *Archives of Physical Medicine and Rehabilitation*, *92*(12), 1967–1973. doi:10.1016/j.apmr.2011.06.026 PMID:22133243

Ostlie, K., Lesjo, I. M., Franklin, R. J., Garfelt, B., Skjeldal, O. H., & Magnus, P. (2012). Prosthesis rejection in acquired major upper-limb amputees: A population-based survey. *Disability and Rehabilitation. Assistive Technology*, *7*(4), 294–303. doi:10.3109/17483107.2011.635405 PMID:22112174

Ostlie, K., Lesjo, I. M., Franklin, R. J., Garfelt, B., Skjeldal, O. H., & Magnus, P. (2012). Prosthesis use in adult acquired major upper-limb amputees: Patterns of wear, prosthetic skills and the actual use of prostheses in activities of daily life. *Disability and Rehabilitation. Assistive Technology*, *7*(6), 479–493. doi:10.3109/17483107.2011.653296 PMID:22315926

Paillard, J. (2013). Body schema and body image: a double dissociation in deafferented patients. In Motor control today and tomorrow. Academic Publishing House.

Paivio, A. (1971). *Imagery and verbal processes*. Holt, Rinehart, and Winston.

Pandilovski, M. (Ed.). (2008). *Art in the Biotech Era*. Experimental Art Foundation.

Panofsky, E. (2000). Ideia: a evolução do conceito de belo. *Nature*, *286*, 74–76.

Pareyson, L. (2001). *Problemas da Estética*. São Paulo: Martins Fontes.

Peirce, C. S. (1977). *Semiótica*. São Paulo: Editora Perspectiva.

Pelbart, P. P. (2003). *Vida capital: ensaios de biopolítica*. São Paulo: Iluminuras.

Pelloso, R. R., & Ferraz, M. G. (2015). *Ética e moral como modos de produção de subjetividade*. Retrieved October 15, 2015 from http://www.scielo.br/pdf/ trans/v82n2/29418.pdf

Perlovsky, L., & Ilin, R. (2012). *Brain. Conscious and Unconscious Mechanisms of Cognition, Emotions, and Language*. Available at http://www.ncbi.nlm.nih.gov/pmc/articles/PMC4061812/

Phelps, E. (2004). A Human emotion and memory: Interactions of the amygdala and hippocampal complex. *Current Opinion in Neurobiology*, *14*(2), 198–202. doi:10.1016/j.conb.2004.03.015 PMID:15082325

Pinker, S. (1995). *The language instinct: The new science of language and mind*. London: Peguin Books.

Pinker, S. (1997). *How the Mind Works*. New York: W. W. Norton & Company.

Pinker, S. (2011). *The Better Angels of Our Nature: Why Violence Has Declined*. New York: Viking Books.

Plaza, J. (2003). *Arte/ciência: uma consciência*. Revista Ars, Universidade de São Paulo.

Portinari, J. C. (2000). Projeto Portinari. *Estudos. Avançados, 14*(38). Retrieved from http://fmri.ucsd.edu/Research/whatisfmri.html

Prata, C. S., Masonb, R. A., & Justb, M. A. (2011). Individual differences in the neural basisi of causal inferencing. *Brain and Language*, *116*(1), 1–13. doi:10.1016/j.bandl.2010.08.004 PMID:21051084

Press, C., Heyes, C., Haggnard, P., & Eimer, M. (2008). Visuotactile learning and body representation: An ERP study with rubber hands and rubber objects. *Journal of Cognitive Neuroscience*, *20*(2), 312–323. doi:10.1162/jocn.2008.20022 PMID:18275337

Purves, D., Augustine, G.J., & Fitzpatrick, D. (n.d.). *Neuroscience*. Available at http://www.ncbi.nlm.nih.gov/books/NBK11066/

Raihani, N. J., & Bshary, R. (2011). Resolving the iterated prisoner's dilemma: Theory and reality. *Journal of Evolutionary Biology*, *24*(8), 1628–1639. doi:10.1111/j.1420-9101.2011.02307.x PMID:21599777

Ramachandran, V.S., & Blakeslee, S. (2012). *Fantasmas da mente: uma investigação dos mistérios da mente humana*. Editora Record.

Ramachandran, V. S. (2011). *The tell-tale brain: A neuroscientist's quest for what make us human*. New York: Norton.

Ramachandran, V. S. (2014). *O que o cérebro tem para contar: desvendando os mistérios da natureza humana*. Rio de Janeiro: Zahar.

Ramachandran, V. S., & Blakeslee, S. (1998). *Phantoms in the brain: probing the mysteries of the human mind*. New York: Willian Morrow and Company.

Ramachandran, V. S., & Hirstein, W. (1999). The science of art: A neurological theory of aesthetic experience. *Journal of Consciouness Studies, 6,* 15–51.

Ramachandran, V. S., & Hubbard, E. M. (2003). The phenomenology of synaesthesia. *Journal of Consciousness Studies*, *10*(8), 49–57.

Ramachandran, V. S., & Hubbard, E. M. (2005). The emergence of the human mind: Some clues from synesthesia. In *Synesthesia - Perspectives from Cognitive Neuroscience. Oxford-NewYork* (pp. 147–190). Oxford University Press.

Rigney, A. (2005). Plenitude, Scarcity, and the Circulation of Cultural Memory. *Journal of European Studies*, *35*(1), 11–28. doi:10.1177/0047244105051158

Rogers, A. (2015, February 26). The Science of why no one agrees on the color of this dress. *Wired.*

Rogers, A. (2016, February 26). A year ago, a Dress murdered the idea of objective color. *Wired.*

Rohlfing, K. J., Rehm, M., & Goecke, K. U. (2010). *Situatedness: The Interplay between Context(s) and Situation*. Technology, Bielefeld University. Retrieved June 27, 2010 from http://mm-werkstatt.informatik.uni-augsburg.de/files/publications/68/situatedness.pdf

Rollins, M. (2003). Pictorial Strategies and Perceptual Content. In H. Hecht, R. Schwartz, & M. Atherton (Eds.), *Looking Into Pictures: An Interdisciplinary Approach to Pictorial Space* (pp. 99–112). Cambridge, MA: MIT Press.

Rose, G. (2007). *Visual Methodologies: An Introduction to the Interpretation of Visual Materials*. London, UK: Sage Publications.

Rubin, D. C. (2005). *A Basic-Systems Approach to Autobiographical Memory*. Retrieved February 12, 2015, from http://dukespace.lib.duke.edu/dspace/bitstream/handle/10161/10105/RubinBasicSystems.pdf?sequence=1

Sacks, O. (1996). *An Anthropologist on Mars*. New York: Vintage.

Sacks, O. (1998). *The Man Who Mistook His Wife for a Hat: and other Clinical Tales*. New York: Touchstone.

Sacks, O. (2007). *Alucinações Musicais*. São Paulo: Companhia das Letras.

Saltzman, A. (2004). *El cuerpo diseñado: sobre la forma em el proyecto de la vestimenta* [The designed body: on the form in a clothing project]. Buenos Aires: Paidós.

Sampaio, E., Maris, S., & Bach-y-Rita, P. (2001). Brain plasticity: 'visual' acuitu of blind persons via the tongue. *Brain Research*, *908*(2), 204–207. doi:10.1016/S0006-8993(01)02667-1 PMID:11454331

Santaella, L. (2001). *Matrizes da Linguagem e Pensamento*. São Paulo: Iluminuras.

Santos, D. R. D. (n.d.). *Contribuições da neurociência à aprendizagem escolar na perspectiva da educação inclusiva* [Contributions of neuroscience to academic learning from the perspective of educational inclusion]. Available at http://www.psicopedagogia.com.br/new1_artigo.asp?entrID=1395#.VMvVIUfF-yU

Sartwell, C. (2014). *Beauty*. The Stanford Encyclopedia of Philosophy. Retrieved from http://plato.stanford.edu/archives/spr2014/entries/beauty/

Schacter, D. L. (2001). The Seven Sins of Memory: How the Mind Forgets and Remembers. New York: Houghton Mifflin.

Schafer, M. R. (2001). *A Afinação do Mundo*. São Paulo: UNESP.

Schafer, M. R. (2003). *O Ouvido Pensante*. São Paulo: UNESP.

Searle, J. R. (1997). *The mystery of consciousness*. New York: The New York Review of Books.

Seeley, W. P. (2011). *What is the Cognitive Neuroscience of Art ... and Why Should We Care?* American Society for Aesthetics – Aesthetics on-line. Available from http://www.aesthetics-online.org/newsletter/31.2.pdf

Seeley, W. P. (2011). What is the cognitive neuroscience of art ... and why should we care? *ASA Newsletter, 21*(2). Retrieved November 18, 2015, from: https://c.ymcdn.com/sites/aesthetics-online.org/resource/resmgr/Newsletters/31.2.pdf

Senior, C. (2003). Beauty in the brain of the beholder. *Neuron*, *38*(4), 525–528. doi:10.1016/S0896-6273(03)00293-9 PMID:12765605

Shapiro, A. (2011). Of Art, Biotech and the Body in the World, by Claudio Cravero. *The Illusion Beyond Art*. Retrieved from http://www.alan-shapiro.com

Shevell, S. K. (2003). Color appearance. In S. K. Shevell (Ed.), *The Science of Color* (2nd ed.; pp. 149–190). Oxford, UK: Elsevier. doi:10.1016/B978-044451251-2/50005-2

Shimamura, A. (2012). *Toward a Science of Aesthetics: Issues and Ideias. In Aesthetic Science connecting minds, brains, and experience* (pp. 4–26). Oxford University Press, Inc.

Shortess, G. K., Clarke, J. C., & Shannon, I. (1997). The shape of things: But not the golden section. *Empirical Studies of the Arts*, *15*(2), 165–165. doi:10.2190/3UBV-2LJU-T2T9-D7XN

Silva, H. F. J. & Marar, J. F. (2010) Emoção, Percepção e Design: Aspectos Teóricos e Práticos da Ciência Cognitiva e Neurociência no Projeto de Produto. *Proceedings of: Congresso Brasileiro de Pesquisa e Desenvolvimento em Design*.

Simperl, E., Michel, M., & Burger, T. (2010). Achieving maturity: The state of practice in ontology engineering. *International Journal of Computer Science and Applications &Technomathematics Research Foundation*, *7*(1), 45–65.

Sjoberg, D. I. K., Dyba, T., & Jorgensen, M. (2007). The Future of Empirical Methods in Software Engineering Research. In *Proceedings of Workshop Future of Software Engineering*. doi:10.1109/FOSE.2007.30

Skov, M. (2009). Neuroaesthetic problems: a framework for neuroaesthetic research. In M. Skov & O. Vartanian (Eds.), *Neuroaesthetics* (pp. 9–26). Amityville, NY: Baywood.

Sloterdijk, P. (2009). *Esferas I: burbujas*. Madrid: Ediciones Siruela.

Smith, G. M. (2003). *Film Structure and the Emotion System*. New York: Cambridge Published in the United States of America by Cambridge University Press. doi:10.1017/CBO9780511497759

Snow, C. P. (1998). *The Two Cultures*. Cambridge University Press.

Snow, C. P. (2012). *The two cultures*. Cambridge, MA: University Press. doi:10.1017/CBO9781139196949

Solso, R. L. (1994). *Cognition and the Visual Arts*. Cambridge, MA: MIT Press.

Sorger, R., & Udale, J. (2009). Fundamentos do design de moda [Fundamental principles of fashion design]. Porto Alegre: Bookman.

Sosnin, P. (2013). Role Intellectual Processor in Conceptual Designing Of Software Intensive Systems. In *Proceedings of the 11-th International conference on Computational Science and Applications* (LNCS), (vol. 7973, pp. 1-16). Springer.

Sosnin, P. (2013). Scientifically Experimental Way-of-Working in Conceptual Designing of Software Intensive Systems. In *Proceedings of the IEEE 12th International Conference on Intelligent Software Methodologies, Tools and Techniques* (pp. 43-51). IEEE.

Sosnin, P. (2015). A Precedent-Oriented Simulation in Designing the templates of Airplane Parts. In *Proceedings of 29th European Conference on Modelling and Simulation*.

Sosnin, P., & Maklaev, V. (2014). Question-Answer Reflections in a Creation of an Experience Base for Conceptual Designing the Family of Software Intensive Systems. Springer International Publishing.

Souriau, E. (2010). *Vocabulaire d'esthétique*. Paris: Presses Universitaires de France.

Southekal, P. H., & Levin, G. (2011). Formulation and Empirical Validation of a GQM Based Measurement Framework. In *Proceedings of 11th International Symposium on Empirical Software Engineering and Measurement*. doi:10.1109/ESEM.2011.59

Souza, W. G. D. (2007). *Modelagem no Design do Vestuário* [Modelling in Clothes Design]. Available at http://fido.palermo.edu/servicios_dyc/encuentro 2007/02_auspicios_publicaciones/actas_diseno/articulos_pdf/A6045.pdf

Spaine, P. A. D. A. (2010). *Modelagem plana do vestuário: diretrizes para a indústria do vestuário e o ensino-aprendizado*[Flat file modelling of clothing; guidelines for the clothing industry and teaching/learning]. Available at https://www.faac.unesp.br/Home/Pos-Graduacao/MestradoeDoutorado/Design/Dissertacoes/patricia-aparecida-de-almeida-spaine.pdf

Spaine. (2013). *Modelagem plana: contribuição para a evolução do vestuário* [Flat file modelling; a contribution to the evolution of clothing]. Available at http://www.cih.uem.br/anais/2013/?l=trabalhos&id=511

Sporns, O. (n.d.). *Network Analysis, Complexity, and Brain Function*. Available at http://csc.ucdavis.edu/~cmg/netdyn/Net_Brain_final.pdf

Stafford, B. (1991). *Body Criticism: Imaging the Unseen in Enlightenment Art and Medicine*. Cambridge, MA: MIT Press.

Stapp, H. O. (1993). *Mind, matter, and quantum mechanics*. Berlin: Springer-Verlag. doi:10.1007/978-3-662-08765-7

Starr, G. (2013). *Feeling Beauty: The Neuroscience of Aesthetic Experience*. MIT Press. doi:10.7551/mitpress/9780262019316.001.0001

Stephen, F. B., & Judith, A. B. (1985). Patterns of Acceptance and Rejection of Upper Limb Prostheses. *Orthotics and Prosthetics*, *39*(2), 40–47.

Stewart, I. (1996). Tales of a neglected number. *Scientific American*, *274*, 102–102.

Struchiner, M., Ricciardi, R. M. V., & Vetromille, V. P. (1998). O painel de especialistas no processo de apreciação analítica de sistemas hipermídia para o ensino de graduação.*IV Congresso RIBIE*, Brasília.

Sturken, M., & Cartwright, L. (2002). *Practices of Looking: An Introduction to Visual Culture* (2nd ed.). Oxford, UK: Oxford University Press.

Sturken, M. (1997). *Tangled Memories: The Vietnam War, the AIDS Epidemic, and the Politics of Remembering*. Oakland, CA: University of California Press.

Symons, D. (1979). *The Evolution of Human Sexuality*. Oxford University Press.

Talon-Hugon, C. (2010). *L'Esthétique*. Paris: Presses Universitaires de France.

Tenenbaum, J., & Tomizu, A. (2008). Narrative Meaning Creation in Interactive Storytelling. *International Journal of Computational*, *1*(2), 3–20.

Thompson, H. (2012). Performance enhancement: Superhuman athletes. *Nature News*. Retrieved May 19, 2014, from http://www.nature.com/news/performance-enhancement-superhuman-athletes-1.11029

Thompson, E. P. (2002). *A voz do passado, História Oral*. Rio de Janeiro: Editora Paz e Terra.

Thornhill, R. (2003). Darwin aesthetics inform traditional aesthetics. In E. Voland & K. Grammer (Eds.), *Evolutionary Aesthetics* (pp. 9–35). Berlin: Springer-Verlag. doi:10.1007/978-3-662-07142-7_2

Tokuhama-Espinosa, T. N. (2008). *The scientifically substantiated art of teaching: a study in the development of standards in the new academic field of neuroeducation (mind, brain and education science).* Doctoral Research Study. Capella University. Available at https://www.researchgate.net/publication/36710537_The_Scientifically_Substantiated_Art_of_Teaching_A_study_in_the_development_of_standards_in_the_new_academic_field_of_neuroeducation_mind_brain_and_education_science

Tokuhama-Espinosa, T. (2010). *Mind, Brain, and Education Science: The new brain-based learning.* New York: W. W. Norton & Company.

Tonkinwise, C. (2014). *Design for Transition - From and to What?* Retrieved April 25, 2015, from https://www.academia. edu/11796491

Tonkinwise, C. (2015). *Committing to the Political Values of Post-Thing-Centered Designing.* Retrieved August 2, 2015, from https://www.academia. edu/14560093

Tooby, J., & Cosmides, L. (2008). The evolutionary Psychology of the emotions and their relationship to internal regulatory variables. In M. Lewis, J. M. Haviland-Jones, & L. F. Barret (Eds.), *Handbook of Emotions* (3rd ed.; pp. 114–137). New York, NY: Guilford Press.

Torabi Nami, M., & Ashayeri, H. (2011). Where Neuroscience and Art Embrace: The Neuroaesthetics. *Basic and Clinical Neuroscience*, *2*(2), 6–11.

Trivers, R. (1971). The evolution of reciprocal altruism. *The Quarterly Review of Biology*, *46*(1), 35–57. doi:10.1086/406755

Trivers, R. (2011). *The Folly of Fools: The Logic of Deceit and Self-Deception in Human Life.* New York: Basic Books.

Trivers, R. L. (1972). Parental investment and sexual selection. In B. Campbell (Ed.), *Sexual selection and the descent of man, 1871-1971* (pp. 136–179). Chicago: Aldine.

Tsakiris, M., Hesse, M. D., Boy, C., Haggard, P., & Fink, G. R. (2007). Neural Signatures of Body Ownership: A Sensory Network for Bodily Self-Consciousness. *Cereb Cortex*, *17*(10), 2235–2244. doi:10.1093/cercor/bhl131 PMID:17138596

Tseng, L. Y., Tseng, P., Liang, W. K., Hung, D. L., Tzeng, O. J. L., Muggleton, N. G., & Juan, C.-H. (2014). The role of superior temporal sulcus in the control of irrelevant emotional face processing: A transcranial direct current stimulation study. *Neuropsychology*, *64*, 124–133. doi:10.1016/j.neuropsychologia.2014.09.015 PMID:25261612

Tsukiura, T., & Cabeza, R. (2011). Shared brain activity for aesthetic and moral judgments: Implications for the Beauty-is-Good stereotype. *Social Cognitive and Affective Neuroscience*, *6*(1), 138–148. doi:10.1093/scan/nsq025 PMID:20231177

Tulving, E. (1972). Episodic and Semantic Memory. In E. Tulving & W. Donaldson (Eds.), *Organization Of Memory* (pp. 381–403). New York, NY: Academic Press.

Tulving, E. (1985). How Many Memory Systems are There? *The American Psychologist*, *40*(4), 385–398. doi:10.1037/0003-066X.40.4.385

Tulving, E., & Craik, F. (Eds.). (2000). *The Oxford Handbook of Memory*. Oxford, UK: Oxford University Press.

Tzschentke, T. M., & Schmidt, W. J. (2000). Functional relationship among medial prefrontal cortex, nucleus accumbens, and ventral tegmental area in locomotion and reward. *Critical Reviews in Neurobiology*, *14*(2), 131–142. doi:10.1615/CritRevNeurobiol.v14.i2.20 PMID:11513242

Uexküll, T. (2014). A teoria da *Umwelt* de Jakob von Uexküll. *Revista Galáxia, 7*, 19-48. Retrieved from: http://revistas.pucsp.br/index. php/galaxia/article/viewFile/1369/852

Umberto, E. (2004). *História da beleza* (E. Aguiar, Trans.). Rio de Janeiro, Brazil: Editora Record.

Van Dijck, J. (2005). *The Transparent Body: A Cultural Analysis of Medical Imaging (In Vivo: the Cultural Mediations of Biomedical Science)*. Seattle, WA: University of Washington Press.

Vanzin, T. (2005). *TEHCo: modelo de ambientes hipermídia com tratametno de erros, apoiado na teoria da cognição situada. UFSC, 2005. Tese (Doutorado em Engenharia de Produção) Programa de Pós-Graduação em Engenharia de Produção*. Florianópolis, SC: Universidade Federal de Santa Catarina.

Varela, F. J., Thompson, E., & Rosch, E. (2003). *A mente incorporada: Ciências cognitivas e a experiência humana*. Porto Alegre: Artmed.

Vartaniana, O., & Skov, M. (2014). Neural correlates of viewing paintings: Evidence from a quantitativemeta-analysis of functional magnetic resonance imaging data. *Brain and Cognition*, *87*, 52–56. doi:10.1016/j.bandc.2014.03.004 PMID:24704947

Vassilieva, J. (2013). Eisenstein/Vygotsky/Luria's project: Cinematic Thinking and the Integrative Science of Mind and Brain. *Screening the Past Publications, 38*. Available from http://www.screeningthepast.com/2013/12/eisenstein-vygotsky-luria%E2%80%99s-project-cinematic-thinking-and-the-integrative-science-of-mind-and brain/

Vieira, J. A. (2007). *Teoria do conhecimento e arte*. Fortaleza: Lab. Estudos da Oralidade.

Vignemont, F. (2003). The Body in the Mind. In *Proceedings of CUNY*.

Vignola, J. (1562). Regola delli cinque ordini d'architettura. Rome: n.p. Facsimile.

Vitruvius, . (1970). *On Architecture, Frank Granger, trans*. Cambridge, MA: Harvard University Press.

Voland, E., & Grammer, K. (2003). Introduction. In E. Voland & K. Grammer (Eds.), *Evolutionary Aesthetics* (pp. 1–5). Berlin: Springer-Verlag. doi:10.1007/978-3-662-07142-7

Wagner, U., Gals, S., Halder, H., Verleger, R., & Born, J. (2004). Sleep inspires insight. *Nature*, *427*(6972), 352–355. doi:10.1038/nature02223 PMID:14737168

Waldert, S., Preissl, H., Demandt, E., Braun, C., Birbaumer, N., Aertsen, A., & Mehring, C. (2008). Hand Movement Direction Decoded from MEG and EEG. *The Journal of Neuroscience*, *28*(4), 1000–1008. doi:10.1523/JNEUROSCI.5171-07.2008 PMID:18216207

Walker, S. (2006). Sustainable by Design: Explorations in Theory and Practice. London: Earthscan.

Wallisch, P. (2015, February 27). Lessons from the Dress – Vision is fundamentally ambiguous but usually not this ambiguous. *Slate*.

Webber, B., & Webb, N. (2010). Question Answering. In Handbook of Computational Linguistics and Natural Language Processing. Blackwells. doi:10.1002/9781444324044.ch22

Webster, M. (2011). *m-w.com*. Encyclopædia Britannica.

Wertheimer, M. (1971). O fenômeno Phi como um exemplo de nativismo na percepção. In *Textos básicos de História da Psicologia*. São Paulo: Herder e EDUSP.

Wessberg, J., Stambaugh, C. R., Kralik, J. D., Beck, P. D., Laubach, M., Chapin, J. K., & Nicolelis, M. A. et al. (2000). Real-time prediction of hand trajectory by ensembles of cortical neurons in primates. *Nature*, *408*(6810), 361–365. doi:10.1038/35042582 PMID:11099043

Wilson, D. H. (2012). Bionic Brains and Beyond, *Wall Street Journal*. Retrieved May 19, 2014, from http://www.wsj.com/articles/SB10001424052702303640104577436601227923924

Wilson, E. O. (1999). *A Unidade do Conhecimento*. Rio de Janeiro: Editora Campus.

Wilson, S. (2002). *Information Arts: Intersections of Art, Science and Technology*. Cambridge, MA: The MIT Press.

Wilson, S. (2002). *Information arts: Intersections of art, science, and technology*. The MIT Press.

Winkler, A. D., Spillmann, L., Werner, J. S., & Webster, M. A. (2015). Asymmetries in blue–yellow color perception and in the color of 'the dress'. *Current Biology*, *25*(13), R547–R548. doi:10.1016/j.cub.2015.05.004 PMID:25981792

Winston, J., O'Doherty, J., Kilner, J., Perrett, D., & Dolan, R. (2007). Brain systems for assessing facial attractiveness. *Neuropsychologia*, *45*(1), 195–206. doi:10.1016/j.neuropsychologia.2006.05.009 PMID:16828125

Witt, P. R. (2013). *O mistério da chaleira*. Retrieved April 18, 2015, from http://surdezsilencioemvoo-deborboleta.com/blog/2013/06/03/o-misterio-da-chaleira/

Xu, S., & Rajlich, V. (2005). Dialog-Based Protocol: An Empirical Research Method for Cognitive Activity in Software Engineering. In *Proceedings of the 2005 ACM/IEEE International Symposium on Empirical Software Engineering*.

Yong, E. (2011, December7). Out-of-body experience: Master of illusion. *Nature*, *480*(7376), 168–170. doi:10.1038/480168a PMID:22158222

Zaro, M. A., Rosat, R. M., Meireles, L. O. R., Spindola, M., Azevedo, A. M. P. D., Bonini-Rocha, A. C., & Timm, M. I. (2010). Emergência da Neuroeducação: a hora e a vez da neurociência para agregar valor à pesquisa educacional [The emergence of neuroeducation: the time and place for neuroscience to aggregate value for educational research]. *Ciências & Cognição, 15*. Available at http://www.cienciasecognicao.org/revista/index.php/cec/article/view/276

Zatorre, R.J., Fields, R.D., & Johansen-Berg, H. (2013, May 22). Plasticity in Gray and White: Neuroimaging changes in brain structure during learning. *Nat Neurosci.*

Zeisel, J. (2006). *Inquiry by design: Environment/behavior/neuroscience in architecture, interiors, landscape, and planning*. New York: Norton.

Zeki, S. (2000). *Abstraction and idealism: From Plato to Einstein: how do we acquire knowledge?* Available at http://www.vislab.ucl.ac.uk/pdf/platotoeinstein.pdf

Zeki, S. (2001). Artistic Creativity and the Brain. *Science Magazine, Essays on Science and Society, 293*(5527). Available from http://www. sciencemag.org/content/293/5527/51.full

Zeki. (2001). *Artistic Creativity and the Brain*. Available at http://www.sciencemag.org/content/293/5527/51.full

Zeki, S. (1999). Art and the brain. *Journal of Consciousness Studies: Art and the Brain, 6*, 76–96.

Zeki, S. (1999). *Inner vision: An exploration of art and the brain*. New York: Oxford University Press.

Zeki, S. (2000). *Inner vision*. New York: Oxford University Press.

Zeki, S., & Lamb, M. (1994). The neurology of kinetic art. *Brain, 117*(3), 607–636. doi:10.1093/brain/117.3.607 PMID:8032869

Zeki, S., Romaya, J. P., Benincasa, D. M., & Atiyah, M. F. (2014). The experience of mathematical beauty and its neural correlates. *Frontiers in Human Neuroscience, 8*(68), 1–12. PMID:24592230

Zwart, H. (2014). *Limitless as a Neuro-pharmaceutical Experiment and as a Daseinsanalyse: on the use of fiction in preparatory debates on cognitive enhancement. Medicine, Health Care and Philosophy*. Dordrecht: Scientific Contribution; doi:10.1007/s11019- 013-9481-5

Zwijnenberg, R. (2009). *Brains*. Art, and the Humanities.

About the Contributors

Rachel Zuanon is the coordinator of the PhD and Master's Design Program of the School of Art, Architecture, Design and Fashion at the University Anhembi Morumbi, Brazil. Researcher and professor in the same Design Program, she is also a designer and media artist. CEO of the All Affective Design for Innovation and Zuannon Integrated Solutions. Head of CNPq research group of design, creation, language and technology and Brazil leader in the research project "Tremors Time Interactive Convergent Project", focused on the development of innovative wearable interfaces for games held in international cooperation with OCAD University (Toronto), Xenophile Media (Toronto) and Zuannon Integrated Solutions (Brazil) and supported by ISTP I FAPESP. Author of several scientific works, she had presented her research in EUA, Canada, China, Japan, Singapore, Taiwan, Germany, Greece, Portugal, Turkey, UK, Vienna, and many other places around the world. She is also an Ad-hoc Assessor for the most important Brazilian Agencies for scientific research support, as FAPESP and CAPES. She is dedicated to research in design of bio-wearable computers since 1998. In 2010, in PhD and Master's Design Program, she founded and since then she coordinates the Sense Design Lab – a laboratory dedicated to support researches focused on Design and Neuroscience relationship. She was granted the privilege of displaying the "Neurobodygame - Co-evolving affective wearable computer #4" artwork as FILE PRIX LUX 2010 finalist at SESI-SP Art Gallery Exhibition. In 2008, featured "BioBodyGame – Co-evolving affective wearable computer #3" artwork at Gameplay exhibition, held by Itaú Cultural. In 2007, she won Rumos Artes Cibernética prize with Biocybernetic Relational Object - Co-evolving affective wearable computer #2. In 2006, she had the artwork "Co-evolving affective wearable computer" exhibited at Primeira Pessoa exhibition, held by Itaú Cultural. In 2003, this very same research was granted an honorable mention from Rumos Itaú Pesquisa. She was a finalist for the Rumos Dance 2003 - Videodance award, granted by Itaú Cultural. She received the APCA award in 2000, granted by Sao Paulo Association of Art Critics. In 1999, she was a VITAE fellow in two residency programs at the American Dance Festival.

* * *

Bruno Araujo has Master (2009) and PhD (2012) in Neuroscience at the Federal University of São Paulo (UNIFESP/EPM). Postdoctoral (2013-2014) in the Sanford Consortium for Regenerative Medicine at the department of pediatrics of the University of California, San Diego (UCSD). He is currently Postdoctoral research at the Federal University of São Paulo (UNIFESP), Department of Neuroscience/ Neurosurgery at the laboratory of Neuroscience, supported by the fellow of Fundação de Amparo a Pesquisa do Estado de São Paulo -FAPESP (2014/08049-1).

Patricia Bieging has PhD in progress in Communication Sciences from University of São Paulo, Master Degree in Education from Federal University of Santa Catarina, postgraduated degree in Advertising and Marketing, undergraduated in Social Communication. Teach in undergraduate and postgraduate programs in Design, Communication and New Media. Bieging is executive director and film producer at Pimenta Cultural.

Raul Inácio Busarello has PhD in progress and Master degree in Engineering and Management of Knowledge from Federal University of Santa Catarina. He has postgraduation in Graphic Design and Corporative Strategy and ungraduated in Social Communication. Busarello is a professor in Design, Communication and New Media in undergraduation and postgraduation programs. Also he is a creative director and filmmaker at Pimenta Cultural.

Ana Teresa Contier pursues interdisciplinary research concerning empathy, theory of mind and technology. She holds a masters degree in electrical engineering from the University of São Paulo and is currently a PhD student at São Paulo Federal University (Unifesp) under its neurology and neurosciences program.

Jean Faber has a Bachelor in Physics (2000) from Federal University of Juiz de Fora (UFJF), Master (2003) and doctorate (2005) in Computational Modeling Program of the National Laboratory for Scientific Computing (LNCC) in the field of Information and Quantum Computation. Postdoctoral (2007-2010) at the International Institute of Neuroscience of Natal - Edmond Lily Safra (IINN-ELS) in the field of Neural Signal Analysis and Sleep-Learning. Postdoctoral (2010-2011) at Commissariat à l'Energie Atomique et aux Energies Alternatives (CEA / LETI / CLINATEC), supported by the Nanoscience Foundation (Grenoble / France) on the project of Neural Processing and Classification for Brain Computer Interface Systems. He is currently Adjunct Professor A I at the Federal University of São Paulo (UNIFESP), Department of Science and Technology / Biomedical Engineering - Nucleus of Neuroengineering and Computation.

Claudia Feitosa-Santana holds a Post Doctorate in Integrative Neuroscience from The University of Chicago, a PhD in Neuroscience and Behavior from the Psychology Institute of the University of São Paulo, a MS in Experimental Psychology from the Psychology Institute of the University of São Paulo, a Specialization in Management from the Escola Politécnica (POLI) of the University of São Paulo, a BA with Honors in Architecture and Urbanism from the Faculdade de Arquitetura e Urbanismo (FAU) of the University of São Paulo, a BS with Honors in Civil Engineering from the Mackenzie University. She is currently a Visiting Professor from the Center of Mathematics, Computation and Cognition of the Federal University of ABC, a researcher from the Image Department of the Albert Einstein Israelite Hospital, and a Senior Lecturer from the Liberal Arts Department of The School of The Art Institute of Chicago (SAIC). A fierce researcher in the studies of the influence of context in perception and behavior, ranging from colors to decision making, including the Fine Arts, she was awarded by the International Color Vision Society, the Pavlov Institute, the Albert Einstein Israelite Hospital, and the University of São Paulo.

Fernando L. Fogliano graduated in Physics, PhD in Communication and Semiotics, Post - Doctorate studies in Art and Technology at the Department of Arts of the Universidade Estadual Paulista, UNESP in Sao Paulo, Brazil. Develops its research in the fields of Complexity and Neuroscience considering issues in Art and Interaction Design. In the Arts field, integrates SCIArts Group, developing projects in Interactive Art.

Mateus Sousa Franco has a Bachelor of Science and Technology from the Federal University of Sao Paulo. Has interdisciplinary training with emphasis in Materials Science and Computation. Worked on researches related to synthesis and characterization of high-performance polymers. Nowadays is studying Computer Engineering at the same institution and works as scientific initiation student at the Nucleus of Neuroengineering and Computation, participating in researches which involve neuroscience, robotics, prosthetics and rehabilitation.

Alexandre Siqueira de Freitas is a pianist and Professor at the Universidade Federal do Sul da Bahia, in the Arts department. Between 2014 and 2015, he held a Postdoctoral internship at Universidade Estadual Paulista, under the supervision of Professor Rosangella Leote. His current research focuses on epistemological questions of arts and sciences, music education, criticism and art history.

Geraldo Coelho Lima Júnior graduated at Design from Universidade do Estado de Minas Gerais (1980) PhD Student on Design Phd and Master's Program. Master's at Design (2008) from the same instituition, acting on the following subjects: fashion, design, culture, sense perception, neuro-education.

Paula Landim graduated at Architecture and Urbanism from Universidade de São Paulo (1987), master's at Geography from Universidade Estadual Paulista Júlio de Mesquita Filho (1994) and ph.d. at Architecture and Urbanism from Universidade de São Paulo (2001). Has experience in Architecture and Urbanism, and Design.

Rosangella Leote is a multimedia artist/researcher. Phd in Communication Sciences by USP, Master of Arts Mestre by UNICAMP and Bachelor in Arts by UFRGS. She is professor at Universidade Estadual Paulista (UNESP), teaching for Doctorate, Master, Specialization and Under Graduation. She is a member of the editorial and scientific board of the ARTECH - International Conference on Digital Arts (Portugal), of the editorial committee of journals CITAR – Journal of Science and Technology of the arts (PT), Revista del BR::AC (Barcelona Investigación: Arte y Creación"); Revista Galáxia (SP), Revista Tecnologia e Sociedade (PR) e Valise (RS), representing the Committee's Visual Poetics of ANPAP and reviewer Ad Hoc from Capes and FAPESP. Was scholar of Research Productivity from CNPq (2007-2011). Is the leader/coordinator of GIIP - "International and Inter-Institutional Group of Research in Convergence Art, Science and Technology" registered in CNPq and certified by UNESP, where there are 5 lines of research. Today her research is about multisensoriality and multimodality based on neuroscience theories, also including handicap technologies for arts. She supervised more them 120 concluding works on all academic levels. She coordinate the Post Graduation Program on Arts at the Art Institute UNESP - São Paulo. Perform in solo and group tecnoperformances (she creates de term). Do sculptures and other interactive works, direct videos (Best Experimental Video at the 2th FESTLATINOBA Festival de Cyne e y Vídeo Latino-americano Argentina 2004). Has been working with Interactive Multimedia Installations with the SCIArts-Interdisciplinary team (Sergio Motta award

2000 and 2005), which is a member since its starting (1995). Coordinator of the Multimedia Communication (PUCSP) until August 2007 and professor until 2009. Taught courses for the Post-Graduation Strictu Senso (Technology of Intelligence and Digital Design) and Graduation (Multimedia). She teaches issues related to the language of digital video 3D (as music videos, animation, and technology). Was professor at the post-graduation Lato Sensu (Specialization) course at PUCSP: Technology Aesthetics. She supervised more them 120 students on all degrees (including postdoctoral), and participated of more them 200 juris. She, also coordinates an international research network connected to GIIP, with national and international partnerships from University of Barcelona (ES), Javeriana University in Bogota (CO), University of VIC (ES), University of Aveiro (PT), and: UFG, UNB, UFSM, UFRJ, UFBa. The total number of researchers, today, in this network is 33.

Maria Manuela Lopes is a visual artist and researcher based in Portugal and the UK. Her current practice is transdisciplinar and based on issues of memory and self-identity informed by life sciences and medical research and presented in the form of time-based installations, occasionally including biological materials. She has been working and showing nationally and internationally and also teaching fine arts in Portugal since 1998. She has studied fine arts - sculpture at the FBAUP-Porto, Portugal and did an MA at Goldsmiths College in London. She has a Doctorate from Brighton University + UCA Farnham, UK (supervised by de Kathleen Rogers, João Lobo Antunes e Judith Williamson), working with representational strategies of Alzheimer's disease in a neuroscience laboratory at Hospital Santa Maria and Molecular Medicine Institute in Lisbon. She is currently involved in a Post Doc project in University of Aveiro ID+ and University of Oporto IBMC (Institute of Molecular and Cellular Biology) – supervised by Kathleen Rogers, Alexandre Quintanilha e Rosa Oliveira-, that extends the PhD project into a wider cultural scenario questioning what it means to be human in a techno-enhanced society. Maria is also assistant-Director of two residency programs: 1) 'artists in Labs' Ectopia - Lisbon, and 2) Cultivamos Cultura, an ecological oriented residency program in a farm in Alentejo. She has concurrently been presenting her work internationally at conferences and also publishing.

Marta de Menezes is a Portuguese artist (b. Lisbon, 1975) with a degree in Fine Arts by the University in Lisbon, a MSt in History of Art and Visual Culture by the University of Oxford, and a PhD candidate at the University of Leiden. She has been exploring the intersection between Art and Biology, working in research laboratories demonstrating that new biological technologies can be used as new art medium. In 1999 de Menezes created her first biological artwork (Nature?) by modifying the wing patterns of live butterflies. Since then, she has used diverse biological techniques including functional MRI of the brain to create portraits where the mind can be visualised (Functional Portraits, 2002); fluorescent DNA probes to create micro-sculptures in human cell nuclei (nucleArt, 2002); sculptures made of proteins (Proteic Portrait, 2002-2007), DNA (Innercloud, 2003; The Family, 2004) or incorporating live neurons (Tree of Knowledge, 2005) or bacteria (Decon, 2007). Her work has been presented internationally in exhibitions, articles and lectures. She is currently the artistic director of Ectopia, an experimental art laboratory in Lisbon, and Director of Cultivamos Cultura in the South of Portugal.

Ibrahim Elias Nasseh has bachelor in Physician and Phylosophy from Universidade de São Paulo - USP. PhD (2004) in Neuroscience at the Federal University of São Paulo (UNIFESP/EPM).

Hosana Celeste Oliveira is artist and researcher in the field of art, science and technology. She got a bachelor degree in Fine Arts and master degree in Multimedia, both at Universidade Estadual de Campinas/ UNICAMP (Brazil). Currently, she is PhD candidate in Arts at Universidade Estadual Paulista / UNESP (Brazil), funded by Capes (Ministry of Education). She accomplished one year of her doctoral studies at the Media Lab - School of Arts, Design and Architecture of Aalto University (Finland). In the same Finnish university she followed the researches conducted by the Laboratory of Computational Engineering - Cognitive Science and Tecnology that investigate the brain and cognition through physiological monitoring, in real time, while the subject watches a film. She was visiting researcher at Köln International School of Design/ KISD (2007 – 2008), funded by DAAD-STIBET, and at Kunsthochschule für Medien Köln/ KHM (2009 – 2010), both in Germany. Also, she collaborated with En-Fer Atelier (The Netherlands, 2005 - 2007) developing web and graphic design projects. Her research agend is built around neuroscientific approaches on perception and emotion, embodied cognition, interactive narratives, and biosensitive computational interfaces.

Tiago Viger Ortiz completed his technical school in informatics on 2010. He graduated in Computer Science in 2015 and is currently doing a Masters Degree in Computer Information System at the Federal University of São Paulo, member of the robotic team and collaborating on a prosthetic arm's scientific research within his university.

Petr Ivanovich Sosnin: 1964 – 1968: Ulyanovsk politechnical institute, Ulyanovsk: Engineer, Computer engineering 1968 – 2013: Ulyanovsk state technical university (assistant, tutor, docent, professor, head of computer department) In 1974 got PhD- degree in Samara politechnical institute In 2003 got Doctor of Sciences degree in Moscow aviation institute Research interests: collaborative activity of designers in computerized technological environments, modeling of professional experience and human-computer interactions, problem-solving activity Author of eleven books and more than 350 scientific publications in Russian and English languages. Scientific advisor: in twenty two PhD-dissertations Member of TPC and Steering Committees of numerous foreign and national conferences in the field of software engineering and Artificial Intelligence. Chairman of the International Conference on Interactive Systems (1995-2013) which were held in Russia (Ulyanovsk). Member: IEEE, Computer Society, ACM, Russian and European Association of Artificial Intelligence Reward: the honored specialist of High Education in Russian Federation.

Laila Torres graduated in biology, PhD and postdoctoral (2011) in Neuroscience at the Federal University of São Paulo (UNIFESP/EPM). Postdoctoral (2013-2014) in the Sanford Consortium for Regenerative Medicine at the department of pediatrics of the University of California, San Diego (UCSD). Research associate (2014-2015) Hospital Albert Einstein- São Paulo-Brazil.

Index

A

B

C

D

CPSIA information can be obtained
at www.ICGtesting.com
Printed in the USA
BVOW10*2329150716
455465BV00008B/126/P